BIORESOURCES AND BIOPROCESS IN BIOTECHNOLOGY FOR A SUSTAINABLE FUTURE

BIORESOURCES AND BIOPROCESS IN BIOTECHNOLOGY FOR A SUSTAINABLE FUTURE

Edited by

Leonardo Sepúlveda Torre, PhD

Juan Carlos Contreras-Esquivel, PhD

Ann Rose Abraham, PhD

A. K. Haghi, PhD

AAP | APPLE ACADEMIC PRESS

First edition published 2024

Apple Academic Press Inc.
1265 Goldenrod Circle, NE,
Palm Bay, FL 32905 USA

760 Laurentian Drive, Unit 19,
Burlington, ON L7N 0A4, CANADA

CRC Press
2385 NW Executive Center Drive,
Suite 320, Boca Raton FL 33431

4 Park Square, Milton Park,
Abingdon, Oxon, OX14 4RN UK

Library and Archives Canada Cataloguing in Publication

CIP data on file with Canada Library and Archives

Library of Congress Cataloging-in-Publication Data

CIP data on file with US Library of Congress

ISBN: 978-1-77491-432-8 (hbk)
ISBN: 978-1-77491-433-5 (pbk)
ISBN: 978-1-00341-004-1 (ebk)

About the Editors

Leonardo Sepúlveda Torre, PhD

Researcher, School of Chemistry, Universidad Autónoma de Coahuila, Saltillo, Coahuila, México

Leonardo Sepúlveda Torre, PhD, is a Researcher at the School of Chemistry at the Universidad Autónoma de Coahuila, Saltillo, Coahuila, México, where he is also a member of the Bioprocesses and Microbial Biochemistry Group. His postgraduate studies were in related topics on food biotechnology in the Food Research Department of the UAdeC. In 2011, he made a research stay at the Institute of Biotechnology and Bioengineering at the University of Minho (Uminho), Braga, Portugal. He worked as a collaborator at the Center of Biological Engineering in Uminho, Braga, Portugal, with the link project Biotechnologies for Regional Food Biodiversity in Latin America. He carried out his postdoctoral program at the same institution, writing on "Assisted Extraction by Fermentation of Polyphenols from Agro-industrial Waste."

Juan Carlos Contreras-Esquivel, PhD

Professor and Head of the Laboratory of Applied Glycobiotechnology, Faculty of Chemistry, Department of Food Research, Universidad Autonoma de Coahuila, Saltillo, Mexico

Juan Carlos Contreras-Esquivel, PhD, is a full Professor and Head of the Laboratory of Applied Glycobiotechnology at the Food Research Department, School of Chemistry, the University of Coahuila, Mexico. He is involved in teaching undergraduate and graduate students with a special focus on food science and food bioscience. His research interests are mainly applied glycol technology for polysaccharide extraction by using ecofriendly technologies (enzymatic, microwave, pulsed electric field, ultrasonic) and oligosaccharide production by degradative or biosynthetic routes. Dr. Contreras-Esquivel earned a bachelor's degree in Biological Chemistry and holds an MSc degree in Food Science and Technology, both from the Universidad Autonoma de Chihuahua, Mexico. He received his PhD from the Universidad Nacional de La Plata, Argentina.

Ann Rose Abraham, PhD
Assistant Professor, Department of Physics Sacred Heart College (Autonomous), Thevara, Kochi, Kerala, India

Ann Rose Abraham, PhD, is currently working as an Assistant Professor at the Department of Physics, Sacred Heart College (Autonomous), Thevara, Kochi, Kerala, India. Her PhD thesis was titled, Development of Hybrid Mutliferroic Materials for Tailored Applications. She has expertise in the field of condensed matter physics, nanomagnetism, multiferroics, and polymeric nanocomposites, etc. She has research experience at various reputed national institutes like Bose Institute, Kolkata, India, SAHA Institute of Nuclear Physics, Kolkata, India, UGC-DAE CSR Centre, Kolkata, India and collaborations with various international laboratories. She is a recipient of a Young Researcher Award in the area of physics and Best Paper Awards–2020, 2021, a prestigious forum for showcasing intellectual capability. She served as assistant professor and examiner, at the Department of Basic Sciences, Amal Jyothi College of Engineering, under APJ Abdul Kalam Technological University, Kerala, India. Dr. Ann is a frequent speaker at national and international conferences. She has a good number of publications to her credit in many peer-reviewed high impact journals of international repute. She has authored many book chapters and edited more than 10 books with Taylor and Francis, Elsevier, etc. Dr. Ann received her MSc, MPhil, and PhD degrees in Physics from School of Pure and Applied Physics, Mahatma Gandhi University, Kerala, India.

A. K. Haghi, PhD
Coimbra University, Portugal

A. K. Haghi, PhD, has published over 250 academic research-oriented books as well as over 1000 research papers published in various journals and conference proceedings. He has received several grants, consulted for several major corporations, and is a frequent speaker to national and international audiences. He is founder and former editor-in-chief of the *International Journal of Chemoinformatics and Chemical Engineering*, published by IGI Global (USA) as well as the *Polymers Research Journal*, published by Nova Science Publishers (USA). Professor Haghi has acted as an editorial board member of many international journals. He has served as a member of the Canadian Research and Development Center of Sciences and Cultures (CRDCSC) and the Research Chemistry Centre, Coimbra, Portugal.

Dr. Haghi holds a BSc in urban and environmental engineering from the University of North Carolina (USA), an MSc in mechanical engineering from North Carolina A&T State University (USA) and an MSc in applied mechanics, acoustics, and materials from the Université de Technologie de Compiègne (France), and a PhD in engineering sciences from Université de Franche-Comté (France).

Contents

Contributors

Rafail A. Afanas'ev
Pryanishnikov All-Russian Scientific Research Institute of Agrochemistry, Moscow, Russia

Cristóbal Noé Aguilar
Bioprocess and Bioproducts Research Group, Food Research Department, School of Chemistry, Autonomous University of Coahuila, Saltillo, Coahuila, México

Nancy Verónica Pérez Aguilar
School of Chemistry, Autonomous University of Coahuila, Saltillo, Coah., México

Candy Carranza Álvarez
School of Professional Studies Huasteca Zone, Romualdo del Campo 501 Fracc. Rafael Curiel, Cd Valles, S.L.P., México

Roberto Arredondo-Valdés
Food Science Research Department, Faculty of Chemical Sciences of the Autonomous University of Coahuila, Saltillo, Coahuila, Mexico; Nanobioscience Group, Faculty of Chemical Sciences of the Autonomous University of Coahuila, Saltillo, Coahuila, Mexico

Juan A. Ascacio-Valdés
Bioproducts Research Group, Food Research Department, School of Chemistry, Autonomous University of Coahuila, Saltillo, Coahuila, México

Nahum Andrés Medellín Castillo
Postgraduate Studies and Research Center, School of Engineering, Autonomous University of San Luis Potosí, Zona Universitaria, San Luis Potosí, S.L.P., México

Mónica L. Chavez-González
Bioprocess and Bioproducts Research Group, Food Research Department, School of Chemistry, Autonomous University of Coahuila, Saltillo, Coahuila, México

Juan Carlos Contreras-Esquivel
Laboratory of Applied Glycobiotechnology, Academic Group of Food Science and Technology, School of Chemistry, Faculty of Chemical Sciences, Universidad Autonoma de Coahuila, Blvd. Venustiano Carranza Esq. Jose Cardenas Valdez s/n, Saltillo, Coahuila, Mexico

Sergio Arturo Coronado-Contreras
Bioprocess and Microbial Biochemistry Group, Food Research Department, School of Chemistry, Autonomous University of Coahuila, Saltillo, Coahuila, México

Yogesh Bharat Dalvi
Scientist, Pushpagiri Research Center, Pushpagiri Institute of Medical Sciences and Research Center, Tiruvalla, Pathanamthitta, Kerala, India

Samir Das
ICAR Research Complex for NEH Region, Umiam, Meghalaya, India

Miriam Desirée Dávila Medina
Bioprocess and Microbial Biochemistry Group, Food Research Department, School of Chemistry, Autonomous University of Coahuila, Saltillo, Coahuila, México

Berni Isaías Esquivel-González
Food Science Research Department, Faculty of Chemical Sciences of the Autonomous University of Coahuila, Saltillo, Coahuila, Mexico; Nanobioscience Group, Faculty of Chemical Sciences of the Autonomous University of Coahuila, Saltillo, Coahuila, Mexico

Lázaro A. González Fernández
Multidisciplinary Postgraduate Program in Environmental Sciences, Autonomous University of San Luis Potosí, Av. Manuel Nava 201, Zona Universitaria, San Luis Potosí, S.L.P., México

Paola Elizabeth Díaz Flores
School of Agronomy and Veterinary Medicine, Carretera San Luis Potosí–Matehuala Km. 14.5, Ejido Palma de la Cruz, Soledad de Graciano Sánchez, S.L.P., México

Amado Enrique Navarro Frómeta
Technological University of Izucar of Matamoros, Prolongación Reforma 168, Santiago Mihuacán, Izúcar de Matamoros, Puebla, México

Soujanya Gadi
Faculty of Chemical Sciences, Universidad Autonoma de Coahuila, Saltillo, Coahuila, Mexico

Mario Alberto Gaitan-Hinojosa
Biochemical Engineering and Bioprocess Laboratory, Chemical Science Faculty, Universidad de Colima, Km 9 Carr. Colima-Coquimatlán s/n. Coquimatlán, Colima, Mexico

Virginia García-Cañas
Molecular Nutrition and Metabolism, Institute of Food Science Research (CIAL, CSIC), Madrid, Spain

Ana Laura García-Pérez
School of Chemistry, Universidad Autonoma de Coahuila, Saltillo, Coahuila, Mexico

Sandeep Ghatak
ICAR Research Complex for NEH Region, Umiam, Meghalaya, India

Rafael Gomes-Araújo
Sustainable Applied Biotechnology Group, Tecnológico de Monterrey, Monterrey, Nuevo León, México

Mayela Govea-Salas
Food Science Research Department, Faculty of Chemical Sciences of the Autonomous University of Coahuila, Saltillo, Coahuila, Mexico; Nanobioscience Group, Faculty of Chemical Sciences of the Autonomous University of Coahuila, Saltillo, Coahuila, Mexico

Rodrigo Guzmán-Pedraza
School of Chemical Engineering, Autonomous University of Yucatan, Merida, Yucatan, Mexico

José Luis Martínez Hernández
Food Science Research Department, Faculty of Chemical Sciences of the Autonomous University of Coahuila, Blvd. V. Carranza e Ing. José Cárdenas V., Col. República, Saltillo, Coahuila, Mexico

Anna Iliná
Food Science Research Department, Faculty of Chemical Sciences of the Autonomous University of Coahuila, Saltillo, Coahuila, Mexico; Nanobioscience Group, Faculty of Chemical Sciences of the Autonomous University of Coahuila, Saltillo, Coahuila, Mexico

C. Jimtha John
Assistant Professor, Department of Integrated MSc Biology, St. Joseph's College, Irinjalakuda, Thrissur, Kerala, India

J. Buenrostro-Figueroa Juan
School of Fundamental Biology and Biotechnology, Siberian Federal University, Krasnoyarsk, Russia; Federal Research Center "Krasnoyarsk Science Center SB RAS," Institute of Biophysics SB RAS, Krasnoyarsk, Russia

Alexey M. Komelin
Mari State University, Yoshkar-Ola, Russia

K. D. Lazareva
Pryanishnikov All-Russian Scientific Research Institute of Agrochemistry, Moscow, Russia

Leonardo Sepúlveda
Bioprocess and Microbial Biochemistry Group, Food Research Department, School of Chemistry, Autonomous University of Coahuila, Saltillo, Coahuila, México

P. Lochana
Department of Chemistry, School of Sciences, Jain Deemed to be University, Bangalore, Karnataka, India

Sonia A. Lozano-Sepúlveda
Department of Biochemistry and Molecular Medicine, School of Medicine, Autonomous University of Nuevo Leon, Monterrey, Nuevo León, México

Rodrigo Macias-Garbett
Laboratory of Applied Glycobiotechnology, Academic Group of Food Science and Technology, School of Chemistry, Universidad Autonoma de Coahuila, Blvd. Venustiano Carranza Esq. Jose Cardenas Valdez s/n, Saltillo, Coahuila, Mexico

T. E. Mantseva
Pryanishnikov All-Russian Scientific Research Institute of Agrochemistry, Moscow, Russia

Emilio Mendez-Merino
Sigma Alimentos, San Pedro Garza Garcia, Nuevo Leon, Mexico

Genrietta E. Merzlaya
Pryanishnikov All-Russian Scientific Research Institute of Agrochemistry, Moscow, Russia

A. Arun Prince Milton
ICAR Research Complex for NEH Region, Umiam, Meghalaya, India

Rafael Minjares-Fuentes
Faculty of Chemical Sciences, Universidad Juárez del Estado de Durango, Gómez Palacio, Durango, Mexico

Julio Montañez
Faculty of Chemical Sciences, Universidad Autonoma de Coahuila, Saltillo, Coahuila, Mexico

Lourdes Morales-Oyervides
Faculty of Chemical Sciences, Universidad Autonoma de Coahuila, Saltillo, Coahuila, Mexico

Sergei I. Novoselov
Mari State University, Yoshkar-Ola, Republic of Mari El, Lenin Square, Russia

Samuel Pérez-Vega
Faculty of Chemical Sciences, Universidad Autonoma de Chihuahua, Chihuahua, Mexico

Armando Pineda-Contreras
Biochemical Engineering and Bioprocess Laboratory, Chemical Science Faculty, Universidad de Colima, Km 9 Carr. Colima-Coquimatlán s/n. Coquimatlán, Colima, Mexico

M. Chendu Bharat Prasad
ICAR Research Complex for NEH Region, Umiam, Meghalaya, India

G. Bhuvana Priya
College of Agriculture, Central Agricultural University (Imphal), Kyrdemkulai, Meghalaya, India

Svetlana V. Prudnikova
School of Fundamental Biology and Biotechnology, Siberian Federal University, Krasnoyarsk, Russia

Rogelio Flores Ramírez
Coordination for the Innovation and Application of Science and Technology, San Luis Potosí, S.L.P., México

Alejandra Solis Ramos
Bioprocesses and Bioproducts Research Group, Food Research Department, School of Chemistry, Autonomous University of Coahuila, Saltillo, Mexico

Fernanda Lizeth Rebolledo-Ramírez
Food Science Research Department, Faculty of Chemical Sciences, Universidad Autónoma de Coahuila, Saltillo, Mexico

Yadira Karina Reyes-Acosta
Bioprocess and Microbial Biochemistry Group, Food Research Department, School of Chemistry, Autonomous University of Coahuila, Saltillo, Coahuila, México

Ana María Rivas-Estilla
Department of Biochemistry and Molecular Medicine, School of Medicine, Autonomous University of Nuevo Leon, Monterrey, Nuevo León, México

Rosa María Rodríguez-Jasso
Food Science Research Department, Faculty of Chemical Sciences of the Autonomous University of Coahuila, Saltillo, Coahuila, Mexico; Biorefinery Group, Food Research Department, Faculty of Chemistry Sciences, Autonomous University of Coahuila, Saltillo, Coahuila, Mexico

Xochitl Ruelas-Chacón
Department of Food Science and Technology, Autonomous Agrarian University Antonio Narro, Saltillo, Coahuila, Mexico

Aidé Saenz-Galindo
School of Chemistry, Universidad Autonoma de Coahuila, Saltillo, Coahuila, Mexico

Carolina Simó-Ruiz
Molecular Nutrition and Metabolism, Institute of Food Science Research (CIAL, CSIC), Madrid, Spain

Michael O. Smirnov
Pryanishnikov All-Russian Scientific Research Institute of Agrochemistry, Moscow, Russia

Nadezhda V. Streltsova
School of Fundamental Biology and Biotechnology, Siberian Federal University, Krasnoyarsk, Russia

Sabu Thomas
School of Fundamental Biology and Biotechnology, Siberian Federal University, Krasnoyarsk, Russia; International and Interuniversity Center for Nanoscience and Nanotechnology, Mahatma Gandhi University, Kottayam, Kerala, India

Leonardo Sepúlveda Torre
School of Chemistry, Autonomous University of Coahuila, Blvd. V. Carranza s/n. Col. República Oriente C. P., Saltillo, Coah., México

Laura Leticia Valdez-Velázquez
Biochemical Engineering and Bioprocess Laboratory, Chemical Science Faculty, Universidad de Colima, Colima, Mexico

Ruby Varghese
Department of Chemistry, School of Sciences, Jain Deemed to be University, Bangalore, Karnataka, India; Pushpagiri Research Center, Pushpagiri Institute of Medical Sciences and Research Center, Tiruvalla, Pathanamthitta, Kerala, India

Oscar Fernando Vázquez-Vuelvas
Laboratory of Applied Glycobiotechnology, Academic Group of Food Science and Technology, School of Chemistry, Universidad Autonoma de Coahuila, Blvd. Venustiano Carranza Esq. Jose Cardenas Valdez s/n, Saltillo, Coahuila, Mexico; Biochemical Engineering and Bioprocess Laboratory, Chemical Science Faculty, Universidad de Colima, Mexico

Namitha Vijay
Pushpagiri Research Center, Pushpagiri Institute of Medical Sciences and Research Center, Tiruvalla, Pathanamthitta, Kerala, India

Tatiana G. Volova
School of Fundamental Biology and Biotechnology, Siberian Federal University, Krasnoyarsk, Russia; Federal Research Center "Krasnoyarsk Science Center SB RAS," Institute of Biophysics SB RAS, Krasnoyarsk, Russia

Abbreviations

AFE	avocado fruit extract
AID	alloxan-induced diabetic rats
AMF	arbuscular mycorrhizal fungi
AMVRT	avian myeloblastosis virus reverse transcriptase
ATR	attenuated total reflection
AZ	azoxystrobin
BGL	blood glucose level
Bmec	bovine mammary epithelial cells
CAMP	competitive annealing mediated isothermal amplification
CO_2	carbon dioxide
CPA	cross-priming amplification
dCPA	double cross-priming amplification
DENV-2	dengue virus
DIF	difenoconazole
EAE	enzyme-assisted extraction
EO	essential oil
FPLC	fast protein liquid chromatography
FTIR	Fourier transform infrared
GHP	good hygiene practices
GPC	gel permeation chromatography
HACCP	hazard analysis and critical control points
HAD	helicase-dependent amplification
HC	high cholesterol
HFD	high-fat diet
HPLC	high-performance liquid chromatography
HTC	hydro-thermal coefficient
IAA	indole-3-acetic acid
IIR	innate immune response
LAMP	loop-mediated isothermal amplification
LDL	low- and high-density lipoproteins
LEAS	lipid-rich extract of avocado seeds
LHS	labile humus substances
MAC	maximum allowable concentration
MAE	microwave-assisted extraction

MSD	microwave steam distillation
MSDf	microwave steam diffusion
NAAT	nucleic acid amplification tests
NASBA	nucleic acid sequence-based amplification
NMR	nuclear magnetic resonance
NPK	nitrogen, phosphorus, and potassium
OHE	ohmic heating extraction
PDA	potato dextrose agar
PEF	pulsed electric field technology
PGPR	plant growth promoting rhizobacteria
PHAs	polyhydroxyalkanoates
PMC	peat-manure compost
PSR	polymerase spiral reaction
RCA	rolling circle amplification
RP	reverse primer
RPA	recombinase polymerase amplification
RPI	research products international
SBBH	sugarcane bagasse hemicellulosic hydrolysate
SC-CO_2	supercritical carbon dioxide
SCFE	supercritical fluid extraction
SCP	single-cell protein
sCPA	single cross-priming amplification
SD	steam distillation
SDA	strand displacement amplification
SDGs	sustainable development goals
SEA	strand exchange amplification
SEC	size exclusion chromatography
SFME	solvent-free microwave extraction
SHDF	steam hydraulic diffusion
SID	streptozotocin
SPIA	single primer isothermal amplification
SRCA	saltatory rolling circle amplification
STZ	streptozotocin
SWE	subcritical water extraction
TC	total cholesterol
TG	triglycerides
TPC	total phenolic content
UAE	ultrasound-assisted extraction
UNGA	United Nations General Assembly
VOC	volatile organic compounds

Preface

This new research-oriented book reports the latest developments in the field of biotechnology and focuses on the different aspects of bio-resources and bioprocesses in biotechnology for sustainable development. This new volume reviews achievements in bioprocess and biosystem engineering, biosynthesis, food, agriculture, and biotechnology-related issues.

The book expounds on the fact that biological alternatives can always replace harmful chemical products and hence maintain the ecosystem for a sustainable future. The book is organized into two parts covering the role of biotechnology in industrial products, environmental remediation, and agriculture, along with updated research case studies.

This new book covers a variety of fields in agricultural biotechnology while focusing on selected commercial crops. It also discusses key parameters in the evaluation of biodegradation in bio-based materials in soil for the better management of natural resources. Case studies on bioremediation and biodegradation are presented that demonstrate ways to improve the condition of polluted soil and water bodies.

This will be a valuable reference source for chemical engineers, practicing agricultural students, environmental engineers, and professionals and academics in other allied fields likewise.

Preface

PART I

Biological Activities, Biomass Utilization, and Bio-Based Chemistry

CHAPTER 1

Biological Activities of Compounds from Avocado (*Persea americana*) Agro-Industrial Waste: Toward a Green and Sustainable Future

FERNANDA LIZETH REBOLLEDO-RAMÍREZ,[1] RAFAEL GOMES-ARAÚJO,[2] ROSA MARÍA RODRÍGUEZ-JASSO,[1] JUAN A. ASCACIO-VALDÉS,[1] ROBERTO ARREDONDO-VALDÉS,[1] CRISTÓBAL NOÉ AGUILAR,[1] ANNA ILINÁ,[1] BERNI ISAÍAS ESQUIVEL-GONZÁLEZ,[1] SONIA A. LOZANO-SEPÚLVEDA,[3] ANA MARÍA RIVAS-ESTILLA,[3] VIRGINIA GARCÍA-CAÑAS,[4] CAROLINA SIMÓ-RUIZ,[4] MÓNICA L. CHAVEZ-GONZÁLEZ,[1] and MAYELA GOVEA-SALAS[1]

[1]*Food Science Research Department, Faculty of Chemical Sciences, Universidad Autónoma de Coahuila, Saltillo, Mexico*

[2]*Institute of Advanced Materials for Sustainable Manufacturing, Tecnológico de Monterrey, Monterrey, México*

[3]*Department of Biochemistry and Molecular Medicine, School of Medicine, Universidad Autónoma de Nuevo Leon, Monterrey, México*

[4]*Molecular Nutrition and Metabolism, Institute of Food Science Research (CIAL, CSIC), Madrid, Spain*

ABSTRACT

Avocado residues, such as seeds, peels, and leaves, are waste generated by the agri-food industry in large quantities since avocado is a fruit of high

Bioresources and Bioprocess in Biotechnology for a Sustainable Future. Leonardo Sepúlveda Torre, Juan Carlos Contreras-Esquivel, Ann Rose Abraham, & A. K. Haghi (Eds.)

nutritional value consumed in much of the world. However, the increase in its waste is an environmental problem, so its use represents a viable alternative to this problem. It has been shown that these residues are a great source of bioactive compounds such as phenols, tannins, anthocyanins, and flavonoids, among others, whose application in areas such as health is of great importance since these compounds have characteristics due to their antioxidant nature that give them biological properties such as anticancer, antiviral, antimicrobial, antioxidant, hypoglycemic, and others that have been studied to prevent or counteract diseases and that will be described in this chapter.

1.1 BIOACTIVE COMPOUNDS OF AVOCADO WASTES

Persea americana (avocado) is a fruit that belongs to the laurel family Lauraceae, this fruit has been cultivated for its high nutritional content since about 8000 B.C., and it is consumed worldwide. Avocado is endemic to tropical and subtropical areas of Mexico and other parts of Central America; nevertheless, nowadays is cultivated in many countries of Africa, Asia, and some parts of Europe [11, 15, 26, 59]. Mexico is the largest avocado producer, followed by the Dominican Republic. In 2018, the world production was 6 million tons, and Mexico produced nearly 35% of it.

Nowadays, avocado is part of the daily diet in many countries, and it is consumed mainly as fresh fruit; however, a few years ago, this fruit started to gain access in the industry by starting to commercialize in a broad diversification of products such as avocado paste, dried or dehydrated avocado, refreshing drinks, frozen products, and oil [35, 37]. It is essential to mention that not only in the food but also in the cosmetic and even pharmaceutical industry, avocado (especially oil) has been used to fabricate different products, for example, soaps, shampoo, body lotions, and cosmetics [18, 21]. The high consumption and production of avocado, no matter if it is consumed as fresh fruit or as an industrialized product, implies a considerable generation of agro-industrial residues such as seeds, peel, and leaves. Most of the time, these residues are discarded with no further applications, representing an environmental problem [5, 65]. Nevertheless, with new technologies and methods, industries and research groups have started to take advantage of avocado waste to obtain other kinds of products [19]. Besides avocado's pulp (edible part) nutritional value, which includes lipids, proteins, fatty acids, vitamins, and polyphenols, among others [4, 20], avocado wastes are a source of bioactive compounds, most of them are polyphenols which include condensed tannins, phenolic acids, and flavonoids, such as procyanidins and flavonols [4, 25, 40, 75]. These compounds could represent an impulse in the alimentary, cosmetic, and pharmaceutical industries [43]. Furthermore, polyphenols have been a target of interest due to their multiple biological effects, such as antioxidant, anticancer, anti-inflammatory, antiviral, antimicrobial, and antifungal activity.

Bioactive compounds in avocado wastes, such as polyphenols, have been primarily found in peels and seeds; however, more detailed information will be discussed in the following paragraphs. A few examples of these compounds are shown in Table 1.1.

1.1.1 AVOCADO PEEL

According to different authors, avocado peel, also known as epicarp, has shown a high content of phenolic compounds. The total phenolic content (TPC) reported in avocado peel varies depending on the extraction method; nevertheless, according to literature, this value could oscillate from 12.52 to 1058 mg GAE g^{-1} [45, 62, 64]. The variation in TPC, not only in peel but also in seeds and leaves, relies on avocado variety, harvest conditions, extraction methods, and an extraction solvent, among others. The most common polyphenols found in the avocado peel are flavonoids such as catechins, some phenolic acids, proanthocyanidins, and procyanidins at different degrees of polymerization [26, 40, 45, 48].

1.1.2 AVOCADO SEED

In the same way as avocado peel, the avocado seed has shown a high polyphenols content. The TPC in this residue has shown values between 1.55 and 1303 mg GAE g^{-1} [64–66]. The most representative polyphenols in the avocado seed are phenolic acids, condensed tannins, and other flavonoids, including procyanidins and catechins [25, 68, 72].

1.1.3 AVOCADO'S LEAVES

As well as peel and seeds, avocado leaves show a content of polyphenols. Nevertheless, compared to peel and seeds, leaves show the lowest concentrations. The TPC reported goes from 1.78 to 43.82 mg GAE g^{-1} [8, 53]. The polyphenols found in avocado leaves mainly include flavonoids and tannins.

TABLE 1.1 Bioactive Compounds in the Avocado Peel, Seed, and Leaves

Compound	Source
Phenolic Acids	
Caffeic acid	Leaves, peel, seeds
Caffeoylquinic acid	
Chlorogenic acid	Peel, seeds
Cinnamic acid	Leaves
Coumaric acid	Leaves, peel
Coumaroylquinic acid	Leaves, seeds
Ferulic acid	Peel, seeds
Sinapic acid	
Syringic acid	
Flavonoids	
Catechin	Leaves, seeds
Epicatechin	Peel, seeds
Epicatechin gallate	
Kaempferol	Leaves, peel, seeds
Quercetin	Peel, seeds
Quercetin-3-O-arabinosyl-glucoside	Leaves, peel
Quercetin-diglucoside	
Isorhamnetin	Seeds
Rutin	Leaves, peel
Vitexin	Leaves
Condensed Tannins	
Proanthocyanidin dimer B1	Seeds
Proanthocyanidin dimer B2	
Procyanidin dimer A	Peel
Procyanidin dimer B1	
Procyanidin tetramer A	Peel, seeds
Procyanidin trimer A1	Leaves, seeds
Procyanidin trimer A2	Seeds
Procyanidin trimer B	
Others	
Hydroxytyrosol glucoside	Leaves, seeds
Tyrosol glucoside	Peel, seeds
Terpenoids	
b-caryophyllene	Leaves
b-pinene	
Monoterpene	

However, they also include phenolic acids and other bioactive compounds not related to the polyphenols family, such as saponins and terpenoids [4, 55, 60, 77]. Furthermore, in avocado leaves, a toxin called persin has shown antifungal [82], antioxidant, and anti-cancer activity.

On the other hand, not specifically in avocado wastes, other phytochemicals such as tocopherols, sterols, and carotenoids have been found [37].

The use of avocado waste for different purposes represents an economic and social impact since their application promotes recycling. Nowadays, recycling has become a new industry that generates new jobs and, thus, boosts the economy of the world. Additionally, bioactive compounds could represent new alternative therapies for the treatment of different diseases since they possess biological activities.

In addition to knowing the properties of the different bioactive compounds present in the seed and peel of avocado, it is also essential to know how they are processed by the human body. Therefore, below is a diagram (Figure 1.1) that explains the general aspects of the metabolism of chlorogenic acids, a family of polyphenolic esters with anti-inflammatory properties whose pharmacological potential is related to their absorption in the body, according to the research presented by different authors [10, 70, 75, 76].

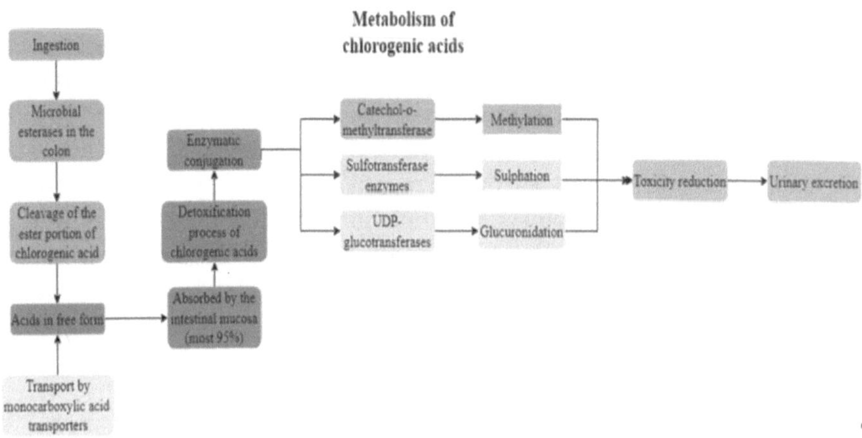

FIGURE 1.1 General aspects of chlorogenic acid metabolism.

1.2 ANTI-CANCER ACTIVITY

More than 20 bioactive compounds have been isolated from avocados related to anti-cancer activities, such as long-chain lipid molecules, long-chain fatty acids, and their derivatives (avocatins, pahuatins, persenins, and polyhydroxy fatty alcohols) [43]. The phytochemical compounds that have been isolated from *P. americana* may have effects on cell cycle arrest, delaying development and activating apoptosis processes in some cancer cell lines. Besides, *P. americana* contains persenones A and B, vitamins that could prevent carcinogenesis [2]. The results show that lipid extracts of *P. americana* (seed and fruit) possess significant inhibition for hepatocellular carcinoma cell line HepG2 and colon cancer cell line HCT116 compared with the drug sorafenib. These findings also revelated that the IC_{50} value of extracts of seeds was less than the fruit extract against both used cell cancer lines as compared with the sorafenib. Lara et al. [43] found apoptotic activity by activating caspase 8 and caspase 9 and cytotoxic activity with $IC_{50} = 28$ μg/mL in lipid extracts obtained from the avocado seed variety *Drymifolia* in colorectal cancer cells. The avocado seed extract is considered a good source of compounds with anti-cancer activity due to the flavonoids, carotenoids, and persin contained in the seed [39]. Among the bioactive compounds that avocado contains are phytosterols, mainly β-sitosterol, which contribute to cancer treatment by suppressing carcinogenesis. Lutein is a carotenoid that is part of the bioactive compounds in avocados and provides protection against prostate cancer [21]. Components of avocado pulp, such as tocopherols, lutein, and other carotenoids, were found in an *in vitro* study to have inhibitory effects against cancer [71]. Bhuyan et al. [12] mention that methanol extracts of avocado seed and peel stand out for their ability to activate caspase-3 and its target protein-PARP in MDA-MB-231 cells (cell line used for *in vitro* study of breast cancer). An ethanol extract of whole avocado seeds activated the transcription factor p53, caspase-3, apoptosis-inducing factor, and oxidative stress-dependent apoptosis in lymphoblastic leukemia cells, as well as acetone extract of avocado pulp induced the arrest of PC-3 prostate cancer cells in the G2/M phase. Based on the above, the agro-industrial residues of avocado are considered a natural source that provides potential biological components with anticancer activity.

1.3 TRADITIONAL APPLICATIONS OR USES IN MEDICINE

Some studies have been carried out about the uses and applications in medicine of compounds obtained from agro-industrial residues of avocado. Avocado content has interesting bioactivities for the medical industry [77]. Recent research carried out by Gallegos et al. [30] identified 10 plant species with applications in the prevention, control, and treatment of skin diseases, including *Persea americana*. It was found that the use of avocado pulp ointment has properties to combat skin blemishes, as well as to accelerate the suppuration process in infected wounds.

Other valuable contributions about the medicinal properties of *P. americana Mill* found that the seeds of this fruit can be helpful in the treatment of hypertension, inflammatory diseases, diabetes, and the reduction of cholesterol [17]. A study in a rural community in Mexico reported that 28% of the population uses avocado as a medicinal treatment for an upset stomach [36]. The test was carried out on 13 healthy adults who, for 6 days, replaced butter in their diet with avocado oil extracted from the pulp of the Mexican creole genotype fruit at 35°C. The results showed benefits in the postprandial profile of insulin, glycemia, total cholesterol, low-density lipoproteins, triglycerides, and anti-inflammatory activity with an effect like the drug ibuprofen. In addition, oil extracted from the avocado pulp was added to cream for skin with vitamin B_{12}, which showed good tolerance and improvements for long-term topical psoriasis treatment [28]. High numbers of plants have been used in traditional medicine to treat various illnesses, including neurodegenerative diseases such as Alzheimer's and other memory-related disorders. Due to this, avocados are considered one of the main foods to help prevent Alzheimer's disease [4]. An investigation by Zhao et al. [83] indicated that the acetogenin found in the avocado pulp showed an inhibitory effect on platelet aggregation, providing a possible preventive effect on thrombus formation and cardiovascular protection.

1.3.1 ANTIOXIDANT ACTIVITY

Avocado extracts contain a considerable polyphenols quantity, and that could be useful in pharmaceutical products, cosmetics, and nutrition. It has been studied that extract seed and peel have better antioxidant properties than pulp due to the concentration of compounds [10]. Bhuyan et al. [12] found that the main antioxidant compounds of avocados come from mostly phenolic compounds, continued monounsaturated and polyunsaturated acids, carotenoids, acetogenins, and tocopherols. According to Ejiofor et al. [23], tannin is a polyphenol with antioxidant activity. Tannins protect against oxidative cell damage. Avocado seed provides a higher content of polyphenols, even more than some synthetic antioxidants. López-Cobo et al. [45] found that the avocado pulp and seed with a higher degree of maturation (overripe) contained a greater phenolic concentration. Egbuonu et al. [22] studied the vitamins that avocado seed produces. The results prove that there is a high concentration of vitamin C (97.8%) and vitamin A; both help with antioxidant capacity in animals and humans. Folasade et al. [29] showed that avocado seed extract contains phenols, flavonoids, and proanthocyanidins. Flavonoids are found in high concentrations in the pulp and leaves. The antioxidant potential of the seed is related to the solubility of the organic solvents used to obtain the polyphenols. Acetone extract has the best performance for this purpose.

Some authors mention that a suitable temperature is needed for it to have a good antioxidant capacity; generally, with a high temperature, the concentration of the antioxidant capacity increases. In recent articles, Figueroa et al. [25, 26] confirm that more phenolic acid subtypes have been studied before, from avocado peel and seed, which were characterized by five and six types of hydroxybenzoic acid.

1.3.2 ANTIVIRAL ACTIVITY

Avocado seed and peel extracts have been shown to have antiviral activity against different types of viruses. Avocado residues contain water-soluble flavonoids; these have been reported to have significant antiviral activity, Athaydes et al. [7] mention that the amount in mg/g of flavonoids in the avocado seed is 0.35 ± 0.01 mg Eq/g. Amado et al. [3] found that avocado skin presents 2.74 mg Eq/g. Figueroa et al. [25, 26] reported that avocado seed contains quercetin-3-β-glucoside; this flavonoid has proven antiviral activity against the Zika virus; it also contains O-glycosidic which has an inhibitory activity against herpes simplex. Furthermore, in their study, they mention 63 tannin groups in the seed coat, and a significant amount of procyanidin type A was found. Vilhelmova-Ilieva et al. [76] indicate that the use of tannins to combat some types of viruses, such as enteroviruses, adenoviruses, influenza, AIDS, and the most critical ones, such as herpes simplex virus type 1 and 2, has been studied in different articles. Ejifor et al. [23] present data where the levels of tannins in the avocado seed are 6.98 ± 0.23 g/100 g tannins. Villa-Rodriguez et al. [78] reported that pulp contains carotenoids, specifically β-carotene and lutein, and discussed the importance of avocado ripening as it decreases β-carotene levels and increases trans-lutein levels [78]. Hegazy et al. [32] mention that carotenoids have been used for replication inhibition of hepatitis C and hepatitis B viruses. Wu et al. [81] described the use of avocado extracts to see the inhibition of dengue virus (DENV-2) replication, and it was found that interferon, together with NF-k beta, has an inhibitory effect on virus replication. With the above data, it is proved that there is a high degree of compounds with antiviral properties that give avocado seed, peel, and pulp residual importance; for this reason, it is necessary to carry out tests to see the effect on different types of viruses.

1.3.3 HYPOGLYCEMIC ACTIVITY

Nardi et al. [49] found a decrease in sugar levels in rats that were treated with ethanol extract from avocado seed at concentrations of 300, 600, and 1200 mg/Kg BW at different times and with 7 repetitions, which were subsequently administered 2 mg/Kg BW of glucose 30 minutes after administering the treatment, obtaining a decrease in the sugar levels in the groups treated with the ethanol extract of the avocado seed, at a concentration of 300 mg/Kg BW the most significant reduction occurred, although the results obtained for each concentration of the extract compared to the positive control (glibenclamide) are not significant. The avocado seed extract, in the results obtained by Nardi et al. [49], contained tannins, saponins, and flavonoids, whose compounds, Oktaria et al. [56] mention that they inhibit the action of S-GLUT 1 (sodium-glucose transporter).

Tests conducted on streptozotocin (STZ)-induced diabetic rats to which a group of rats was administered avocado fruit extract (AFE) as treatment, indicating a decrease in glucose and glycosylated hemoglobin levels being significant to the control group (treated with gliclazide); In addition to the activities of hexokinase, pyruvate kinase, glucose-6-phosphatase, fructose-1, 6-bisphosphatase, glucose-phosphate dehydrogenase, and glycogen phosphorylase returned almost to normal; it also restored glycogen and glycogen synthase levels. All results were compared to the control group. Therefore, according to the results obtained, the administration of AFE to rats improved the activity of key enzymes in carbohydrate metabolism, giving information on the mechanism of action of AFE in the decrease of glucose levels.

In a study conducted in diabetes-induced albino rats that were treated with aqueous extract and methanolic extract of avocado seeds, Ejiofor et al., [23] found that in all rats treated with the extract of the avocado seed, either aqueous or methanolic, glucose levels decreased by a greater percentage than rats treated with insulin, but the methanolic extract in the concentrations studied (200 and 300 mg/b. wt.) obtained a greater decrease in blood glucose levels (BGLs); there was a significant increase in the level of total bilirubin and conjugated bilirubin. Vítek, [79] mentions that high levels of bilirubin could protect against diabetes. Cheriyath et al., [16] mention that high levels of bilirubins lead to more efficient glucose utilization thanks to the increased mobilization of glucose to cells. Broadhurst, [14] mentions that the hypoglycemic effect of the avocado seed extract could be due to calcium, magnesium, potassium, sodium, zinc, and chromium, which are important in the regulation of gluconeogenesis enzymes blocking this pathway and thus improving glucose utilization.

1.3.4 IMMUNOMODULATORY ACTIVITY

Lara et al., [44] discovered that the lipid-rich extract of avocado seeds (LEAS) modulates the inflammatory response in Caco-2 human colon cancer cells, evaluating the cytokines secreted by Caco-2 cells treated with 28 μg/mL of LEAS for 6, 12, and 24 hours showing an increase in IL-6 and IL-10 secretion, a decrease in IL-1b secretion after 6 hours of treatment with the extract, secretion of IL-8 in Caco-2 cells treated with LEAS at different times, while cells treated with 0.1% DMSO did not secrete said chemokine. One of the roles of IL-6 described by Fisher et al., [27] includes eliciting a favorable antitumor response that, in turn, promotes T-cell proliferation and survival, infiltration to lymph nodes, and mobilization toward tumors. Overexpression of IL-10 favors the proliferation of tumor cells and generates immunosuppression of the anti-tumor response. Lara et al., [43] mention that conflicting data indicate that IL-10 decreases inflammation and increases the anti-tumor response, activating regulatory T cells.

Báez et al., [9] analyzed the effects of LEAS on the regulation of the innate immune response (IIR) in bovine mammary epithelial cells (bMEC) according to the levels of mRNA of proinflammatory cytokines (TNF-a, IL-1 b, and IL-6) and in the anti-inflammatory cytokine IL-10, in cells pretreated with LEAS before and after *S. aureus* infection, the results showed an increase in the levels of IL-10 mRNA, a decrease in the levels of TNF-a mRNA, both results in infected back. Also, there was a decrease in TNF-a secretion and an increase in IL-1b secretion, both resulting in infected cells. bMEC, as mentioned by Oviedo et al., [57], act as danger sensors and have the capacity to generate a defense response; therefore, they are considered an immunological defense barrier.

1.4 CHOLESTEROL-LOWERING ABILITY OF *PERSEA AMERICANA*

Cardiovascular accidents are one of the main causes of mortality in the world, considering high cholesterol levels, dyslipidemia, and platelet aggregation as risk factors [69]. Cholesterol levels are measured with different parameters such as total cholesterol (TC), triglycerides (TG), and low- and high-density lipoproteins (LDL and HDL) [69]. The avocado (pulp, seed, and peel) contains different phenolic compounds, phytosterols, triterpenes, fatty acids, and proanthocyanidins, among others, which is why different studies place *Persea americana* avocado as nutritious food, including its seed in the diet because its toxicity is low [5] and it has important biological activities, one of them is its ability to reduce cholesterol in the blood of an organism, especially by the phytosterol β-sitosterol [80].

Tabeshpour [71] shows the study of two extracts of avocado seed and pulp administered to rats fed a high-fat diet (HFD) for 6 weeks; the results were more significant with the extract of the seed at decreased serum levels of TC, TG, and LDL-C, there was also a decrease in TC and TG in the liver of rats. In addition, Tabeshpour [71] mentions an aqueous extract of the seed of *P. americana* administered to rabbits for two months in doses of 100 and 200 mg/kg. As a result, a reduction in the number of total lipids was observed, including TC, TG, and LDL-C.

An experiment with hypertensive rats (SHR) and healthy rats (SR) was carried out by Uzukwu [73], to which he added different amounts of *P. americana* seed powder to his diet. The results describe that the maximum blood cholesterol reduction was obtained with those SHR rats with the highest percentage of extract (4%); this decrease may be due to phytosterols such as β-sitosterol that induce it by decreasing absorption and cholesterol solubility in the intestine. Phytosterols bind more easily with bile salts due to their hydrophobicity resulting in the excretion of LDL-C that was not absorbed [73].

It is important to know that the presence of saponins in avocado seeds also has an essential role in the ability to reduce cholesterol in the blood since these bind to cholesterol, forming insoluble complexes, thus preventing its reabsorption [33].

A study with seed extracts of *P. americana* and *Citrullus lanatus* with alloxan-induced diabetic rats (AID rats) presented an excess of fatty acids in the plasma because alloxan promotes their transformation into triglycerides and cholesterol. Ukpabsi [74] shows that the treatment with avocado extract caused a decrease in TC, TG, and LDL-C and increased HDL levels; however, it just represents this it normalized the lipid profile that was affected when applying alloxan, compared with the positive control Atorvastatin [74].

Manal [47] made an experiment comparing two extracts, one of avocado pulp and another of the seed, in mice with high cholesterol (HC); the treatment consisted of 30% of the avocado pulp, reducing the TC by 22.48%,

TG 38.21%, and LDL-C by 29.27% and increasing HDL-c by 31.33%, this because the avocado is rich in monounsaturated fatty acids. Treatment with 30% avocado seed extract in HC rats reduced TC by 31.97%, TG by 43.19%, and LDL-C by 47.77%, increasing HDL-C by 34.33% compared to the control group. The more significant effect of the seed is due to its high content of soluble dietary fiber, flavonoids, and phenolic compounds [47].

An ethanolic extract of the seed of *P. americana* was administered to male Wistar rats with a high lipid profile at different doses for two weeks. Fidrianny [69] found that the lowest dose of 10 mg/kg/day reduced TC and LDL-C, while the highest doses of 20 and 40 mg/kg/day decreased the number of triglycerides in the blood. Due to the results shown, avocado and its components, *P. americana*, maybe a medical strategy to lower the serum lipid level and thus prevent the risk of suffering a cardiovascular accident [69].

1.5 *PERSEA AMERICANA*, LIKE A GLUCOSE-LOWERING AGENT

The increase in the amount of glucose in the blood, either due to a deficiency in insulin secretion or because the body has a resistance to it, is diagnosed as Diabetes mellitus.

Avocado is a fruit with a minimum content of sugar in the form of D-mannoheptulose but with a large number of secondary metabolites such as tannins, saponins, flavonoids, and alkaloids [58] that are synthesized during its development as a response of unfavorable conditions [74] that give it different biological activities, such as being a hypoglycemic agent [80].

Oboh [54] experimented with methanolic extracts of the different components of the avocado in rat pancreatic cells to evaluate the inhibition of enzymes that participate in the production and absorption of glucose. It shows that the extract of the avocado peel from *P. americana* has the highest inhibition percentage of α-amylase, and in turn, the extract from the leaves of the plant better inhibits the enzyme α-glucosidase. It considers that its hypoglycemic activity, inhibiting these essential enzymes, could be a therapeutic target for type 2 diabetes mellitus because it slows glucose metabolism, keeping BGLs low [54].

Different studies in diabetic rats, induced by streptozotocin (SID rats), describe those rats were administered 300 mg/kg/day for 4 weeks an ethanolic extract of the pulp of *P. americana*, observing as a result of the decrease in BGL and glycated hemoglobin, increasing plasma insulin levels. In addition, Tabeshpour [71] mentions the evaluation of an aqueous extract of the avocado seed, administered in two doses of 300 and 600 mg in AID (rats), having a decrease in BGL, dose-dependent, the extract also has a protective effect of the pancreatic islets, which were damaged by inducing the disease.

An aqueous extract of *P. americana* seed achieves effects similar to glibenclamide, considered by Ezejiofor [24], who, in his experiment, administered 200, 300, and 400 mg/kg for three weeks to diabetic albino rats, observed a reduction of 58.9% at day 21 of rat BGLs.

Patala [58] used an aqueous-ethanolic extract of the avocado seed, which was obtained after drying and spraying 1 kg of *P. americana* seed and administered to 30 diabetic rats (*Rattus novergicus*) (SID rats). At first, they showed hyperglycemia after 42 days of treatment, with doses of 250, 300, and 350 mg/kg BGL of the rats was between 97.6 and 232.6 mg/dL, compared with metformin which decreases BGL by 67%, the highest doses reached this percentage too. It should be noted that the extract begins to reduce BGLs from day 14. Patala [58] considers that this is because the extract can restore the damage that pancreatic β cells present; likewise, they found that the saponins present can stimulate insulin secretion, which transports glucose from the blood to the tissues to decrease its concentration [58].

An analysis of the extracts of the avocado *P. americana* and one of the watermelons *Citrullus lanatus* was administered to AID rats, which presented, in addition to hyperglycemia, partial damage to the pancreas (100 mg/kg). The results of the avocado extract showed that the extract significantly reduced the amount of glucose in blood compared to the positive control used, insulin, mentioned Ukpabi [74], who attributes hypoglycemic activity to the increased need for glucose in the tissues or the inhibition of its absorption in the intestine [74].

Nardi [49] used an ethanolic extract of avocado seed *P. americana* to monitor the BGL of rats. Three different doses were administered, 300, 600, and 1200 mg/kg, having a negative group with distilled water; after 30 minutes, they were given 2 mg/kg of glucose. The amount of glucose in the blood was examined, and it found that the most significant decrease was the result of the group of rats administered with the lowest dose of the extract (–29.39 g/dL) [49].

A clinical study was carried out with 26 people with obesity who added half of *P. americana* avocado to their diet; as a result, blood insulin was decreased. Sabaté [66] concluded that D-mannoheptulose, present in the pulp of the avocado, inhibits hexokinase, an essential enzyme in the glycolysis process. In addition to this, the avocado intake provides an early feeling of satiety.

According to the studies above, avocado and its components could be added to the daily diet because it acts as a hypoglycemic agent, in addition to presenting other benefits for human health, such as those shown in Figure 1.2.

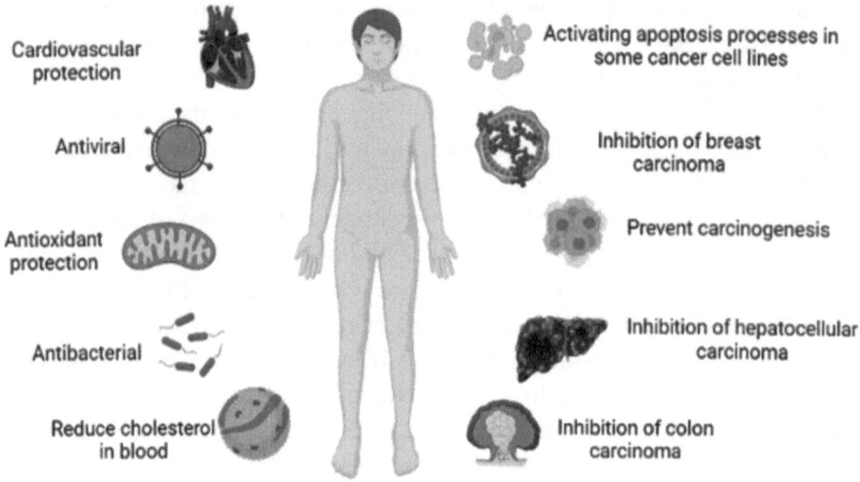

FIGURE 1.2 Biological activities of avocado extracts in human health.

1.6 ANTIMICROBIAL ACTIVITY

The avocado seed has shown huge potential as an antimicrobial. Currently, there is a great interest in studying the different properties of avocado extracts being applied as a treatment for different ailments. In another review, the antimicrobial efficacy of extracts on *Pseudomonas aeruginosa* and *Aspergillus niger*, demonstrating the efficacy of the antibacterial activity of avocado seed extracts, thus presenting another usefulness of these residues that should be exploited. In 2021, Cid-Perez and collaborators verified in their research those certain fatty acids present in avocado seed acetone extracts, such as linoleic acid, have antimicrobial activity, especially against *S. aureus*; however, a significant reduction was observed with the application of extracts on this same microorganism. It happens that these fatty acids alter the cell membrane increasing its oxidative stress and thus inhibiting the process of synthesis of the fatty acids of the bacteria. A study conducted by Villarreal-Lara et al. [76] demonstrated the antimicrobial capacity of acetogenins, lipid derivatives of avocado, on *Listeria monocytogenes* in a frozen food matrix; it indicated an efficient inhibition of spore-forming bacteria such as *C. perfringens* and *Bacillus subtilis*, comparing their effect with other commercial additives.

1.7 INSECTICIDAL ACTIVITY IN AVOCADO EXTRACTS

One of the most recent discoveries of avocado seed extracts is being used as a larvicide in pests of *Ae. Aegypti* in Venezuela, a place where the DENV is very persistent thanks to this vector; however, an investigation was carried out on how the extracts of avocado seed and leaf could be cytotoxic in the larvae of the third and fourth stages of maturation, giving. As a result, the avocado seed had a more toxic effect on the larvae.

Past studies mention that the seed behaves as a reserve organ, where active ingredients that cause cytotoxicity in larvae are possibly stored [13], while other studies show that the use of avocado seed extracts has cytotoxic effects on the intestine of *Anopheles larvae. Gambiae* is another vector that carries the malaria virus in Africa [41]. However, although it is not known which substance causes cytotoxicity in larvae, some researchers point out that the seed extracts contain steroids, triterpenes, sesquiterpene lactones, and fatty acids, among them which stand out 1,2,4-trihydroxy-nonadecane and 1,2,4-trihydroxy-heptadecane [52].

Meanwhile, other researchers concluded that larvicidal activity is mainly related to triterpenes and sesquiterpene lactones present in the avocado seed [61]. However, larvicidal activity may not be measured by a single molecule. The extracts may have molecules that interact, causing the death of the larva.

Briefly, the use of avocado extracts is a possible future option to produce cheaper and less polluting larvicides for nature since, currently, most of these synthetic insecticides cause damage to human health [38] in addition to Excessive use of these has generated resistance to organophosphates and pyrethroids [31].

In conclusion, the uses and applications of bioactive molecules from avocado residues have a wide range of effects in the study mentioned above, models with significant results in various areas, compared to residues from other fruits. Mainly, the effects of the biological activities described here can be attributed to its phenolic nature compounds, which generate action mechanisms related to antioxidant activities, favoring the trapping of free radicals and reducing cellular oxidative stress. All these results can benefit the population in terms of health and biological control. In addition, it generates a positive impact on the environment by using agro-industrial waste that generates significant pollutants.

ACKNOWLEDGMENTS

To the National Laboratory BIOBANCO UANL-UAdeC of CONAHCYT for the support provided.

KEYWORDS

- **anticancer**
- **anti-cancer activity**
- **antiviral**
- **avocado wastes**
- **bioactive compounds**
- **biological activities**
- **insecticidal activity**
- ***Persea americana***

REFERENCES

1. Agrela, I. F., Hidalgo, Y., & Herrera, F., (2014). Larvicide effect of methanolic extracts from seeds and leaves *Persea americana* (Laurales: *Lauraceae)* (avocado) on Aedes aegypti (*Diptera*: *Culicidae*). *Boletín de Malariología y Salud Ambiental, 54*(2), 199–207.
2. Alkhalaf, M. I., Alansari, W. S., Ibrahim, E. A., & Elhalwagy, M. E., (2019). Anti-oxidant, anti-inflammatory and anti-cancer activities of avocado (*Persea americana*) fruit and seed extract. *Journal of King Saud University-Science, 31*(4), 1358–1362.
3. Amado, D. A. V., Helmann, G. A. B., Detoni, A. M., Carvalho, S. L. C. D., Aguiar, C. M. D., Martin, C. A., & Cottica, S. M., (2019). Antioxidant and antibacterial activity and preliminary toxicity analysis of four varieties of avocado (*Persea americana* Mill.). *Brazilian Journal of Food Technology, 22.*
4. Ameer, K. (2016). Avocado is a major dietary source of antioxidants and its preventive role in neurodegenerative diseases. *The Benefits of Natural Products for Neurodegenerative Diseases*, 337–354.
5. Araújo, R. G., Rodriguez-Jasso, R. M., Ruiz, H. A., Govea-Salas, M., Pintado, M. E., & Aguilar, C. N., (2020). Process optimization of microwave-assisted extraction of bioactive molecules from avocado seeds. *Industrial Crops and Products, 154*, 112623.
6. Araújo, R. G., Rodriguez-Jasso, R. M., Ruiz, H. A., Pintado, M. M. E., & Aguilar, C. N., (2018). Avocado by-products: Nutritional and functional properties. *Trends in Food Science & Technology, 80*, 51–60.

7. Athaydes, B. R., Alves, G. M., De Assis, A. L. E. M., Gomes, J. V. D., Rodrigues, R. P., Campagnaro, B. P., & Gonçalves, R. D. C. R., (2019). Avocado seeds (*Persea americana* Mill.) prevent indomethacin-induced gastric ulcers in mice. *Food Research International, 119*, 751–760.
8. Awala, S. I., Ajayi, O. E., & Alabi, O. A., (2017). Evaluation of the *in vitro* antioxidant activities of leaves extracts of *Persea americana* and *Ficus exasperata* collected from Akure, Nigeria. *Evaluation, 17*(1).
9. Báez-Magaña, M., Ochoa-Zarzosa, A., Alva-Murillo, N., Salgado-Garciglia, R., & López-Meza, J. E., (2019). Lipid-rich extract from Mexican avocado seed (*Persea americana* var. drymifolia) reduces *Staphylococcus aureus* internalization and regulates innate immune response in bovine mammary epithelial cells. *Journal of Immunology Research, 2019*, 7083491. https://doi.org/10.1155/2019/7083491.
10. Bahru, T. B., Tadele, Z. H., & Ajebe, E. G., (2019). A review on avocado seed: Functionality, composition, antioxidant and antimicrobial properties. *Chemical Science International Journal, 27*(2), 1–10.
11. Benitez, J., Sánchez, A., Bolaños, C., Bernal, L., Ochoa, C., Velez, C., & Sandoval, A., (2021). Cambios fisicoquímicos del aguacate hass durante el almacenamiento frio y la maduración acelerada. *Biotecnología En El Sector Agropecuario y Agroindustrial, 19*(2), 41–56.
12. Bhuyan, D. J., Alsherbiny, M. A., Perera, S., Low, M., Basu, A., Devi, O. A., & Papoutsis, K. (2019). The odyssey of bioactive compounds in avocado (*Persea americana*) and their health benefits. *Antioxidants, 8*(10), 426.
13. Bobadilla, M., Zavala, F., Sisniegas, M., Zavaleta, G., Mostacero, J., & Taramona, L., (2005). Evaluación larvicida de suspensiones acuosas de *Annona muricata* Linnaeus «guanábana» sobre *Aedes aegypti* Linnaeus (Diptera, Culicidae). *Revista peruana de Biología, 12*(1), 145–152.
14. Broadhurst, C. L., (1997). Nutrition and non–insulin diabetes mellitus form by the combined administration of streptozotocin or alloxan and poly (adenosine diphosphate ribose). *Engg., 2*, 125.
15. Castro-López, C., Bautista-Hernández, I., González-Hernández, M. D., Martínez-Ávila, G. C., Rojas, R., Gutiérrez-Díez, A., & Aguirre-Arzola, V. E., (2019). Polyphenolic profile and antioxidant activity of leaf purified hydroalcoholic extracts from seven Mexican *Persea americana* cultivars. *Molecules, 24*(1), 173.
16. Cheriyath, P., Gorrepati, V. S., Peters, I., Nookala, V., Murphy, M. E., Srouji, N., & Fischman, D., (2010). High total bilirubin as a protective factor for diabetes mellitus: An analysis of NHANES data from 1999-2006. *Journal of Clinical Medicine Research, 2*(5), 201.
17. Chil-Núñez, I., Molina-Bertrán, S., Ortiz-Zamora, L., Dutok, C. M. S., & Souto, R. N. P., (2019). Estado del Arte de la especie *Persea americana* Mill (aguacate). *Amazonia Investiga, 8*(21), 73–86.
18. Cowan, A. K., & Wolstenholme, B. N., (2016). Avocado. *Encyclopedia of Food and Health*, 294–300.
19. Dabas, D., Shegog, R. M., Ziegler, G. R., & Lambert, J. D., (2013). Avocado (*Persea americana*) seed is a source of bioactive phytochemicals. *Current Pharmaceutical Design, 19*(34), 6133–6140.
20. Dreher, M. L., & Davenport, A. J., (2013). Hass avocado composition and potential health effects. *Critical Reviews in Food Science and Nutrition, 53*(7), 738–750.

21. Duarte, P. F., Chaves, M. A., Borges, C. D., & Mendonça, C. R. B., (2016). Avocado: Characteristics, health benefits, and uses. *Ciência Rural, 46*, 747–754.
22. Egbuonu, A. C. C., et al., (2017). Vitamins composition and antioxidant properties in normal and monosodium glutamate-compromised rats' serum of *Persea americana* (avocado pear) seed. *Open Access Journal of Chemistry, 1*, 19–24.
23. Ejiofor, N. C., Ezeagu, I. E., Ayoola, M., & Umera, E. A. (2018). Determination of the chemical composition of avocado (*Persea americana*) seed. *Adv. Food Technol. Nutr. Sci. Open J.*
24. Ezejiofor, A. N., Okorie, A., & Orisakwe, O. E., (2013). Hypoglycaemic and tissue-protective effects of the aqueous extract of *Persea americana* seeds on alloxan-induced albino rats. *The Malaysian Journal of Medical Sciences: MJMS, 20*(5), 31.
25. Figueroa, J. G., Borrás-Linares, I., Lozano-Sánchez, J., & Segura-Carretero, A., (2018). Comprehensive characterization of phenolic and other polar compounds in the seed and seed coat of avocado by HPLC-DAD-ESI-QTOF-MS. *Food Research International, 105*, 752–763.
26. Figueroa, J. G., Borrás-Linares, I., Lozano-Sánchez, J., & Segura-Carretero, A., (2018). Comprehensive identification of bioactive compounds of avocado peel by liquid chromatography coupled to ultra-high-definition accurate-mass Q-TOF. *Food Chemistry, 245*, 707–716.
27. Fisher, D. T., Appenheimer, M. M., & Evans, S. S., (2014). The two faces of IL-6 in the tumor microenvironment. In: *Seminars in Immunology* (Vol. 26, No. 1, pp. 38–47). Academic Press.
28. Flores, M., Saravia, C., Vergara, C. E., Avila, F., Valdés, H., & Ortiz-Viedma, J., (2019). Avocado oil: Characteristics, properties, and applications. *Molecules, 24*(11), 2172.
29. Folasade, O. A., Olaide, R. A., & Olufemi, T. A., (2016). Antioxidant properties of *Persea americana* M. seed as affected by different extraction solvent. *Journal of Advances in Food Science & Technology, 3*(2), 101–106.
30. Gallegos-Zurita, M., & Gallegos, D., (2017). Plantas medicinales utilizadas en el tratamiento de enfermedades de la piel en comunidades rurales de la provincia de Los Ríos Ecuador. In: *Anales de la Facultad de Medicina* (Vol. 78, No. 3, pp. 315–321). UNMSM. Facultad de Medicina.
31. Grisales, N., Poupardin, R., Gomez, S., Fonseca-Gonzalez, I., Ranson, H., & Lenhart, A., (2013). Temephos resistance in *Aedes aegypti* in Colombia compromises dengue vector control. *PLoS Neglected Tropical Diseases, 7*(9), e2438.
32. Hegazy, G. E., Abu-Serie, M. M., Abo-Elela, G. M., Ghozlan, H., Sabry, S. A., Soliman, N. A., & Abdel-Fattah, Y. R., (2020). *In vitro* dual (anticancer and antiviral) activity of the carotenoids produced by haloalkaliphilic archaeon *Natrialba* sp. M6. *Scientific Reports, 10*(1), 1–14.
33. Henry, L. N., Mtaita, U. Y., & Kimaro, C. C., (2015). Nutritional efficacy of avocado seeds. *Glob. J. Food Sci. Technol., 3*(5), 192–196.
34. Hossain, M. B., Rai, D. K., Brunton, N. P., Martin-Diana, A. B., & Barry-Ryan, C., (2010). Characterization of phenolic composition in *Lamiaceae* spices by LC-ESI-MS/MS. *Journal of Agricultural and Food Chemistry, 58*(19), 10576–10581.
35. Hurtado-Fernández, E., Fernández-Gutiérrez, A., & Carrasco-Pancorbo, A., (2018). Avocado fruit—*Persea americana. Exotic Fruits*, 37–48.
36. Jiménez, C. P. A., Hernández, J. M., Espinosa, S. G., Mendoza, C. G., & Bell, T. A. M., (2015). Los saberes en medicina tradicional y su contribución al desarrollo rural:

Estudio de caso Región Totonaca, Veracruz. *Revista Mexicana de Ciencias Agrícolas, 6*(8), 1791–1805.
37. Jiménez, P. P., García, C. P., Quitral, V., Vásquez, K., Parra, R. C., Reyes, F. M., & Soto, C. J., (2020). Pulp, leaf, peel, and seed of avocado fruit: a review of bioactive compounds and healthy benefits. *Food Review International, 37*(6), 619–655.
38. Kamaraj, C., Bagavan, A., Elango, G., Zahir, A. A., Rajakumar, G., Marimuthu, S., & Rahuman, A. A., (2011). Larvicidal activity of medicinal plant extracts against *Anopheles subpictus* & *Culex tritaeniorhynchus*. *The Indian Journal of Medical Research, 134*(1), 101.
39. Karasawa, M. M. G., & Mohan, C., (2018). Fruits as prospective reserves of bioactive compounds: A review. *Natural Products and Bioprospecting, 8*(5), 335–346.
40. Kosińska, A., Karamać, M., Estrella, I., Hernández, T., Bartolomé, B., & Dykes, G. A., (2012). Phenolic compound profiles and antioxidant capacity of *Persea americana* Mill. peels and seeds of two varieties. *Journal of Agricultural and Food Chemistry, 60*(18), 4613–4619.
41. Koua, H. K., Han, S. H., & D'Almeida, M. A., (1998). Histopathology of *Anopheles gambiae* s.l Giles, 1902 (*Diptera, Culicidae*) subjected to the larvicidal activity of the aqueous extract of *Persea americana* Miller, 1768 (*Lauraceae*). *Bulletin de la Societe de Pathologie Exotique (1990), 91*(3), 252–256.
42. Lamacchia, C., (2015). An innovative method for the detoxification of gluten proteins from grains of cereals. *Journal of Food Processing & Technology, 6*(5), 72–83.
43. Lara-Flores, A. A., Araújo, R. G., Rodríguez-Jasso, R. M., Aguedo, M., Aguilar, C. N., Trajano, H. L., & Ruiz, H. A., (2018). Bioeconomy and biorefinery: Valorization of hemicellulose from lignocellulosic biomass and potential use of avocado residues as a promising resource of bioproducts. In: *Waste to Wealth* (pp. 141–170). Springer, Singapore.
44. Lara-Márquez, M., Báez-Magaña, M., Raymundo-Ramos, C., Spagnuolo, P. A., Macías-Rodríguez, L., Salgado-Garciglia, R., & López-Meza, J. E., (2020). Lipid-rich extract from Mexican avocado (*Persea americana* var. drymifolia) induces apoptosis and modulates the inflammatory response in Caco-2 human colon cancer cells. *Journal of Functional Foods, 64*, 103658. https://doi.org/10.1016/j.jff.2019.103658.
45. López-Cobo, A., Gómez-Caravaca, A. M., Pasini, F., Caboni, M. F., Segura-Carretero, A., & Fernández-Gutiérrez, A., (2016). HPLC-DAD-ESI-QTOF-MS and HPLC-FLD-MS as valuable tools for the determination of phenolic and other polar compounds in the edible part and by-products of avocado. *LWT, 73*, 505–513.
46. Mahadeva, R. U. S., (2017). Salutary potential of ethanolic extract of avocado fruit on anomalous carbohydrate metabolic key enzymes in hepatic and renal tissues of hyperglycaemic albino rats. *Chinese Journal of Integrative Medicine*, 1–7.
47. Manal, S., & Sahar, S., (2013). Effects of bioactive component of kiwi fruit and avocado (fruit and seed) on hypercholesterolemic rats. *World Journal of Dairy & Food Sciences, 8*(1), 82–93.
48. Melgar, B., Dias, M. I., Ciric, A., Sokovic, M., Garcia-Castello, E. M., Rodriguez-Lopez, A. D., & Ferreira, I. C., (2018). Bioactive characterization of *Persea americana Mill.* by-products: A rich source of inherent antioxidants. *Industrial Crops and Products, 111*, 212–218.
49. Nardi, L., Lister, I. N., Girsang, E., & Fachrial, E., (2020). Hypoglycemic effect of avocado seed extract (*Persean americana Mill*) from analysis of oral glucose tolerance test on

Rattus norvegicus L. *American Academic Scientific Research Journal for Engineering, Technology, and Sciences, 65*(1), 49–56.

50. Niogret, J., Epsky, N. D., Schnell, R. J., Boza, E. J., Kendra, P. E., & Heath, R. R., (2013). Terpenoid variations within and among half-sibling avocado trees, *Persea americana Mill. (Lauraceae). PLoS One, 8*(9), e73601.
51. Noyko, S. M. D., Tashko, D. M. D., & Jovcho, J. M. D. (2010). *Advanced Colorectal Cancer Is Associated with Enhanced Il-23 and IL-10 Serum Levels. Laboratory Medicine, 41*(3), 159–163.
52. Oberlies, N. H., Rogers, L. L., Martin, J. M., & McLaughlin, J. L., (1998). Cytotoxic and insecticidal constituents of the unripe fruit of *Persea americana. Journal of Natural Products, 61*(6), 781–785.
53. Oboh, G., Adelusi, T. I., & Akinyemi, A. J., (2013). Inhibitory effect of phenolic extract from leaf and fruit of avocado pear (*Persea americana*) on Fe^{2+} induced lipid peroxidation in rats' pancreas *in vitro. FUTA J. Res. Sci., 2*, 276–286.
54. Oboh, G., Isaac, A. T., Akinyemi, A. J., & Ajani, R. A., (2014). Inhibition of key enzymes linked to type 2 diabetes and sodium nitroprusside induced lipid peroxidation in rats' pancreas by phenolic extracts of avocado pear leaves and fruit. *International Journal of Biomedical Science: IJBS, 10*(3), 208–216.
55. Oboh, G., Odubanjo, V. O., Bello, F., Ademosun, A. O., Oyeleye, S. I., Nwanna, E. E., & Ademiluyi, A. O., (2016). Aqueous extracts of avocado pear (*Persea americana* Mill.) leaves and seeds exhibit anti-cholinesterases and antioxidant activities *in vitro. Journal of Basic and Clinical Physiology and Pharmacology, 27*(2), 131–140.
56. Oktaria, Y. E., Azizah, T., & Sutrisna, E. M., (2015). The hypoglycemic effect of avocado seed (*Persea Americana Mill*) and histopathologic profile. *Journal Pharm. Bio. Sci., 6*(4), 136–141.
57. Oviedo-Boyso, J., Valdez-Alarcón, J. J., Cajero-Juárez, M., Ochoa-Zarzosa, A., López-Meza, J. E., Bravo-Patiño, A., & Baizabal-Aguirre, V. M., (2007). The innate immune response of the bovine mammary gland to pathogenic bacteria is responsible for mastitis. *Journal of Infection, 54*(4), 399–409.
58. Patala, R., Dewi, N. P., & Pasaribu, M. H., (2020). Efektivitas ekstrak etanol biji alpukat (*Persea americana* Mill.) terhadap kadar glukosa darah tikus putih jantan (*Rattus novergicus)* model hiperkolesterolemia-diabetes: Effectiveness of ethanol extract of avocado seeds (*Persea americana* mill.) on blood glucose levels of male white rat (*Rattus novergicus)* hipercolesterolemia-diabetes model. *Jurnal Farmasi Galenika (Galenika Journal of Pharmacy) (e-Journal), 6*(1), 7–13. https://doi.org/10.22487/j24428744.2020.v6.i1.13929.
59. Pérez, Á. S., Ávila, Q. G., & Coto, A. A., (2015). Avocado (*Persea americana* Mill). *Cultivos Tropicales, 36*(2), 111–123.
60. Rahman, N., & Dewi, N. U. B., (2018). Phytochemical and antioxidant activity of avocado leaf extract (*Persea americana* Mill.). *Asian Journal of Scientific Research, 11*(3), 357–363.
61. Ramos-Jerz, M. D. R., Villanueva, S., Jerz, G., Winterhalter, P., & Deters, A. M., (2013). *Persea americana* Mill. seed: Fractionation, characterization, and effects on human keratinocytes and fibroblasts. *Evidence-Based Complementary and Alternative Medicine, 2013*.
62. Rodríguez-Carpena, J. G., Morcuende, D., Andrade, M. J., Kylli, P., & Estévez, M., (2011). Avocado (*Persea americana* Mill.) phenolics, *in vitro* antioxidant and antimicrobial

activities, and inhibition of lipid and protein oxidation in porcine patties. *Journal of Agricultural and Food Chemistry, 59*(10), 5625–5635.

63. Rodríguez-López, C. E., Hernández-Brenes, C., Treviño, V., & Díaz De La, G. R. I., (2017). Avocado fruit maturation and ripening: Dynamics of aliphatic acetogenins and lipidomic profiles from mesocarp, idioblasts and seed. *BMC Plant Biology, 17*(1). doi: 10.1186 /s12870-017-1103-6.
64. Rosero, J. C., Cruz, S., Osorio, C., & Hurtado, N., (2019). Analysis of phenolic composition of byproducts (seeds and peels) of avocado (*Persea americana* Mill.) cultivated in Colombia. *Molecules, 24*(17), 3209.
65. Saavedra, J., Córdova, A., Navarro, R., Díaz-Calderón, P., Fuentealba, C., Astudillo-Castro, C., & Galvez, L., (2017). Industrial avocado waste: Functional compounds preservation by convective drying process. *Journal of Food Engineering, 198*, 81–90.
66. Sabaté, J., Wien, M., & Haddad, E., (2015). Post-ingestive effects of avocados in meals on satiety and gastric hormone blood levels. *Human Health Nut., 2015*, 459–461.
67. Segovia, F. J., Hidalgo, G. I., Villasante, J., Ramis, X., & Almajano, M. P., (2018). Avocado seed: A comparative study of antioxidant content and capacity in protecting oil models from oxidation. *Molecules, 23*(10), 2421.
68. Setyawan, H. Y., Sukardi, S., & Puriwangi, C. A., (2021). Phytochemicals properties of avocado seed: A review. In: *IOP Conference Series: Earth and Environmental Science* (Vol. 733, No. 1, p. 012090). IOP Publishing.
69. Soemardji, A. A., Umar, M. H., & Fidrianny, I., (2016). Lipid profile and platelet aggregation of ethanolic seed extract of avocado (*Persea americana* mill.) in hyperlipidemic male Wistar rat. *Asian Journal of Pharmaceutical and Clinical Research*, 143–147.
70. Soledad, C. P. T., Paola, H. C., Enrique, O. V. C., Israel, R. L. I., Guadalupe Virginia, N. M., & Raúl, Á. S., (2021). Avocado seeds (*Persea americana* cv. Criollo sp.): Lipophilic compounds profile and biological activities. *Saudi Journal of Biological Sciences, 28*(6), 3384–3390.
71. Tabeshpour, J., Razavi, B. M., & Hosseinzadeh, H., (2017). Effects of avocado (*Persea americana*) on metabolic syndrome: A comprehensive systematic review. *Phytotherapy Research, 31*(6), 819–837.
72. Tremocoldi, M. A., Rosalen, P. L., Franchin, M., Massarioli, A. P., Denny, C., Daiuto, É. R., & Alencar, S. M. D., (2018). Exploration of avocado by-products as natural sources of bioactive compounds. *PloS One, 13*(2), e0192577.
73. Uchenna, U. E., Shori, A. B., & Baba, A. S., (2017). Inclusion of avocado (*Persea americana*) seeds in the diet to improve carbohydrate and lipid metabolism in rats. *Revista Argentina de Endocrinología y Metabolismo, 54*(3), 140–148.
74. Ukoabi, C. F., Minaseidiema, R., Anyaogu, E. J., Njoku, C. I., & Esihe, T. E. (2020). Amelioration of hyperglycemia and dyslipidemia in alloxan-induced diabetic rats using *Citrullus lanatus* and *Persea Americana*. *Journal of Life Sciences Research, 7*(1), 1–7.
75. Velderrain-Rodríguez, G. R., Quero, J., Osada, J., Martín-Belloso, O., & Rodríguez-Yoldi, M. J., (2021). Phenolic-rich extracts from avocado fruit residues as functional food ingredients with antioxidant and antiproliferative properties. *Biomolecules, 11*(7), 977.
76. Vilhelmova-Ilieva, N., Galabov, A. S., & Mileva, M. (2019). Tannins as antiviral agents. *Tannins-Structural Properties, Biological Properties and Current Knowledge.*
77. Villa-Rodriguez, J. A., Yahia, E. M., González-León, A., Ifie, I., Robles-Zepeda, R. E., Domínguez-Avila, J. A., & González-Aguilar, G. A., (2020). Ripening of 'Hass'

avocado mesocarp alters its phytochemical profile and the *in vitro* cytotoxic activity of its methanolic extracts. *South African Journal of Botany, 128*, 1–8.

78. Villarreal-Lara, R., Rodríguez-Sánchez, D. G., Díaz De La, G. R. I., García-Cruz, M. I., Castillo, A., Pacheco, A., & Hernández-Brenes, C., (2019). Purified avocado seed acetogenins: Antimicrobial spectrum and complete inhibition of *Listeria monocytogenes* in a refrigerated food matrix. *CyTA-Journal of Food, 17*(1), 228–239.
79. Vítek, L., (2012). The role of bilirubin in diabetes, metabolic syndrome, and cardiovascular diseases. *Frontiers in Pharmacology, 3*, 55.
80. Weschenfelder, C., Dos Santos, J. L., De Souza, P. A. L., De Campos, V. P., & Marcadenti, A., (2015). Avocado and cardiovascular health. *Open Journal of Endocrine and Metabolic Diseases, 5*(7), 77.
81. Wu, Y. H., Tseng, C. K., Wu, H. C., Wei, C. K., Lin, C. K., Chen, I. S., & Lee, J. C., (2019). Avocado (*Persea americana*) fruit extract (2R, 4R)-1, 2, 4-trihydroxyheptadec-16-yne inhibits dengue virus replication via upregulation of NF-κB–dependent induction of antiviral interferon responses. *Scientific Reports, 9*(1), 1–10.
82. Yasir, M., Das, S., & Kharya, M. D., (2010). The phytochemical and pharmacological profile of *Persea americana* Mill. *Pharmacognosy Reviews, 4*(7), 77.
83. Zhao, C. N., Meng, X., Li, Y., Li, S., Liu, Q., Tang, G. Y., & Li, H. B., (2017). Fruits for prevention and treatment of cardiovascular diseases. *Nutrients, 9*(6), 598.

CHAPTER 2

Valorization of Brown Algae Biomass and By-Products

LÁZARO A. GONZÁLEZ FERNÁNDEZ,[1]
NAHUM ANDRÉS MEDELLÍN CASTILLO,[2]
AMADO ENRIQUE NAVARRO FRÓMETA,[3]
CANDY CARRANZA ÁLVAREZ,[4] ROGELIO FLORES RAMÍREZ,[5]
PAOLA ELIZABETH DÍAZ FLORES,[6] LEONARDO SEPÚLVEDA TORRE,[7]
and NANCY VERÓNICA PÉREZ AGUILAR[7]

[1]*Multidisciplinary Postgraduate Program in Environmental Sciences, Autonomous University of San Luis Potosí, Av. Manuel Nava 201, Zona Universitaria, San Luis Potosí, S.L.P., México*

[2]*Postgraduate Studies and Research Center, School of Engineering, Autonomous University of San Luis Potosí, Av. Manuel Nava 8, Zona Universitaria, San Luis Potosí, S.L.P., México*

[3]*Technological University of Izucar of Matamoros, Prolongación Reforma 168, Santiago Mihuacán, Izúcar de Matamoros, Puebla, México*

[4]*School of Professional Studies Huasteca Zone, Romualdo del Campo 501 Fracc. Rafael Curiel, Cd Valles, S.L.P., México*

[5]*Coordination for the Innovation and Application of Science and Technology, Av. Sierra Leona 550, Col. Lomas 2a. Sección, San Luis Potosí, S.L.P., México*

[6]*School of Agronomy and Veterinary Medicine, Carretera San Luis Potosí–Matehuala Km. 14.5, Ejido Palma de la Cruz, Soledad de Graciano Sánchez, S.L.P., México*

[7]*School of Chemistry, Autonomous University of Coahuila, Blvd. V. Carranza s/n. Col. República Oriente C. P., Saltillo, Coah., México*

Bioresources and Bioprocess in Biotechnology for a Sustainable Future. Leonardo Sepúlveda Torre, Juan Carlos Contreras-Esquivel, Ann Rose Abraham, & A. K. Haghi (Eds.)

ABSTRACT

The valorization of brown algae biomass and its by-products is a topic of growing interest due to its potential as a source of renewable energy and high value-added products. Brown algae are a type of marine algae found in cold and temperate waters around the world. The biomass of brown algae can be exploited in a variety of ways. One of the most promising applications is the production of biofuels from the sugars and lipids found in the algae. These biofuels can be used as a sustainable alternative to fossil fuels in sectors such as transportation and electricity generation. In addition to biofuels, brown algae can also be used as a feedstock to produce high value-added chemicals. For example, the polysaccharides present in brown algae have gelling and thickening properties and are therefore used in the food and cosmetics industry. Other brown algae by-products include bioactive compounds with antioxidant, antibacterial and antiviral properties, which have applications in the pharmaceutical and functional food industry. In addition to their potential as a source of energy and high value-added chemicals, brown algae also offer significant environmental benefits. The most important of these is that they are bioindicator species for environmental pollution. Because of their ability to bind heavy metals and other chemical species in water, they have been widely used as an indicator of pollution in the environments in which they grow. Additionally, they have been used as adsorbents of these pollutants due to the properties of their cell wall. For all these reasons, brown algae are at the center of novel environmental applications that are currently being developed. Both the biomass itself and its by-products are materials of great economic and environmental value and will continue to be so in the coming years.

2.1 INTRODUCTION

Although the term algae are used generically to refer to aquatic plants and give the impression of defining a homogeneous group of plants, the truth is that it comprises the most varied, complex, and plastic group (morphologically, biochemically, and physiologically) of the Vegetal kingdom. The macroalgae and seagrass beds have been used for centuries as green manure (or semi-composted) in almost all coastal agricultural areas and, above all, island areas, where they ensure that their use exempts them from practicing crop rotation. In some North Sea islands, they have even formed the basis of the existence of agriculture since the agricultural land has been (and continues) manufacturing man mixing sand and silt with the arriving macroalgae [44].

Despite being a relatively unknown group, algae offer many possibilities in terms of their use. Perhaps the best known is the one they have as food, and although this culture is not extended, in some countries, especially on the Asian coast, they are a daily ingredient in their dishes. They have also been used as fertilizers, and in many coastal towns, it is possible to see people collecting algae to give them these uses. With the arrival of new technologies, they have become used industrially as a source of chemical products or in the production of shampoos or creams, while the food industry has incorporated them into everyday foods such as custards and ice cream or jams. Its use also covers the field of medicine, in which, since ancient times, they have been used to combat diseases and all kinds of illnesses [1].

Marine algae are innocuous species with an inorganic content rich in calcium, magnesium, sodium, and potassium, which are identified in cellular processes. Consequently, algae-based biosorbents can be accepted when applied to water cleaning or wastewater treatment.

In coastal regions, seaweeds are fundamental primary producers, and several species are being considered as raw material for their diversity of products of economic importance, which has resulted in an increase in their demand. Therefore, it is necessary to monitor the bioaccumulation of certain xenobiotics in them because, for example, there are some that are used for direct or indirect human consumption or for livestock consumption [17].

2.2 BROWN ALGAE AND FOOD PRODUCTS

There is scientific evidence showing that incorporating seaweed and/or seaweed isolates in food matrices can positively affect the nutritional, organoleptic, textural, healthy, and even improved preservation characteristics of foods and beverages. Seaweeds that are consumed directly can belong to several groups of algae. Among them, the most important are red algae (*Rhodophyta*, Figure 2.1(a)), brown algae (*Phaeophyta*), and green algae (*Chlorophyta*). All of them live in very diverse environments with extreme changes in salinity, temperature, lighting, and nutrients, with an extraordinary ability to adapt [41].

Specifically, brown algae are a rich source of iodine. This is an aspect of great interest due to the recent importance of iodine deficiency in European population groups. In addition, they have interesting antioxidant, antimicrobial, and anti-inflammatory properties that can affect lowering cholesterol and blood pressure or help with digestion and weight control. Despite its multiple advantages, seaweed also poses some limitations at a nutritional level, so food technologists must work on advances that allow overcoming the current obstacles for its incorporation as a valuable ingredient in more fortified foods and healthy [20].

FIGURE 2.1 The most important algae groups: (a) red algae; (b) brown algae; and (c) green algae.

For example, the iodine content of raw brown seaweed may be too high for daily intake. In addition, they have a high salt content, which limits the amounts that can be incorporated into food products and dishes. It should also be noted that depending on the location of the algae culture, they can also accumulate certain heavy metals. Another important aspect to consider in the application of algae as an ingredient in food is the bioavailability of its proteins or other nutrients present in its composition. This bioavailability is closely related to its structural properties and the possible presence of antinutrient compounds, thus limiting its absorption after ingestion [4].

All these elements show the need for technological innovations in cultivation, processing, and subsequent technological adaptations to optimize this raw material as a feed ingredient. These innovations seek to improve the food safety, digestibility, nutritional value, and flavor of brown algae.

The incorporation of algae as an ingredient in foods favors their texture and palatability, improves its physical properties, and performance in certain products, and enables the substitution of certain additives. Thus, proteins and fibers favor solubility, water retention capacity, emulsifying activity, foam stability, viscosity, and gelation [28].

For example, the functionality of algae in processed meat products, such as hamburgers, sausages, and meat emulsions, has multiple benefits from a technological point of view. Algae can influence pH due to acidic components such as fucoidans and alginic acid; in stability thanks to the antioxidants that retard the rancidity of these products; its high content of mineral salts allows to reduce the addition of salt and improve the properties of water retention in the food, in addition to reducing the percentage of saturated fats [32]. Its application in cereal-based products, such as bread, flour or pasta, is also notable, where it improves the interaction between starch granules and protein matrix [37].

On the other hand, the use of algae in food has been more focused on incorporating its extracts, rich in certain components, into food for human consumption. The most obvious example is the *hydrocolloids*, which are the structural component of the cell walls of algae. It is a set of polysaccharides that contribute to stabilizing emulsions, suspensions, and foams, in addition to controlling the growth of crystals [19].

2.3 BROWN ALGAE AS BIOINDICATORS

In aquatic ecosystems, pollution by organic or inorganic sources causes a series of physicochemical changes in the water, which affect the composition and distribution of communities [27]. In aquatic organisms, the effects of being subjected to a toxic discharge occur over time, from individual responses (biochemical and physiological) to population, community, and ecosystem responses; and the magnitude of the changes registered in the organisms depends on the duration of the disturbance of the initial conditions of the aquatic system, its intensity and nature [31].

In recent years, a large number of environmental agencies around the world use methods for evaluating water quality based on the use of biological communities [2]. The reasons for the use of living organisms to monitor water quality are mainly the low cost and ease of implementing this type of study, compared to expensive chemical or toxicity analyses.

Bioindicators constitute a large group of plant, fungal, or animal species whose presence or status in a given ecosystem provides information on certain ecological characteristics of the ecosystem or the possible environmental impact of some practices on it. These are mainly used for the evaluation of the environmental quality of ecosystems. All bioindicators must meet a series of requirements for their use, such as dispersion and abundance in the territory, sedentary lifestyle, and tolerating pollutants in concentrations similar to those of the contaminated ecosystem without lethal effects [23, 29, 39, 40].

Algae are frequently used as bioindicators of water quality since they are easily sampled and indicate the presence of contaminants quite clearly, due to their sensitivity to changes in the characteristics of the environment they inhabit [23].

The most used are diatoms, which are part of the phytoplankton of both marine and freshwater environments. They show the trophic state of the environment according to its abundance (from ultraoligotrophic to hyper-eutrophic), as well as the mineralization of the water or its salinity. They are also very sensitive to water acidification and bioaccumulate a wide variety of contaminants. Being microorganisms, they detect short-term changes and present a rapid response to those changes [30].

Macroalgae are also used as bioindicators, generally in the marine environment. Like diatoms, they bioaccumulate various pollutants, such as heavy metals or PAHs. This property means that species of the *Fucus*, *Ulva*, *Ascophyllum*, or *Enteromorpha* genera are used as sentinel organisms against certain pollutants. Being phototrophic organisms, their greater or lesser abundance can also indicate the transparency of the water in the environment [35].

In a work carried out by Méndez-Rodríguez et al. [24] on the accumulation of metals in macroalgae reported that Fe, Mn, and Zn were the most abundant elements in all species analyzed. Next came Cu, Ni, Co, and Cd. However, the levels of most of the metals analyzed varied widely, depending on the collection sites and the particular species. Also, Haritonidis & Malea [16] carried out a study on the bioaccumulation of metals in the macroalgae Ulva rigida in the Thermaikos Gulf (Greece). The temporal and spatial trend of contamination provided by the results on the detected concentrations of metals in macroalgae may not correspond to that evaluated in other bioindicator species, such as mollusks and barnacles.

2.4 HEAVY METALS BIOSORPTION WITH BROWN ALGAE BIOMASS

The great diversity of marine algae allows it to increase its selectivity and efficiency. Different adsorption capacities and selectivity have been discovered by red, green, and brown algae against various heavy metals [25]. The chemical composition and presence of different adsorption centers (fucanoids, alginates, phosphated proteins, etc.) [34] allow greater adsorption of certain metals due to their size, degree of solvation, presence of chelating ions, molecular sieves, ion exchange with species present in the algae, etc. [11, 34].

The mechanisms by which heavy metals are fixed in brown algae have been studied by Crist et al. [7] and Ragan [33]. These state that they incorporate these metals in high concentrations through polyanionic polysaccharides present in the cell wall and by the affinity between metals and polyphenols. Finally, the high content of alginates (Figure 2.2) in seaweeds (compared to the other functional groups identified as adsorption centers) makes them ideal models to identify the biosorption mechanism, specially to investigate metal-algae interactions at the molecular level [36].

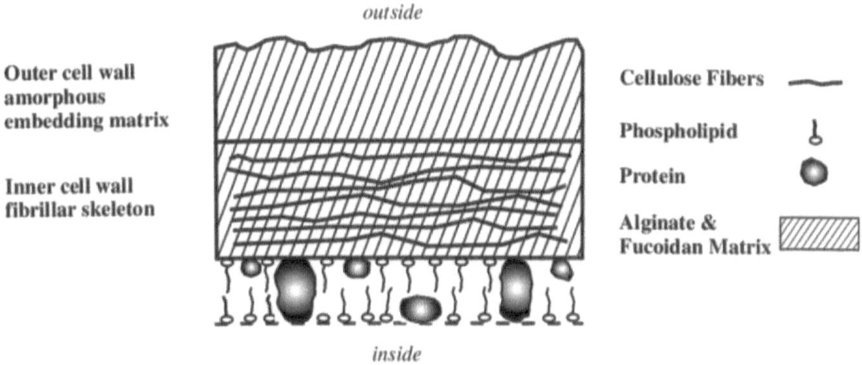

FIGURE 2.2 Cell wall structure in brown algae.

Source: Reprinted from Ref. []. Copyright © 2003 Published by Elsevier Ltd.

The existence of a single functional group responsible for the adsorption of heavy metals allows its mechanism to be clearly elucidated by means of different techniques, such as the determination of the ionization constant [26] of algae, ionic strength effect, desorption of adsorbed metals by chelating species and acids, ion exchange, infrared spectroscopy, scanning electron microscopy, etc. Marine algae are perhaps the only adsorbents whose adsorption capacity is due exclusively to alginates in more than 90% [9, 10, 15, 21, 42].

Marine algae are a promising biosorbent of heavy metals and various pollutants and, due to their intrinsic characteristics, have received increasing attention in recent decades. Unfortunately, despite its recent development, biosorption is reduced to discontinuous processes at the laboratory level. The biosorption mechanism is being elucidated, but the type of metal ion-adsorbent interaction and its factors are unknown. Likewise, the optimal conditions reached are not completely adjusted to the conditions of conventional wastewater (ionic strength, interfering ions, detergents, acidity, organic content, etc.).

Algae arouse superior interest in developing new biosorbent materials, not only because of their adsorption capacity but also because they are present in abundant quantities and are easily accessible in seas and oceans. Still, there are only a few publications on biosorption using algae in relation to the investigations using other biomaterials; moreover, there are a few regarding multimetallic systems and dynamical systems. According to the literature, brown algae have a higher metal adsorption capacity than red and green algae [6].

In the near future, many aspects should be considered regarding the adsorption mechanism, such as studying it more intensively using various techniques and combinations of them, as well as the factors that influence the equilibrium state, such as pH, temperature, and the role that the co-ion (anion) plays in the biosorption process.

Better mathematical models of equilibrium and kinetics are also required that are adapted to real conditions, including parameters that conventional models neglect. This could be strengthened by a better application of biosorption technology, such as improving the physical-chemical conditions with pH and ionic content similar to those of real wastewater, and finally analyzing the possibility of recovering the heavy metal after being removed from the wastewater solution and adsorbed on the biosorbent.

Even though the biosorption of heavy metals by marine algae has been extensively studied during the last decades, there are still many unanswered questions and aspects to be determined for its complete optimization, but what can be sure is that the use of marine algae is one of the best ecological weapons we have for the decontamination of our planet.

Algae bind only the free metal ions through two physical-chemical processes. The first process is rapid and reversible and involves adsorption of the metal ion on the outer surface of the cell wall. This process can be ionic or by complex formation with cell wall ligands. The polymers that make up the cell wall are rich in carboxylic, phosphoryl, hydroxyl, and aromatic groups that can bind cations or produce organic complexes that can influence metal absorption. The second mechanism of metal incorporation is slower, it is regulated by cellular metabolism, and metals are stored in the cytoplasm in vacuoles rich in polyphenols [14].

As a product of this accumulation, the algae can reach trace element contents several orders of magnitude higher than in the water. Another version of how heavy metals are bound to algae was studied by Crist et al. [8] and Ragan [33], who state that these metals are incorporated in large concentrations through polyanionic polysaccharides present in the cell wall, and by affinity between metals and polyphenols (Figure 2.3).

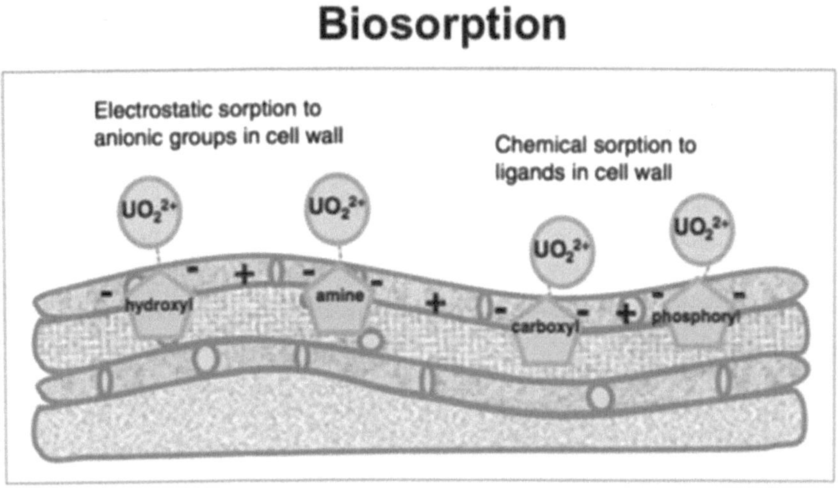

FIGURE 2.3 Mechanisms by which contaminants bind to adsorbents during the biosorption process.

Source: Adapted from Wikipedia. Image by Xavier Dengra. https://creativecommons.org/licenses/by-sa/3.0/deed.en

2.5 ALGAL BIO-OIL

The great dependence on energy and the high pollution generated by fossil fuels has led us to promote the development of renewable energy sources with a low environmental impact. This is how biofuels are presented as a great alternative to fossil fuels since fuels such as biodiesel, biogas, and bioethanol can be produced from biomass [38].

One of the alternatives that have attracted much interest in recent decades is the use of algae biomass to produce bioethanol. Algae have high concentrations of carbohydrates in the form of polysaccharides (sugar polymers), which are released after fragmentation using enzymes and can be fermented to bioethanol [22].

Algae can be processed differently to obtain a broad spectrum of products. For some time now, its use as an alternative to current biomass raw materials for the generation of biofuels has been gaining increasing interest among the scientific community, businesspeople, and the general public. In fact, algae are recognized as a potential source for biodiesel production due to their high oil content and rapid growth. However, this is not the only application [43].

Biofuel extraction technology is relatively simple. It includes a first pressing stage with which approximately 70% of the oil is extracted and a second stage with an organic solvent reaches up to 99%, although the latter makes the process more expensive [5]. Given the high viscosity that the original virgin oil reaches, it can be used directly in diesel engines once they have been adapted to operate with this highly viscous fuel. The triglycerides that makeup vegetable oils can also be transformed into monoesters and glycerin by means of the trans-esterification reaction with methanol. The molecules that make up the biodiesel resulting from this last process have a lower molecular weight, and its viscosity is substantially lower, so it can be used as fuel in compression engines. Obviously, the biodiesel obtained by either of the two routes does not contain sulfur, is not toxic, and is easily biodegradable [12].

2.6 BROWN ALGAE AS FERTILIZERS

The use of algae as a fertilizer date back to the 19th century, when the inhabitants of the coasts collected the large brown algae dragged by the tide and contributed them to their land. At the beginning of the 20th century, a small industry based on the drying and grinding of algae developed but was weakened by the arrival of synthetic chemical fertilizers [13].

Brown algae and its derivatives improve the soil and invigorate the plants, increasing the yields and quality of the crops; so as this practice spreads, it will replace the use of synthetic chemical products with organic ones, thus favoring sustainable agriculture. Algae have better properties than fertilizers because they release nitrogen more slowly, and they are also rich in microelements and do not generate adventitious seeds [18].

Thanks to its high fiber content, macro-, and micronutrients, amino acids, vitamins, and plant phytohormones, algae act as a soil conditioner and contribute to moisture retention. In addition, due to their mineral content, they are a useful fertilizer and a source of trace elements [10].

Algae such as *Ascophyllum nodosum* (Figure 2.4(a)), *Fucus serratus* (Figure 2.4(b)), *Sargassum* spp. (Figures 2.4(c) and 2.4(d)), and *Laminaria* (Figure 2.4(e)) are used to cultivate potatoes, artichokes, citrus, orchids, and grasses. *Corallines* (Figure 2.2(f)), a calcified red alga, has a high carbonate content and are used as well as soil conditioners to correct the pH in acid soils. Brown algae are also capable of activating the immune system of the crops, generating higher production of higher quality and more resistant to diseases and environmental stress [3].

(a) ***Ascophyllum nodosum*** **(b)** ***Fucus serratus*** **(c)** ***Sargassum*** **spp.**

(d) ***Sargassum*** **spp.** **(e)** ***Laminaria*** **(f) Corallines**

FIGURE 2.4 Main algae species used as fertilizers for food crops.

KEYWORDS

- **algal bio-oil**
- **bioindicators**
- **biomass**
- **brown algae**
- **fertilizers**
- **food products**
- **heavy metals biosorption**

REFERENCES

1. Baghel, R. S., Suthar, P., Gajaria, T. K., Bhattacharya, S., Anil, A., & Reddy, C. R. K., (2020). Seaweed biorefinery: A sustainable process for valorizing the biomass of brown seaweed. *Journal of Cleaner Production, 263*, 121359.
2. Barbour, M. T., (2001). Rapid bioassessment protocols for use in streams and wadeable rivers: Periphyton. *Benthic Macroinvertebrates and Fish.*
3. Baroud, S., Tahrouch, S., El Mehrach, K., Sadki, I., Fahmi, F., & Hatimi, A., (2021). Effect of brown algae on germination, growth, and biochemical composition of tomato

leaves (*Solanum lycopersicum*). *Journal of the Saudi Society of Agricultural Sciences, 20*(5), 337–343.

4. Blikra, M. J., Henjum, S., & Aakre, I., (2022). Iodine from brown algae in human nutrition, with an emphasis on bioaccessibility, bioavailability, chemistry, and effects of processing: A systematic review. *Comprehensive Reviews in Food Science and Food Safety, 21*(2), 1517–1536.
5. Brar, P. K., Örmeci, B., & Dhir, A., (2022). Algae: A cohesive tool for biodiesel production along with wastewater treatment. *Sustainable Chemistry and Pharmacy, 28*, 100730.
6. Brinza, L., Dring, M. J., & Gavrilescu, M., (2007). Marine micro and macroalgal species as biosorbents for heavy metals. *Environmental Engineering & Management Journal (EEMJ)*, *6*(3).
7. Crist, R. H., Oberholser, K., McGarrity, J., Crist, D. R., Johnson, J. K., & Brittsan, J. M., (1992). Interaction of metals and protons with algae. 3. Marine algae, with emphasis on lead and aluminum. *Environmental Science & Technology, 26*(3), 496–502.
8. Crist, R. H., Oberholser, K., Schwartz, D., Marzoff, J., Ryder, D., & Crist, D. R., (1988). Interactions of metals and protons with algae. *Environmental Science & Technology, 22*(7), 755–760.
9. Cuizano, N. A., Llanos, B. P., Chang, L., & Navarro, A. E., (2007). Equilibrio ácido-base de algas marinas del litoral peruano elucida su alta afinidad por contaminantes ambientales. *Revista de La Sociedad Química Del Perú, 73*(2), 85–93.
10. Davis, T. A., Volesky, B., & Mucci, A., (2003). A review of the biochemistry of heavy metal biosorption by brown algae. *Water Research, 37*(18), 4311–4330.
11. De La Rocha, S. R., Sánchez-Muniz, F. J., Gómez-Juaristi, M., & Marín, M. T. L., (2009). Trace elements determination in edible seaweeds by an optimized and validated ICP-MS method. *Journal of Food Composition and Analysis, 22*(4), 330–336.
12. Deshpande, G., Shrikhande, S., Patle, D. S., & Sawarkar, A. N., (2022). Simultaneous optimization of economic, environmental, and safety criteria for algal biodiesel process retrofitted using dividing wall column and multistage vapor recompression. *Process Safety and Environmental Protection.*
13. Farnham, W., Murfin, C., Critchley, A., & Morrell, S. (2019). Distribution and control of the brown alga *Sargassum muticum. International Seaweed Symposium (X*[th]*)*, 277–282.
14. Garnham, G. W., Codd, G. A., & Gadd, G. M., (1992). Kinetics of uptake and intracellular location of cobalt, manganese, and zinc in the estuarine green alga *Chlorella salina. Applied Microbiology and Biotechnology, 37*(2), 270–276.
15. González, F., Romera, E., Ballester, A., Blázquez, M. L., Muñoz, J. Á., & García-Balboa, C., (2011). Algal biosorption and biosorbents. In: *Microbial Biosorption of Metals* (pp. 159–178). Springer.
16. Haritonidis, S., & Malea, P., (1999). Bioaccumulation of metals by the green alga Ulva rigida from Thermaikos Gulf, Greece. *Environmental Pollution, 104*(3), 365–372.
17. Hashemian, M., Ahmadzadeh, H., Hosseini, M., Lyon, S., & Pourianfar, H. R. (2019). Production of microalgae-derived high-protein biomass to enhance food for animal feedstock and human consumption. In: *Advanced Bioprocessing for Alternative Fuels, Biobased Chemicals, and Bioproducts* (pp. 393–405). Elsevier.
18. Islek, C., (2022). Green fertilizer and algal biofertilizer. *Introduction and Application of Organic Fertilizers as Protectors of Our Environment,* 182.
19. Jensen, A., (1993). Present and future needs for algae and algal products. *Fourteenth International Seaweed Symposium*, 15–23.

20. Küpper, F. C., & Carrano, C. J., (2019). Key aspects of the iodine metabolism in brown algae: A brief critical review. *Metallomics, 11*(4), 756–764.
21. Lodeiro, P., Cordero, B., Grille, Z., Herrero, R., & Sastre De, V. M. E., (2004). Physico-chemical studies of cadmium (II) biosorption by the invasive alga in Europe, *Sargassum muticum. Biotechnology and Bioengineering, 88*(2), 237–247.
22. Maity, S., & Mallick, N., (2022). Trends and advances in sustainable bioethanol production by marine microalgae: A critical review. *Journal of Cleaner Production,* 131153.
23. Markert, B. A., Breure, A. M., & Zechmeister, H. G., (2003). Definitions, strategies, and principles for bioindication/biomonitoring of the environment. In: *Trace Metals and Other Contaminants in the Environment* (Vol. 6, pp. 3–39). Elsevier.
24. Méndez-Rodríguez, L., Piñón-Gimate, A., Casas-Valdez, M., Cervantes-Duarte, R., & Arreola-Lizárraga, J. A., (2022). Macroalgae from two coastal lagoons of the Gulf of California as indicators of heavy metal contamination by anthropogenic activities. *Journal of the Marine Biological Association of the United Kingdom,* 1–13.
25. Murphy, V., Hughes, H., & McLoughlin, P., (2007). Cu (II) binding by dried biomass of red, green and brown macroalgae. *Water Research, 41*(4), 731–740.
26. Navarro, A. E., (2006). Propiedades ácido-básicas de lentinus edodes y cinética de biosorción de cadmio (II). *Revista Latinoamericana de Recursos Naturales, 2*(2), 47–54.
27. Ospina, A. N., & Peña, E. J., (2004). Alternativas de monitoreo de calidad de aguas: Algas como bioindicadores. *Acta Nova, 2*(4), 513–517.
28. Palasí, M. J. T., (2015). *Caracterización Físico-Química y Nutricional de Algas en Polvo Empleadas Como Ingrediente Alimentario.* Universitat Politècnica de València.
29. Phillips, D. J. H., (1977). The use of biological indicator organisms to monitor trace metal pollution in marine and estuarine environments—A review. *Environmental Pollution (1970), 13*(4), 281–317.
30. Piña Leyte-Vidal, J. J., González-Fernández, L. A., Gutiérrez-Artiles, O., Márquez-Llauger, L., & Del, C. T. A., (2019). Caracterización de tres bioindicadores de contaminación por metales pesados. *Revista Cubana de Química, 31*(2), 293–308.
31. Pinilla, G., (2000). *Indicadores Biológicos en Ecosistemas Acuáticos Continentales de Colombia.* Bogotá: Universidad Jorge Tadeo Lozano.
32. Quitral, V., Jofré, M. J., Rojas, N., Romero, N., & Valdés, I., (2019). Algas marinas como ingrediente funcional en productos cárnicos. *Revista Chilena de Nutrición, 46*(2), 181–189.
33. Ragan, M. A., (1986). Phlorotannins, brown algal polyphenols. *Progress in Phycological Research, 4,* 177–241.
34. Rojas, G., Silva, J., Flores, J. A., Rodriguez, A., Ly, M., & Maldonado, H., (2005). Adsorption of chromium onto cross-linked chitosan. *Separation and Purification Technology, 44*(1), 31–36.
35. Russo, D., Salinas-Ramos, V. B., Cistrone, L., Smeraldo, S., Bosso, L., & Ancillotto, L., (2021). Do we need to use bats as bioindicators? *Biology, 10*(8), 693.
36. Schiewer, S., & Volesky, B., (2000). Biosorption by marine algae. In: *Bioremediation* (pp. 139–169). Springer.
37. Ścieszka, S., & Klewicka, E., (2019). Algae in food: A general review. *Critical Reviews in Food Science and Nutrition, 59*(21), 3538–3547.
38. Singh, A., Prajapati, P., Vyas, S., Gaur, V. K., Sindhu, R., Binod, P., Kumar, V., et al., (2022). A comprehensive review of feedstocks as sustainable substrates for next-generation biofuels. *BioEnergy Research,* 1–18.

39. St-Cyr, L., & Campbell, P. G. C., (1994). Trace metals in submerged plants of the St. Lawrence River. *Canadian Journal of Botany, 72*(4), 429–439.
40. St-Cyr, L., & Campbell, P. G. C., (2000). Bioavailability of sediment-bound metals for *Vallisneria americana* Michx, a submerged aquatic plant, in the St. Lawrence River. *Canadian Journal of Fisheries and Aquatic Sciences, 57*(7), 1330–1341.
41. Stevenson, J., (2014). Ecological assessments with algae: A review and synthesis. *Journal of Phycology, 50*(3), 437–461.
42. Tapia, P., Santander, M., Pavez, O., Valderrama, L., Guzman, D., & Romero, L., (2011). Biosorption of copper ions with biomass of algae and dehydrated waste of olives. *Revista de Metalurgia, 47*(1), 15–28.
43. Yaashikaa, P. R., Devi, M. K., Kumar, P. S., & Pandian, E., (2022). A review on biodiesel production by algal biomass: Outlook on lifecycle assessment and techno-economic analysis. *Fuel, 324*, 124774.
44. Zuli, W. U., & Shouyu, Z., (2019). Effect of typhoon on the distribution of macroalgae in the seaweed beds of Gouqi Island, Zhejiang Province. *Journal of Agricultural Science and Technology, 21*(9), 159.

CHAPTER 3

Effectiveness of Bio-Based Fertilizer Systems for a Sustainable Future

RAFAIL A. AFANAS'EV, GENRIETTA E. MERZLAYA, and MICHAEL O. SMIRNOV

Pryanishnikov All-Russian Scientific Research Institute of Agrochemistry, Moscow, Russia

ABSTRACT

The purpose of a multifactorial long-term field experiment, carried out by conventional methods on sod-podzolic light loamy weakly acidic humus soil, was a comparative assessment of three fertilization systems: mineral, organomineral, and organic. In a long field experiment conducted in the "nonchernozem" zone of Russia for 30 years of effect and seven years of aftermath of three fertilizer systems were studied. It was found that the highest yield of feed units in the process of the action of fertilizers achieved for the mineral system–40.0 C/ha, the lowest–for organic–34.2 C/ha. In the aftermath, the organomineral system took the forefront in terms of crop rotation productivity (32.1 C/ha), the last place was involved by the mineral system–28.2 C/ha. Depending on the fertilizer systems, the agrochemical properties of the soil converted with a negative balance of basic nutrients; such as Nitrogen, Phosphorus, and Potassium (NPK). In action, the content of topsoil has decreased for all fertilizer systems, but the content of mobile phosphorus has increased. In the aftereffect of fertilizer systems in the soil, the humus and phosphorus content continued to decrease, with the exception of the variant with a mineral system. The mineral fertilizer system during the period of effect was superior to other systems. This system had the greatest negative impact on the ph $_{soil}$. The organic system can meet the requirements

Bioresources and Bioprocess in Biotechnology for a Sustainable Future. Leonardo Sepúlveda Torre, Juan Carlos Contreras-Esquivel, Ann Rose Abraham, & A. K. Haghi (Eds.)

of livestock enterprises in the dairy sector. Mineral and organo-mineral fertilization systems due to the higher yield of crop rotations allow, in addition to feed, to grow agricultural products for food and other purposes.

3.1 INTRODUCTION

Organic (biological, biodynamic, ecological) agriculture in the modern sense is a system of cultivation of agricultural crops without the use of synthetically produced mineral fertilizers and chemical means of plant protection [1]. The history of organic farming is hidden in the depths of centuries, because at the beginning of agriculture, which arose about 10,000 years ago, mankind tried to increase crop yields in various ways, including the use of manure and other organic substances. As you know, the lack of organic farming or its limited capabilities caused the death of powerful civilizations and States of Mesopotamia, the Easter Islands, the highlands of Eastern Greece, and huge areas of arable land in Central America, where the Mayan civilization flourished, due to the gradual depletion of the soil and inexorable decline in yields [2].

The question of the modern civilization development was radically resolved with the emergence of the industry of mineral fertilizers, primarily nitrogen. In the very beginning of the 20th-century Fritz Haber first isolated nitrogen from the air, creating the first artificial fertilizers-nitrates, which replaced the Chilean saltpeter. The emergence of synthetic fertilizers and the beginning of their mass production met the need to increase the production of agricultural products in the world. According to the conclusion of D. N. Pryanishnikov [3], in a number of Western European countries (Belgium, England, etc.) due to the development of organic agriculture, the yield of grain crops, in particular winter wheat, increased from 7 to 15–17 C/ha, but then stabilized and for about half a century (until 1890) remained almost constant at the achieved level. Further increase in productivity in Western Europe was associated with the beginning of mass application of mineral fertilizers at the beginning of 20th-century. Since that time, in just two decades, the yield of wheat has almost doubled, reaching, for example, in 1910–1913. In the Netherlands, 26 C/ha on average for the whole country, in Belgium–25 C/ha, in Germany-up to 22 C/ha.

Modern agricultural production is also mainly based on the active use of mineral fertilizers. With an annual use of 175.5 million tons of NPK, mineral fertilizers compensate for more than 52% of the removal of crops, i.e., more than 90 million tons of nutrients. All the above indicators indicate the significance of using mineral fertilizers in agricultural production, as well as Supplement with organic fertilizers, and chemical plant protection products. It is due to modern organomineral fertilizer systems that the average annual yield of grain crops in a number of European countries has increased to 60–80 C/ha, and in some farms, including in Russia, to 90–100 C/ha or more. However, in recent decades, due to the current environmental situation in a number of countries in America and Europe, a trend has emerged that revives the principles and technologies of organic farming, based on the rejection of the use of both chemical fertilizers and chemical plant protection agents [1, 2, 4, 5]. The formal date of the emergence of the modern organic movement in agriculture, or more precisely its first version, is chronologically considered to be 1924, but the main provisions of organic agriculture have not changed in its almost century-long history [1]. However, the advantage of organic farming systems over organomineral ones is still based on short-term experiments and currently has more than modest information of this type. Only long-term, methodically sustained field experiments can serve as a source of objective information about comparing the effectiveness of organic and organomineral fertilizer systems. Abroad, in particular in Switzerland, such a long field experience was established in 1978. The purpose of this experience was a comprehensive study of technologies in three fertilizer systems: mineral, organomineral, and organic. On average, over 21 years, the productivity of the organic system was 20% lower than that of the mineral system, with large fluctuations in individual crops.

In our country, the results of a multi-factor long-term field experiment conducted in the non-Chernozem zone near Smolensk may be of interest for a comparative assessment of various fertilizer systems.

3.2 METHODS

The experiment was conducted from 1979 to 2015, i.e., for 37 years. The studies were performed using generally accepted methods [6–9]. The soil of the investigational site is sod-podzolic light loam with the following initial agrochemical indicators: humus content (according to Tyurin) 1.4%C, total nitrogen 0.13%, mobile phosphorus (P_2O_5) (according to Kirsanov) 159 mg/kg, exchange potassium (according to Maslova) (K_2O) 139 mg/kg, pH_{kcl} 5.6. According to the level of fertility, the soil was characterized as weakly acidic, sufficiently humusized, with an increased content of mobile phosphorus and exchangeable potassium.

Repeatability in the experiment is 3-fold, the area of the experimental plot is 112 m^2 (7×16 m). The scheme of the experiment included options: 1-control (without fertilizers), 2-manure at an average annual dose of 9 t/ha of natural humidity, 3-mineral fertilizers ($N_{76,5}$ $P_{76,5}$ $K_{76,5}$), 4-manure at an average annual dose of 3 t/ha + $N_{25,5}$ $P_{25,5}$ $K_{25,5}$. As an organic fertilizer, we used bedding manure of cattle with a humidity of 70% with an organic content of 60%. With one ton of manure, $N_{4,8}P_{2,2}K_{6,9}$ was applied to the soil. Consequently, in the organic system the average annual dose of nutrients was $N_{43,4}P_{19,8}K_{62,1}$ (in the amount of 125.3 kg of NPK), in the mineral system $N_{76,5}P_{76,5}K_{76,5}$ (in the amount of 229.5 kg of NPK), in the organomineral system–$N_{39,8}P_{32,1}K_{46,2}$ (in the amount of 118.1 kg of NPK). Thus, in terms of the active substance, the largest amount of nutrition elements was applied in the mineral system variant, while in the organic and organomineral systems, the difference in total NPK doses was generally insignificant–7.2 kg/ha.

During the 37-year period of experience, five rotations of crop rotations identical in composition and alternation took place. The alternation of crops in the first rotation (1979–1989) was as follows: potatoes, barley, winter rye, oats, pea-oat mixture, winter wheat, barley, perennial grasses of two years of use, winter rye, oats, in the second (1990–1995) and third (1996–2001) rotations: potatoes, barley, perennial grasses of two years of use, winter wheat, oats, in the fourth (2002–2008) and fifth (2009–2015) rotations: annual grasses, winter rye, barley, perennial grasses of two years of use, spring wheat, oats. During 4 rotations from 1979 to 2008, the effect of fertilizers was studied, in the last, fifth rotation from 2009 to 2015, i.e., for seven years, the aftereffect of previously applied fertilizers was studied.

3.3 DISCUSSION

The applied fertilizers had a corresponding effect on crop productivity (Table 3.1). Thus, on average, for 4 rotations, the highest productivity of crop rotations–40.0 C/ha of feed units-was obtained in the variant of the mineral fertilizer system, which was 44.6% higher than the control. The beneficial impact of mineral fertilizers was also shown in the variant of the organomineral system, where the productivity was 36.0 C/ha of feed units, or 30.0% higher than the control one. On average, 34.2 C/ha of feed units were obtained for the manure fertilizer system, which was 23.5% higher than the control. By the least significant difference (LSD_{05}), the variants of all fertilizer systems differed from the control variant, and the mineral system

TABLE 3.1 Crop Rotation Productivity Depending on the Effect and Aftereffect of Various Fertilizer Systems

Variant	Fertilizer System	Effect							Aftereffect		
		Rotations of Crop Rotation				Average of 4 Rotations C/ha f.u.	Increment		For 5 Rotation C/ha f.u	Increment	
		1	22	33	4		C/ha f.u	%		C/ha f.u	%
1	Control	33.7	26.6	29.3	21.1	27.7	–	–	22.7	–	–
2	Organic	37.3	34.4	36.1	28.8	34.2	6.5	23.5	30.5	7.8	34.4
3	Mineral	40.3	45.5	41.4	32.6	40.0	12.3	44.6	28.2	5.5	24.3
4	Organomineral	39.7	38.4	40.1	25.7	36.0	8.3	30.0	32.1	9.4	41.4
LSD_{05}						5.2			3.0		

was distinguished between them, which had a significant difference from the organic and organomineral systems.

The use of fertilizer systems has had a mixed effect on the payback of NPK. On average, for 4 rotations of crop rotations, the lowest payback of 1 kg of NPK–5.2 kg of feed units was noted in the organic system variant, the average–5.4 kg of mineral and the highest–7.1 kg of organic-mineral fertilizer.

In the aftereffect of fertilizers, the difference in yield between the fertilizer variants significantly decreased. With a general decrease in crop yield in the 5th rotation, when the aftereffect of fertilizers was tested, compared with four rotations of their action, it decreased on average by almost 1.2 times, and its greatest decrease was noted for the mineral system–by 1.4 times. At the same time, between the variants with different fertilizer systems, the largest difference in crop yield was 3.9 C/ha of feed units, while in action (on average for rotation) it was equal to 5.8 C/ha of feed units. However, in comparison with the control variant the aftereffect of the fertilizers increased the productivity of crop rotation depending on the previously used systems of fertilization: mineral system by 24.3%, organic–by 34.4% and organomineral–by 41.4%, i.e., the least aftereffect of fertilizers on crop yields was shown in the mineral system variant, and the greatest–in the organomineral system.

The biochemical composition of agricultural crops depended both on their biological characteristics and on the fertilizer, systems used (Table 3.2). Based on dry matter, the highest nitrogen content in the first rotation of the crop rotation was observed in barley and winter wheat grains: 2.2% each, and the lowest–1.4%–in potato tubers. The average position for this indicator was occupied by oats and winter rye (1.57% and 1.85%). The same ratio for these crops is typical for phosphorus. In terms of potassium content, potatoes differed in a positive direction, the tubers of which contained 2.9% of this element on dry matter, while the grain of spring barley was 0.67%, and oats and winter rye were 0.47% each. Accordingly, the total NPK content in potatoes was 5.3%, which is significantly higher than the content of these substances in cereals. Perennial grasses also had a high nitrogen and potassium content–2.3 and 2.8%, while low phosphorus content–0.6%. In total, the content of these elements in the dry mass of herbs was 5.7% and exceeded even their content in potatoes. The content of raw protein in barley, winter wheat and winter rye grains in the action of fertilizers was most affected by the mineral system, where it averaged 13.2%, and the lowest-by the organic system with a value of 11.6%. Oats in comparison with other grain crops had the lowest content of protein substances. The starch content in potato tubers was clearly negatively affected by fertilizer systems, especially mineral and organomineral systems, due to a grow in the yield of this crop, i.e., due to the negative correlation of quantitative and qualitative indicators of potato yield. In the aftereffect, the NPK content in all crops decreased, for example, in barley-from 3.99 to 2.53%, in oats–from 2.96 to 2.34%, in winter rye–from 3.07 to 2.99%, and in perennial grasses-from 5.7 to 3.92%.

TABLE 3.2 Biochemical Composition of Crop Rotations under Different Fertilizer Systems

Crop	Fertilizer System	N	P_2O_5	K_2O	Starch* Dry Protein
		Percentage of Dry Matter (%)			
Potato	Control	1.5	1.0	2.7	20.2*
	Organic	1.2	1.2	3.1	20.1*
	Mineral	1.5	1.0	2.7	19.9*
	Organomineral	1.4	1.1	3.0	19.5*
Barley	Control	2.2	1.2	0.7	12.5
	Organic	2.1	1.1	0.7	12.0
	Mineral	2.4	1.1	0.7	13.6
	Organomineral	2.1	1.1	0.6	12.0
Winter rye	Control	1.6	0.7	0.4	9.1
	Organic	1.8	0.8	0.5	10.2
	Mineral	2.1	0.7	0.5	11.9
	Organomineral	1.9	0.8	0.5	10.8
Oats	Control	1.5	0.9	0.5	8.6
	Organic	1.5	1.0	0.5	8.6
	Mineral	1.7	0.9	0.5	9.7
	Organomineral	1.6	0.9	0.4	9.1
Winter wheat	Control	2.0	1.0	0.5	11.4
	Organic	2.2	1.0	0.5	12.5
	Mineral	2.5	0.9	0.5	14.2
	Organomineral	2.2	1.0	0.5	12.5
Perennial herbs one year of use	Control	2.3	0.6	2.4	–
	Organic	2.3	0.6	3.2	–
	Mineral	2.3	0.6	2.9	–
	Organomineral	2.3	0.6	2.7	–

Calculation of the average annual NPK nutrient balance for 4 crop rotations (Table 3.3) showed that for almost all elements, except for phosphorus in the mineral system variant, it was negative and of course, the highest in

TABLE 3.3 NPK Balance on Average per Year Depending on the Operation of Fertilizer Systems

Fertilizer System	Applied (kg/ha)			NPK Applied in Total (kg/ha)	Removed (kg/ha)			Removal of NPK in Total (kg/ha)	Balance (kg/ha. +/–)			
	N	P_2O_5	K_2O		N	P_2O_5	K_2O		N	P_2O_5	K_2O	NPK
Control	–	–	–	–	69.2	30.4	78.9	178.5	–69.2	–30.4	–78.9	–178.5
Organic	43.4	19.8	62.1	125.3	77.9	37.7	113.9	229.5	–34.5	–17.9	–51.8	–104.2
Mineral	76.5	76.5	76.5	229.5	94.9	42.7	119.3	257.0	–18.4	33.8	–42.8	–27.5
Organo-mineral	39.8	32.1	46.2	118.1	83.4	37.7	111.8	232.9	–43.6	–5.6	–65.6	–114.8

the control variant (–178.5 kg/ha), in which fertilizers were not used. The mineral fertilizer system had the lowest negative NPK balance (–27.5 kg/ha). The intermediate position in the balance of the NPK amount was occupied by the organic and organomineral systems, respectively –104.2 and –114.8 kg ha.

The decrease in crop productivity observed in the 4th rotation of the crop rotation compared to the first three rotations is largely due to a general decrease in soil fertility due to the manifestation of a negative balance of nutrients and their transition, apparently, to a less accessible state for plants.

The long-term effect of fertilizers, as well as the effect taking into account the aftereffect, had a significant impact on the coefficients of NPK use by agricultural crops (Table 3.4). In the variant of the organic system manure nitrogen was used least effectively–by 20.1 and 22.2%, while manure phosphorus and potassium were used more significantly: phosphorus, respectively, by 36.9 and 44.4% and potassium–by 56.4 and 63.5%.

TABLE 3.4 Coefficients of Nutrient Use by Agricultural Crops under Different Fertilizer Systems (%)

Fertilizer System	For 4 Rotations			For 5 Rotations		
	N	P_2O_5	K_2O	N	P_2O_5	K_2O
Organic	20.1	36.9	56.4	22.2	44.4	63.5
Mineral	33.6	16.1	52.9	4.5	18.3	60.2
Organomineral	35.7	22.8	71.3	42.6	30.0	97.4

In the other two fertilizer systems nitrogen use ranged from 33.6% to 42.6%, phosphorus from 16.1% to 30.0%, and potassium from 52.9% to 97.4%. The coefficients of use of phosphorus and potassium obtained in the experiment significantly exceeded the previously established ones, and the utilization coefficient for nitrogen generally corresponded to these indicators [10, 11].

The increased use of phosphorus by plants is mainly related to the system of organic fertilizers. Which is obviously due to the long-term effect and aftereffect of organic substances applied with manure due to its decomposition by soil microflora and activation of biochemical processes in the soil. This is partially confirmed by a biological test on the decomposition of linen fabric, conducted according to the variants of the experiment. So, according to data for 2015, in the variants with the use of manure linen fabric decomposed by 67.3–73.3% under the influence of soil microflora, and in the variant of the mineral system–only by 49.6%. It is possible that the moderate use of nitrogen by plants in the variant of the organic system is due to the activity of certain strains of soil microflora that intercept the nitrogen of manure and turn it into a microbiological form.

As a result of long-term application of fertilizers, the agrochemical composition of the soil changed (Table 3.5).

TABLE 3.5 Agrochemical Properties of the Soil During Long-Term Use of Various Fertilizer Systems

Fertilizer System	pH_{kcl}			Humus (%C)			P_2O_5 (mg/kg)			K_2O (mg/kg)		
	1	2	3	1	2	3	1	2	3	1	2	3
Control	5.0	4.9	5.1	1.40	1.01	0.99	170	75	75	145	55	79
Organic	5.7	5.1	5.4	1.23	1.04	1.04	143	160	141	138	145	70
Mineral	6.0	4.7	5.2	1.53	1.19	1.17	149	174	183	135	100	970
Organo-mineral	5.7	4.9	5.4	1.46	1.28	1.20	174	202	165	139	130	82
On average	5.6	4.9	5.3	1.40	1.13	1.10	159	178	141	139	108	82

Note: 1: initial content; 2: end of 4th rotation; and 3: end of 5th rotation.

The mineral fertilizer system had the greatest negative impact on the soil pH. For all the studied fertilizer systems (effect), the content of humus in the soil decreased, but the content of mobile phosphorus increased. Fertilizer systems had a mixed effect on the content of exchangeable potassium. Increasing it in the organic system and reducing it in the organomineral and mineral systems. In the aftereffect of fertilizer systems, the humus and phosphorus content in the soil continued to decrease, with the exception of the mineral system. Also, the exchange of potassium continued to decrease for all studied systems.

Results of a long-term field experiment with the study of 30-year effect and seven-year aftereffect of fertilizer systems-organic, mineral, and organomineral on sod-podzolic light loamy soil in the conditions of the non-Chernozem zone of the Russian Federation have identified certain patterns of their influence on plants and soil. It was found that the organic system based on the use of bedding manure of cattle, in terms of the productivity of crops of grain-grass and following grain-grass crop rotation, was inferior in the effect of the mineral fertilizer system by 17% and organomineral-by 5.3%. In the aftereffect, it exceeded the mineral system by 8.2% and was inferior to the organomineral system by 5.3%.

Results of a long-term field experiment with the study of 30-year effect and seven-year aftereffect of fertilizer systems-organic, mineral, and organomineral on sod-podzolic sandy loam soil in the conditions of the non-chernozem zone of the Russian Federation have identified certain patterns of their influence on plants and soil. It was found that the organic system based on the use of bedding manure of cattle, in terms of the productivity of crops of grain-grass and following grain-grass tilled crop rotation, was inferior in the action of the mineral fertilizer system by 17% and organomineral-by 5.3%. In the aftereffect, it exceeded the mineral system by 8.2% and was inferior to the organomineral system by 5.3%. On average, taking into account the effect and aftereffect of fertilizers, the yield of crops in the organic system was lower than in the mineral and organomineral systems by 5.3–5.6%. The mineral system of fertilizer during the period of effect exceeded the productivity of organic and organomineral crop rotations, but because of the low aftereffect, it was inferior to these systems. The organo-mineral system of fertilizer in effect was superior to the organic one, but inferior to the mineral one, and in the aftereffect of fertilizers it surpassed both of these systems.

Under the influence of the studied fertilizer systems, the agrochemical properties of the soil changed. The mineral system had the greatest negative impact on the pH of the soil. The content of exchangeable potassium in the soil under the influence of fertilizers significantly decreased in the variants of mineral and organomineral systems. For all the studied fertilizer systems, the content of humus in the soil decreased under direct action of fertilizers, but the content of mobile phosphorus increased. In the aftereffect of fertilizer systems in the soil at the trend level, the content of humus and more significantly phosphorus continued to decrease, with the exception of the mineral system, as well as exchange potassium for all studied fertilizer systems.

According to the conducted research, the amount of nutrients used from the applied fertilizers was determined. Over the seven-year period of experience, the coefficients of manure nitrogen use in the organic system were at the level of 20.1 and 22.2%, which is generally lower than the standard indicators. In mineral and organomineral fertilizer systems, nitrogen utilization rates were 1.5–1.6 and 1.8–1.9 times higher than in organic fertilizer systems and met the standards. The use of phosphorus in the organic system variant exceeded this indicator for the mineral system by more than 2 times, for the organomineral system–by 1.5–1.6 times. The coefficients of potassium use from fertilizers were 56.4–63.5% in the organic system, 52.9–60.2% in the mineral system, and 71.3–97.4% in the organomineral system, which is consistent with or significantly exceeds the known, including regulatory, indicators [10–12].

Based on the data obtained (without evaluating the economic aspects of fertilizer application systems), the organic system can meet the requirements of livestock enterprises in the dairy sector. With a crop yield of 34 c/ha of feed units, which provides an average annual milk yield per cow of at least 4,000 liters of milk without the use of compound feeds [13], and the output of litter manure of at least 8 tons per stall (200 days) period [14], this system is able to ensure balanced production of dairy products. The mineral fertilizer system, due to the higher yield of crop rotations, allows, in addition to feed, to grow agricultural products for food and other purposes. The same conclusion applies to the organic-mineral system of fertilizer. Thus, the choice of a particular system of application of fertilizers will mainly depend on the direction of economic activity of enterprises and economic indicators of production of milk and other agricultural products.

3.4 CONCLUSIONS

1. As a result of long-term experience with the study of the effect and the aftereffect on the soil and plants of three fertilizer systems (organic, mineral, and organomineral), it was found that the organic system in terms of productivity of crops of grain-grass and the following grain-grass-row crop rotation, was inferior to the mineral and organomineral fertilizer system.
2. The mineral fertilizer system during the period of effect was superior to other systems. The organo-mineral fertilizer system in the aftereffect of fertilizers was superior to both of these systems.
3. In the aftereffect, humus, and more significantly phosphorus content continued to decrease in the soil at the trend level, with the exception of the mineral system, as well as decreased exchangeable potassium content for all studied fertilizer systems.
4. During the studied period of experience, the coefficients of manure nitrogen use in the organic system were lower than the standard indicators. In the mineral and organomineral fertilizer systems, the coefficients of nitrogen use met the standards.
5. The use of phosphorus in the organic system variant exceeded this indicator for the mineral system by more than 2 times, for the organomineral system by 1.5 times.
6. The organic system can meet the requirements of livestock enterprises in the dairy sector.
7. Mineral and organo-mineral fertilization systems due to the higher yield of crop rotations allow, in addition to feed, to grow agricultural products for food and other purposes.

KEYWORDS

- **biological**
- **bioprocess**
- **bioresources**
- **biotechnology**
- **standard and durable field experience**
- **standard organomineral fertilizer systems**
- **standard plant products mineral**

REFERENCES

1. Gorchakov, Ya. V., & Durmanov, D. N., (2002). *World Organic Agriculture of the XXI Century* (p. 402). Moscow: Publishing House of the Russian University of Peoples' Friendship. In Russian.
2. Bachin, S., (2016). Manure. *Myths and Reality* (p. 128). Moscow: LLC «Hleb-Sol» («Bread -Salt» in Rus). In Russian.
3. Pryanishnikov, D. N., (1965). *Selected Works* (Vol. 1). Moscow: "Kolos" ("Ear" in Rus.). in Russian.
4. Mineev, V. G., Debrecen, B., & Mazur, T., (1993). *Biological Agriculture and Mineral Fertilizers* (p. 415). Moscow: «Kolos» («Ear» in Rus). In Russian.
5. Dovban, K. I., (2009). *Green Fertilizer in Modern Agriculture: Questions of Theory and Practice* (p. 404). Minsk: «Belorusskaya nauka» («Belarusian science» in Rus.). In Russian.
6. Dospekhov, B. A., (1979). *Methodology of Field Experience* (p. 416). Moscow: «Kolos» («Ear» in Rus.). In Russian.
7. Peregudov, V. N., et al., (1976). *Conducting Multi-Factor Experiments with Fertilizers and Mathematical Analysis of Their Results (Methodological Guidelines)* (p. 111). Moscow: Publishing house of the All-Union Agricultural Academy. in Russian.
8. Yagodin, B. A., Deryugin, I. P., & Zhukov, Yu. P., (1987). *Workshop on Agrochemistry* (p. 517). Moscow: Agropromizdat. In Russian.
9. Arinushkina, E. V., (1961). *Guide to Chemical Analysis of Soils* (p. 491). Moscow: Publishing House of the Moscow University. In Russian.
10. Derzhavin, L. M., (1992). *Application of Mineral Fertilizers in Intensive Agriculture* (p. 212). Moscow: «Kolos» («Ear» in Rus. In Russian.
11. Kirikoi, Ya. T., Merzlaya, G. E., Polunin, S. F., et al., (1991). *Normative Indicators of Nitrogen, Phosphorus and Potassium Content in Organic Fertilizers and Their Use by Agricultural Crops* (p. 56). Moscow: Pryanishnikov All-Russian Scientific Research Institute of Fertilizers. In Russian.

12. Kidin, V. V., (2012). *System of Fertilizer* (p. 534). Moscow: Publishing House of the Russian state University-Moscow agricultural Academy. In Russian.
13. Smurygin, M. A., Iglovikov, V. G., & Tashilin, V. A., (1985). *Handbook of Feed Production* (p. 413). Moscow: Agropromizdat. In Russian.
14. Vasiliev, V. A., & Filippova, N. V., (1988). *Handbook of Organic Fertilizers* (p. 255). Moscow: Rosagropromizdat. In Russian.

CHAPTER 4

Innovative Technologies of Bio-Based Fertilizers for a Sustainable Future

GENRIETTA E. MERZLAYA,[1] K. D. LAZAREVA,[1] T. E. MANTSEVA,[1] MICHAEL O. SMIRNOV,[1] and SERGEI I. NOVOSELOV[2]

[1]*Pryanishnikov All-Russian Scientific Research Institute of Agrochemistry, Moscow, Russia*

[2]*Mari State University, Yoshkar-Ola, Republic of Mari El, Lenin Square, Russia*

ABSTRACT

The results of research are presented and the agroecological justification of innovative technologies for applying fertilizers on a biological basis-manure from cattle and pig farms is given. It is shown that the intra-soil application of manure, while optimizing its doses, improves plant nutrition conditions, increases crop yields by 10–28% compared to surface application, and by 30–78% in relation to control, improves the quality of crop production, and contributes to the effective solution of environmental protection issues. It has been established that the greatest efficiency of biologically based fertilizers is achieved when they are combined with mineral fertilizers in optimal doses.

4.1 INTRODUCTION

In ensuring the food security of the Russian Federation, an important role belongs to the optimization of crop cultivation technologies using organic and mineral fertilizers in order to increase the productivity of agrocenoses,

---

Bioresources and Bioprocess in Biotechnology for a Sustainable Future. Leonardo Sepúlveda Torre, Juan Carlos Contreras-Esquivel, Ann Rose Abraham, & A. K. Haghi (Eds.)

reproduction of soil fertility with effective environmental protection. Of great importance, in this case, belongs to the biologization of agricultural technologies, including through saturation with organic fertilizers [1–3]. It is known that, in recent years, the availability of organic fertilizers in agriculture of Russia has been characterized as extremely low. No more than 1.1–1.3 tons are applied to 1 hectare of sown area in terms of litter manure, which is 15–18% of the need for organic fertilizers to achieve a deficiency-free balance of humus in sod-podzolic soils. Taking into account the growth trend of livestock by 2030, the yield of manure in agricultural organizations in physical weight can reach 300 million tons, and in terms of litter manure about 170 million tons. In connection with the construction of new livestock complexes and large farms, the main increase in the volume of organic fertilizers will occur due to litterless manure and bird droppings, the share of which in the structure of the yield of organic fertilizers reaches 50–70% or more.

It is important to take into account that liquid organic fertilizers produced at large modern enterprises-litterless manure and bird droppings differ, on the one hand, in the presence of a complex of nutrients necessary for plants, and on the other – are characterized by high humidity, which creates certain difficulties in their storage and can cause excessive intake of nutrients, polluting the environment [4–8].

A number of scientific institutions: the Pryanishnikov Research Institute of Agrochemistry, the Mari State University, have developed innovative biologized technologies using intra-soil application of organic fertilizers of high humidity in the form of manure from cattle and pig farms, which make it possible to successfully apply these fertilizers in the fields without risk pollute adjacent territories.

Positive results in this direction were obtained in the long-term research of the Pryanishnikov Institute of Agrochemistry in the Moscow region when testing the effectiveness of litterless manure in increasing doses. A single dose of mature in terms of nitrogen content corresponded to 120 kg/ha rounded off.

4.2 METHODS

The field experiment was carried out on sod-podzolic soil of heavy granulometric composition with a humus content of 1.4–1.5%, poorly provided with mobile compounds of phosphorus and potassium. The area of the experimental plot was 96 m^2, which allowed mechanized application of litterless manure. In the experiment, the following crops were cultivated in the fodder crop rotation: corn (during the first two years) to obtain green feed, vetch-oat mixture, perennial grasses of two years of use.

Litterless semi-liquid cattle manure had a humidity of 86–92%. Around 1 ton of manure contained up to 10 kg of NPK. In the experiment, in addition to increasing doses of semi-liquid manure, mineral, organo-mineral, and organic fertilizer systems with an equalized amount of basic nutrients were used. Fertilizers in all variants of the experiment were applied annually during all 15 years of research. Agrochemical soil parameters and grain quality characteristics were determined according to generally accepted methods [3, 4].

4.3 DISCUSSION

According to the results obtained, in the above experiment (Figure 4.1), with an increase in the level of application of litterless manure with its prolonged use, the average annual productivity of crop rotation increased from 6.07 tons of f.u. in the variant with one dose to 8.50 tons of f.u. at a 5-fold dose. A significant increase in crop rotation productivity was observed at a 3-fold dose of manure corresponding to 345 kg/ha of nitrogen, where 5.03 tons of f.u. per 1 ha was obtained, or 92% higher than in the control. At the same time, the payback of the applied manure by increasing the yield of f.u. with an increase in its dose decreased from 5.9 to 2.7 kg.

Depending on the amount of manure applied, the quality of feed changed. With an increase in the dose of manure in nitrogen content from 115 to 575 kg/ha, the amount of nitrates in the vegetative mass of corn increased from 140 to 220 mg/kg. It is characteristic that even with the maximum dose of manure, the nitrate content in the corn biomass did not exceed the permissible level–500 mg/kg of NO_3, while when applying mineral fertilizers, nitrates in corn accumulated two times higher than normal.

Litterless manure when applied in moderate doses had a positive effect on soil fertility, improving its humus state and availability of mobile compounds of phosphorus and potassium. Despite the annual application of manure during three rotations of the 5-field crop rotation, there was no contamination of groundwater with nitrates.

Based on experimental data on the yield of forage crops and its significant increase with an increase in the amount of manure applied, the quality of the feed obtained, as well as agrochemical properties of the soil and the state of groundwater, effective doses ranged from 200 to 230 kg/ha of nitrogen. At the same time, it should be taken into account that the average annual dose of litterless manure the crop, according to the Methodological Recommendations [4], should not exceed 200–300 kg/ha of nitrogen, depending on the conditions of humidification.

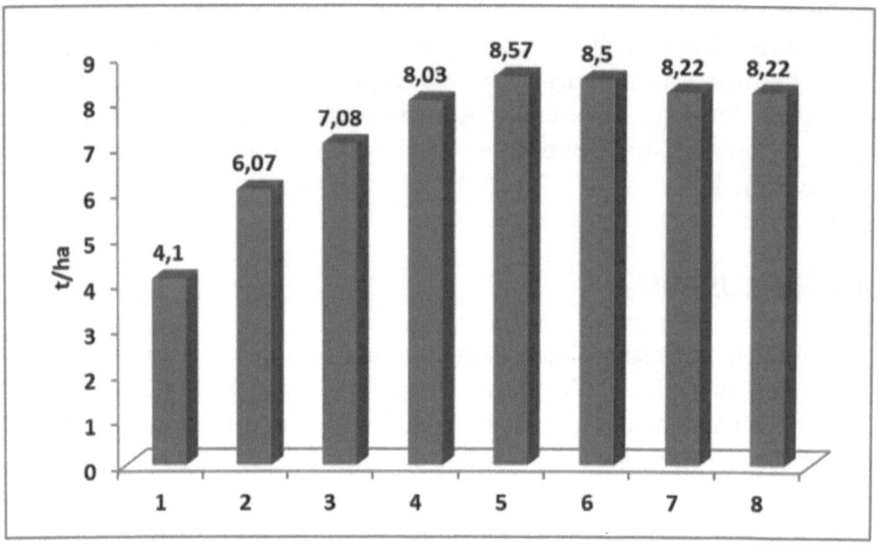

FIGURE 4.1 Average annual productivity (t/ha of feed units) of feed five-field crop rotation with annual application of organic and mineral fertilizers.

Note: 1: Control; 2: 1 dose of manure; 3: 2 doses of manure; 4: 3 doses of manure; 5: 4 doses of manure; 6: 5 doses of manure; 7: NPK equivalent to 2 doses of manure; and 8: NPK equivalent to 1 dose of manure + 1 dose of manure.

The test of the in-soil application of litterless manure showed a positive effect in the work of the Mari State University, conducted jointly with JSC "Shoybulaksky." Litterless pig manure before sowing was applied at a dose of 60 t/ha, and in the top dressing in spring tillering at a dose of 20 t/ha. The machine MZHU-20-1 was used for both surface and subsurface application of litterless manure. The object of research was winter wheat of the Moskovskaya 56 variety. The predecessor was a vetch-oat mixture grown for green fodder. The experiment was carried out on sod-podzolic medium loamy soil, well provided with phosphorus and medium provided with potassium.

It is known that the traditional technology of applying liquid organic fertilizers provides for their spilling over the surface of the field. The main disadvantages of this technology are low economic and environmental efficiency due to large losses of nitrogen applying as well as due to the threat of environmental pollution [5]. Field and laboratory studies have shown that the use of an intra-soil method of litterless pig manure into the soil had a positive effect on the nitrogen regime, increasing the content of ammonium, nitrate, and as a consequence, mineral nitrogen (Table 4.1). During the germination period, the main form of nitrogen in the soil was ammonium nitrogen. In 2018, during the period of winter wheat germination, the content of mineral nitrogen in untreated soil was 13.8 mg/kg. With the surface application of liquid organic fertilizer, the content of mineral nitrogen in the soil was 15.5 mg/kg. Intra-soil application of liquid organic fertilizer provided a significant increase in the content of mineral nitrogen in the soil, up to 21.9 mg/kg. In 2019, the content of mineral nitrogen was higher and amounted to 32.4 mg/kg. On average, over two years of research, the content of mineral nitrogen in the winter wheat germination phase in unfertilized soil was 15.8 mg/kg. With surface application of liquid organic fertilizer, it increased to 22.4 mg/kg, and with intra-soil application to-27.2 mg/kg.

TABLE 4.1 The Influence of Methods of Liquid Pig Manure Application on the Mineral Nitrogen Content in the Soil During the Winter Wheat Germination Phase (mg/kg)

Variant	**$N\text{-}NH_4$**			**$N\text{-}NO_3$**			**$N_{мин}$**		
	2018	**2019**	**On Average for 2 Years**	**2018**	**2019**	**On Average for 2 Years**	**2018**	**2019**	**On Average for 2 Years**
Control (without fertilizers)	10.1	12.4	11.8	3.7	5.4	4.6	13.8	17.8	15.8
Surface application	12.0	21.4	16.7	3.5	7.9	5.2	15.5	29.3	22.4
Intra-soil application	15.6	24.1	19.9	6.3	8.3	7.3	21.9	32.4	27.2

Improving the conditions of plant mineral nutrition had a positive effect on the yield of winter wheat grain. Grain yield increases depended on the method of liquid organic fertilizer application. On average, for three years, when growing winter wheat without the use of fertilizers, grain yield was 2.75 t/ha (Table 4.2). With surface application of liquid organic fertilizer, grain yield increased to 3.30 t/ha. The maximum grain yield of 3.58 t/ha was obtained by growing winter wheat in a variant with an intra-soil application of 60 t/ha of liquid manure. The effect of the intra-soil method in comparison with the surface application of litterless pig manure was significant and amounted to 10.4% (with an increase in grain of 0.28 t/ha).

TABLE 4.2 The Influence of Methods of Liquid Pig Manure Application on the Yield of Winter Wheat Grain (t/ha)

Variant	Top Dressing 20 t/ha	2018	2019	2020	On Average	Yield Increase	
						t/ha	%
1	–	2.80	3.19	2.26	2.75	–	–
	+		3.51	2.58	–		
2	–	3.30	3.88	2.73	3.30	0.55	20.0
	+		4.41	2.99	–		
3	–	3.70	4.10	2.93	3.58	0.83	30.2
	+		4.62	3.30	–		
LSD_{05}	0.20						

Note: 1: Control (without fertilizers); 2: Surface application; and 3: Intra-soil application.

Carrying out top dressing with liquid fertilizer provided a significant increase in grain yield. As can be seen from Table 4.2, the increase in grain yields due to fertilizing with pig manure on a non-fertilized background was 0.32 t/ha, against the background of surface application of organic fertilizer 0.40 t/ha, and intra-soil – 0.45 t/ha (with LSD_{05} = 0.20).

The application of liquid organic fertilizer had a positive effect on the quality of winter wheat grain. Its surface application provided an increase in the content of raw protein in the grain by 0.8%, and the intra-soil – by 1.0% (Table 4.3).

The use of liquid organic fertilizer in top dressing increased the crude protein content in the grain by 0.4–0.6% and the mass of 1000 grains by 0.1–0.5%. At the same time, the protein content in wheat in the variant of intra-soil manure application corresponded to grain of the second class according to SAUS-RF 9353–2016. The nature of grain depended on the weather conditions of the growing season and averaged 801–806 g/l over the years of research, which exceeds the indicators according to SAUS-RF 9353–2016.

A high effect from the intra-soil application of litterless cattle manure was obtained in the scientific and production experience in the farm "Korobovsky" of the Shatura district (the Moscow region) on pasture grassland (the fourth year of use). Manure was applied after grazing animals in the autumn with an intra-soil ABV-f-2.8-RZHT-8 aggregate by surface application. The

TABLE 4.3 The Influence of Methods of Applying Liquid Pig Manure on the Quality of Winter Wheat Grain

Variant	Top Dressing 20 t/ha	Raw Protein (%)			Weight of 1000 Grains (g)			In Natural Terms (g/l)		
		2019	2020	On Average	2019	2020	On Average	2019	2020	On Average
1	–	12.5	9.8	11.2	46.8	42.0	44.4	827	782	804
	+	12.9	10.6	11.8	47.2	42.6	44.9	836	777	806
2	–	13.5	12.5	13.0	46.8	42.1	44.5	826	779	803
	+	14.2	12.8	13.5	47.2	42.8	45.0	836	765	801
3	–	13.8	12.7	13.2	47.0	42.1	45.1	826	778	802
	+	14.3	13.0	13.6	47.5	42.9	45.2	839	768	803

Note: 1: Control (without fertilizers); 2: Surface application; and 3: Intra-soil application.

soil of the site is peat-gley sandy loam with a phosphorus content (P_2O_5) of 133 mg/kg, with a potassium content (K_2O) of 481 mg/kg, pH_{sol} – 6.3; litterless manure (in a dose of 230 kg/ha of nitrogen) with a moisture content of 93.1%, total nitrogen 0.23%, potassium (K_2O) – 0.2%.

As can be seen from Table 4.4, the effect of the intra-soil method of manure application was clearly manifested on the pasture. The yield increase of pasture grass was 78% compared to the control and 28.2% relative to the surface application. The advantage of the intra-soil method of manure application was determined not only in increasing pasture productivity, but also in reducing environmental pollution. So, according to the Saratov Research Institute of Hygiene, the complete self-cleaning of the soil layer on this pasture with the application of manure to a depth of 17 cm was completed in 3 months. Similar sanitary results were obtained during helminthological studies.

TABLE 4.4 The Yield of the Perennial Grass Dry Mass in Pasture Use, Depending on the Application Method of Litterless Manure

Variant	Dry Weight Yield (t/ha)	Yield Increase			
		t/ha		%	
		To Control	To Surface Application	To Control	To Surface Application
1	5.06	–	–	–	–
2	7.03	1.97	–	38.0	–
3	9.01	3.95	1.98	78.0	28.2

Note: 1: Control (without fertilizers); 2: Surface application; and 3: Intra-soil application.

It is known that the greatest effect is achieved with the combined use of organic and mineral fertilizers. This is clearly evidenced by the data obtained in the long-term field experience in the conditions of the western part of the Russian Non-Chernozem zone.

The soil of the experimental site was sod-podzolic, light-loamy cultivated, before the experiment began, contained humus (according to Tyurin) 1.4% C, mobile phosphorus (P_2O_5 – according to Kirsanov) 160 mg kg and potassium (K_2O-according to Kirsanov) 130 mg/kg of soil at a pH of 5.9.

Manure from cattle farms with a small amount of litter was used from organic fertilizers, which was applied in the first rotation of the crop rotation for potatoes and winter wheat, in the second and third rotations–for potatoes. Organic fertilizers had an average humidity of 70% and contained 0.5% total nitrogen, 0.2% P_2O_5 and 0.7% K_2O. The organic matter content was 60%, the C:N ratio is 20. The gross content of heavy metals is low: Cd–0.1, Sg–1, Ni–1, Cu–0.6, Zn–7 mg/kg of dry weight. Ammonium nitrate, superphosphate, and potassium chloride were used as mineral fertilizers in the experiment. Single doses of nitrogen, phosphorus, and potassium of mineral fertilizers were rounded up to 30 kg a.s/ha, organic fertilizers (manure)–3 tons/ha.

Starting from the fifth rotation of the crop rotation, i.e., since 2009, in the variants of the experiment, the aftereffect of previously applied fertilizers was tested when applied under crops against the background of supportive spring fertilizing with nitrogen fertilizers at a dose of 45 kg/ha.

The climate in the research region (Smolensk region) is moderately continental with relatively warm and humid summers and moderately cold winters. The average annual precipitation for May-September is 419 mm, the average annual air temperature is 14.8°C, the hydrothermal coefficient is 1.6 (Figure 4.2).

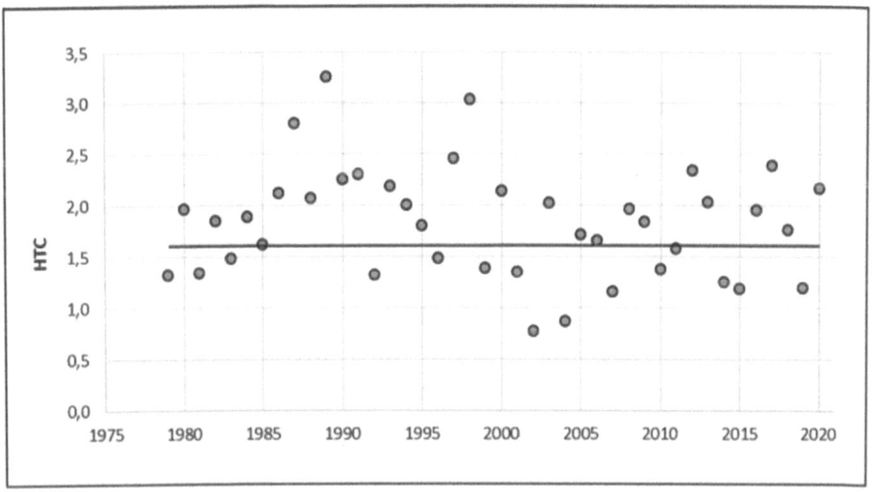

FIGURE 4.2 Hydro-thermal coefficient of Selyaninov (HTC) during the growing season (April–September) in the years of research.

Note: —: average age.

Around 50% of the years were characterized by satisfactory humidification conditions, but even during these years, either excessively wet or dry periods were observed during the vegetation of plants. In the aftereffect of fertilizers, i.e., in the fifth rotation of the crop rotation, the growing seasons of 2010 and 2015 were dry. In general, it should be noted that, despite some unfavorable meteorological periods, agricultural crops in the experiment mainly formed yields sufficient for the research region.

When analyzing the effect of fertilizers, it was found that the use of an organic system in a three-fold dose of manure of 9 tons/ha (variant 6–the interpretation of the variants is given in the legend of Table 4.5) allowed to obtain annual crop rotation productivity for 30 years at the level of 3.61 tons of fodder units per 1 ha, or 21.6% higher than the control (variant 1). In the variant of the organomineral system, also with triple doses (variant 9), crop rotation productivity increased to 4.15 tons of feed units/ha, i.e., in relation to the control by 1.18 tons of feed units/ha (or by 39.8%), and in relation to the organic system by 0.54 tons of feed units/ha (or by 15%). When using the mineral system (variant 5), crop rotation productivity increased to 4.25 tons of feed units/ha, i.e., it was higher than the control by 1.28 tons of feed units/ha (43.1%) and compared with the organic system (option 6) higher by 0.64 tons of feed units/ha (17.8%). With unilateral fertilization, the lowest effect was obtained from phosphorus fertilizers (superphosphate-variant 3) –8.5%, and the greatest–from nitrogen fertilizers (variant 2)–18.7% relative to the control. It is characteristic that in the aftereffect of fertilizers (for the fifth rotation of the crop rotation), the productivity of crops from unilateral application of phosphorus fertilizers (variant 3) increased by more than 3 times, potash fertilizers (variant 4)–by 2 times with a reduced effect of nitrogen fertilizers alone (variant 2), where the increase in relation to the control was 17.6%.

Of the fertilizer systems for crop rotation productivity increases, the organic system prevailed in the aftereffect (variant 6), where the increase reached 33.2% (versus 21.6% in action). It was inferior to the mineral system (variant 5) with an increase of 29.5% (versus 43.1% in action). When using the organomineral system, the increase in productivity in the aftereffect in relation to the increase in action decreased not so significantly (these values were 34.1 and 39.8%, respectively).

The productivity of crop rotation in the variants of the experiment in the aftereffect of fertilizers (in the fifth rotation) compared with their effect (on average for four rotations) decreased markedly: for the organic system (variant 6) by 24%, for the organomineral (variant 9)–by 42% and very sharply for the mineral system (variant 5)–by 87%.

Drawing a conclusion based on the results of the entire experiment, it should be noted that the highest productivity of crop rotation, taking into account by-products, on average over the years of fertilizer action (for 30 years) was obtained in organomineral (variant 8) and mineral (variant 5) variants, amounting to 4.2–4.25 t/ha of grain units, which is 42–43% higher control. Over the seven-year period of aftereffect, the organomineral fertilizer system in maximum doses (variant 11) gave a higher effect, with a yield of grain units of 3.71 t/ha, or 71% higher than the control.

When analyzing the productivity of individual crops, the best results were obtained from applying moderate doses (Table 4.5). So, when cultivating winter wheat, according to the data for an average of five years (1984, 1985, 1995, 2000, 2001) the maximum grain yield was achieved in the organomineral variant with two-fold doses of fertilizers (variant 8), where it amounted to 3.57 t/ha, significantly exceeding the control (by 1.73 t/ha) and a variant with single doses (variant 7) (by 0.72 t/ha) with LSD_{05} = 0.53 t/ha). A higher effect when applying double doses of fertilizers in organomineral variants (variant 8) was also obtained when cultivating spring wheat (on average for 2007–2008) and winter rye (on average for 2009–2010). At the same time, when growing oats, a crop that closes the crop rotation, higher doses of fertilizers were required to form a high grain yield. In particular, a high yield of oats (4 t/ha of grain – on average for 2008–2009) was obtained by applying three-fold doses of fertilizers in the organomineral version (variant 9). It should be noted at the same time that the same level of oat yield was noted in the variant of the mineral system with triple doses of NPK (variant 5).

The experiment conducted on sod-podzolic light loamy soil found out the dependence of the yield of crops cultivated on it on the level of fertilizers applied. In the control variant, where fertilizers were not used, at the end of the experiment, a sharp decrease in humus in the soil was noted compared to its initial content (by 28 and 29% in the fourth and fifth rotations of the crop rotation, respectively). At the same time, the availability of mobile phosphorus and potassium compounds in the soil also worsened. With the use of organomineral fertilizer systems, by the end of the experiment, the availability of mobile phosphorus in the soil improved, but the content of potassium in the soil decreased, with the exception of the variant with maximum doses – N150P150K150+15 t/ha of manure.

4.4 CONCLUSIONS

1. As a result of the conducted research, a scientific justification is given from the agronomic and ecological positions of innovative technologies for the use of biologically based fertilizers, represented by cattle and pig manure.

TABLE 4.5 The Effect of Organic and Mineral Fertilizers in Action on the Yield of Grain Crops During Cultivation in Crop Rotation

Variant	Winter Wheat			Winter Rye			Spring Wheat			Oats		
	Yield (t/ha)	Yield Increase		Yield (t/ha)	Yield Increase		Yield (t/ha)	Yield Increase		Yield (t/ha)	Yield Increase	
		t/ha	%		t/ha	%		t/ha	%		t/ha	%
1	1.84	–	–	2.82	–	–	1.92	–	–	2.25	–	–
2	2.48	0.64	35	3.11	0.29	10	3.05	1.13	60	3.49	1.24	55
3	2.12	0.28	15	3.71	0.89	32	3.06	1.14	59	3.94	1.69	75
4	2.57	0.73	40	3.57	0.75	27	2.76	0.84	44	2.96	0.71	31
5	3.25	1.41	77	3.82	1.00	36	3.45	1.53	79	4.07	1.82	81
6	2.51	0.67	36	3.72	0.90	32	3.02	1.10	57	3.34	1.09	48
7	2.86	1.02	55	3.38	0.56	20	2.85	0.93	48	3.30	1.05	46
8	3.57	1.73	94	3.46	0.64	23	3.34	1.42	74	3.70	1.45	64
9	3.25	1.41	77	3.46	0.64	23	3.53	1.61	84	4.06	1.81	80
10	2.82	0.98	53	3.41	0.59	21	3.23	1.31	68	3.76	1.51	67
11	2.35	0.51	28	3.35	0.53	19	3.96	2.04	106	3.88	1.63	72
LSD_{05}		0.53			0.15			0.57			0.34	

Note: Variants – 1: Control; 2: N90; 3: P90; 4: K90; 5: N90P90K90; 6: Manure 9 t/ha; 7: N30P30K30 + Manure 3 t/ha; 8: N60P60K60 + Manure 6 t/ha; 9: N90P90K90 + Manure 9 t/ha; 10: N120P120K120 + Manure 12 t/ha; and 11: N150P150K150 + Manure 15 t/ha.

2. Agronomically effective and environmentally safe method of applying fertilizers on a biological basis is intra-soil, thanks to which, when optimizing the application doses, plant nutrition conditions improve, and crop yields increase by 10–28% compared with the surface method and by 30–78% compared with the control without fertilizers.
3. When using the intra-soil method of fertilization, the quality of crop production is improved, and environmental protection is ensured.
4. The greatest efficiency of biologically based fertilizers is achieved when they are applied together with mineral fertilizers. Thus, the highest productivity on average over 30 years of fertilizer action was obtained in the organomineral variant with a combination of N60P60K60 with 6 t/ha of manure, amounting to 4.2 t/ha of grain units, which is 42–43% higher than the control.

KEYWORDS

- **bio-based fertilizers**
- **environmental protection**
- **grain units**
- **intra-soil method**
- **manure**
- **organic and mineral fertilizers**
- **plant products quality**
- **yield**

REFERENCES

1. Sychev, V. G., (2019). *Soil Fertility in Russia: The State and Opportunities (to the 100th Anniversary of the Birth of Tamara Nikandrovna Kulakovskaya)* (p. 240). Moscow; Pryanishnikov All-Russian Scientific Research Institute of Agrochemistry, In Russian.
2. Merzlaya, G. E. (2012). *Highly Efficient Systems for the Use of Organic Fertilizers and Renewable Biological Resources*, (p. 216) Vladimir: All-Russian Research Institute of Organic Fertilizers. In Russian.
3. Lazareva, K. D. (1990). *Doses and Terms of Application of Litterless Manure (Methodical Recommendations)*, p. 23. Moscow: Pryanishnikov All-Russian Scientific Research Institute of Agrochemistry, in Russian.

4. Merzlaya, G. E., Volodarskaya, N. V., Ovchinnikova, T. G., Marchenko, N. M., & Rodin, V. M., (1990). *The Intra-Soil Method of Applying Litterless Manure* (p. 9). Moscow: Exhibition of achievements of the National economy. In Russian.
5. Novoselov, S. I. (1978). *Efficiency of Intra-soil Application of Liquid Organic Fertilizers Based on Pig Manure. Waste, the Reasons for Their Formation and Prospects for Use* (pp. 550–552). Collection of scientific papers based on the materials of the International Scientific Ecological Conference. Moscow. In Russian.
6. Tarasov, S. I., & Merzlaya, G. E., (2018). The use of bespodstilochny manure. *Priority Areas of Research. Fertility, 6*, 53–56.
7. Koriat, H., (1972). Verfahren zur aufbereitung und ausbringung der gulle und ihr einsatz in der pflanzenproduktion. *Monatshefte fur Med. Jena., 27*, S. 247–252.
8. Lobl, F., (1990). Vynziti procesv kompostovani pri spracovani organickych hnojiv s velkochovn svirat. Ecologicke aspekty vynziti, upravy a zpracovani kejdy. *Kostelec nad Orlici.*, S. 24–26.

CHAPTER 5

Agro-Industrial Wastes from Fruits, Vegetables, and Cereals: Potential Substrates for the Production of Value-Added Products for a Sustainable Future

ALEJANDRA SOLIS RAMOS,[1] MÓNICA L. CHÁVEZ-GONZÁLEZ,[1,2] ANNA ILINÁ,[2] JOSÉ LUIS MARTÍNEZ HERNÁNDEZ,[2] and CRISTÓBAL NOÉ AGUILAR[1]

[1]*Bioprocesses and Bioproducts Research Group, Food Research Department, School of Chemistry, Autonomous University of Coahuila, Saltillo, Mexico*

[2]*Food Science Research Department, Faculty of Chemical Sciences of the Autonomous University of Coahuila, Blvd. V. Çarranza e Ing. José Cárdenas V., Col. República, Saltillo, Coahuila, Mexico*

ABSTRACT

From the production to the consumption of fruits, vegetables, and cereals, agro-industrial waste is generated in large quantities. These wastes are often disposed of improperly, seriously affecting the environment. The utilization of these wastes is a promising approach to their possible uses. The utilization of these wastes allows the production of various high-value-added products. Through agro-industrial waste, the production of unicellular protein is possible. Single-cell protein is one of the food alternatives in the face of current food safety concerns. They are defined as dried cells obtained from microorganisms. Due to factors such as environmental pollution and population growth, it is essential to search for new food sources sustainably. Lately, the interest in the production of unicellular protein has leaned towards

Bioresources and Bioprocess in Biotechnology for a Sustainable Future. Leonardo Sepúlveda Torre, Juan Carlos Contreras-Esquivel, Ann Rose Abraham, & A. K. Haghi (Eds.)

the use of alternative technologies. The use of agro-industrial residues and alternative technologies allows for the promotion of product development, economizing, balancing, and benefiting environmental concerns. This chapter has focused on the use of agro-industrial residues from fruits, vegetables, and cereals to obtain different high-value-added products. It summarizes the quantity of some of the residues that are generated in greater quantity annually in the world. In addition, it reviews the different methodologies for obtaining various products with different applications. Highlighting its use for the production of microbial biomass and unicellular protein. It addresses the use of different microorganisms using fermentation technology. Challenges in single-cell protein production are also discussed. This chapter is considerable for the concept of biotechnology, where there is interest in converting agro-industrial wastes into new food sources and producing high-value-added products.

5.1 INTRODUCTION

One of the main concerns nowadays is environmental pollution; Factors such as the growing population demand and the growing industries cause serious problems such as the increase of waste [4]. One of the industries that have expanded is agriculture [45]. In the process, consumption, or harvesting of the product of interest, the waste generates agro-industrial residues. These wastes are disposed of through combustion, accumulate in landfills or are dumped in vacant lots. This causes soil and water contamination, gas emissions and attracts rodents and insects. Affects the environment and human health [21, 63]. About 155 billion tons of organic matter are produced annually. Derived from photosynthetic processes and only a minimal part is consumed directly by humans or animals [18]. The volume of waste is so large that collection and disposal are becoming increasingly costly. Interest in utilizing these wastes has been increasing in scientific communities worldwide [5]. In Asian countries, around 4.4 billion tons of solid waste are produced annually. In India, 960 million tons are currently produced by the agribusiness industry [34]. South America is projected to produce 900 million tons of agro-industrial waste by 2025. These agro-industrial wastes are mostly solid and organic in nature [18]. The biomass consists mainly of lignocellulose, which is one of the most abundant resources on Earth [8]. Lignocellulose is a renewable organic material and the main component of all plants [11]. It consists of cellulose, hemicellulose, and lignin. Cellulose is a linear polysaccharide biopolymer of anhydroglucopyranose molecules. These molecules are connected by β-1,4-glycosidic bonds. Hydrogen bonds wrap the cellulose chains in fibrils. Hemicelluloses link these fibrils together. Hemicellulose is an amorphous polymer composed mainly of five different sugars [1]. Hemicellulose is a heterogeneous polymer of pentoses including xylose and arabinose, hexoses mainly mannose, less galactose, glucose, and other sugar acids. Lignin generally has three aromatic alcohols in its structure such as coniferyl, sinapyl, and *p*-coumaryl alcohol. Lignin provides structural strength to biomass fibers by holding together both celluloses and hemicelluloses. This biomass, besides containing sugars also has compounds such as dietary fibers, starch, proteins, micronutrients, and bioactive compounds [73]. Due to the high stability in the structure of the lignocellulosic components, it must be pretreated to produce the depolymerization of carbohydrates [15] and thus obtain high value-added products through different methodologies; physical, chemical, physicochemical, and biological. Fermentation is one of the biological processes that allow the bioconversion of these wastes. The microorganisms degrade using these as substrate for obtaining products [1]. That can be modified and thus used as a source of energy and raw material in different applications. Such as the production of biofuels, phytochemicals, nanomaterials, enzymes, unicellular protein, etc. [21]. They become potential and low-cost resources. Allowing them to be used by scientific communities in conjunction with the development of sustainable technologies [4, 21]. Their application can be for different areas and simultaneously contribute to reducing their environmental impact [26]. Therefore, this chapter presents an overview of some of the different areas of utilization of agro-industrial residues from fruits, vegetables, and cereals to obtain value-added products. Emphasis was made on their use to produce unicellular protein and the importance of being used as fermentation substrates to produce microbial biomass.

5.2 UTILIZATION OF AGRO-INDUSTRIAL WASTES: AN OVERVIEW

Much of the contribution to the economy comes from agriculture [21]; however, a large amount of waste it generates seriously damages the environment, mainly in the emission of greenhouse gases. Production in the agricultural industry has increased significantly over the last 50 years, reaching 23.7 million tons of food per day [22]. With the increase in population, agricultural productivity increases and with it agro-industrial waste [22]. Approximately 28% of the world's land area, equivalent to 1.4 billion hectares of fertile land, is used annually to produce lost or wasted foo [52].

Dey et al. [21] defined agro-industrial waste as the part of the plant that is not consumed by man and divided them into four different sources (Table 5.1), as it mentions that these are not restricted only to the crop but include all those that are produced by the different agricultural operations, from its preparation to its products.

TABLE 5.1 Different Sources of Agricultural Residues

Sources of Agricultural Residues	
Field residues	Leaves, stalks, seeds, straws, stubble, manure, etc.
Solid waste from industrial processing	Molasses, bagasse, husks, pulps, etc.
Livestock wastes	Animal carcasses, urine, wastewater from animal bathing, etc.
Hazardous chemical wastes	Pesticides, herbicides, and insecticides.

Wastes generated from industrial processing products and those intended for human or animal use are considered agricultural solid wastes from industrial processing, including molasses, bagasse, fruit, and vegetable peels and pulps, etc. Waste biomass is derived from different agrarian sources, about 140 billion metric tons, with a growth per year of 7.5% of agricultural solid waste [13]. At least 1300 million tons of food and one-third of its production is discarded per year, where Russia tops the list [20]. About 100 billion tons of food waste generated worldwide comes from cereals, while from vegetables and fruits from 80 billion to about 60 billion tons per year, respectively [52]. Fruit and vegetable waste accounts for more than 50% of the total product [61]. In countries such as North America, Europe, and Asia, the amount of cereal waste represents about 18% [29]. Most of these come to possess a higher nutritional or functional content. These wastes can be reduced with proper management and reuse practices and policies [20]. Table 5.2 shows the utilization of some such wastes for obtaining different products and metabolites, through different methodologies.

5.3 USE OF AGRO-INDUSTRIAL WASTES FOR DIFFERENT APPLICATIONS

Billions of agro-industrial wastes are generated, rich in materials that should be used to reduce their environmental impact. Much of the waste from the agricultural industry that is discarded in the form of stems, pulps, peels, and seeds of fruits, vegetables, and cereals, contains substrates that have various

TABLE 5.2 Use of Agro-Industrial Wastes to Obtain Different Value-Added Products/Metabolites

Waste	Tons/Year	Composition on a Dry Basis	Product	Pre-Treatment
		Cereals		
Corn cob	60 million	Cellulose 35–45%, Hemicellulose from Xylan 30–40%, Lignin 5–20%	Xylooligosaccharides	Enzymatic hydrolysis with endoxylanases
Rice husk	134 million	Cellulose 25–35%, Hemicellulose 18–21%, Lignin 26–31%, Silica 15–17%	Fluoride adsorbent in water	Acid surface modification
Rice straw	600–900 million	Cellulose 39.04%, Hemicellulose 20.91%, Lignin 5.71%	Biofuel	Alcothermal liquefaction
Wheat straw	747 million	Cellulose 34–45%, Hemicellulose 20–30%, Lignin 8–15%	Biogas	Anerobic digestion
Pineapple peel, stem, core	13.5 million	Cellulose 20–25%, Hemicellulose 6–19%, Lignin 4%, Wax 4%, Ash 1–5%	Glycosides	Autohydrolysis
Mango seed	1.0 million	Cellulose 55%, Hemicellulose 20.6%, Lignin 23.85%	Cellulose nanocrystals	Acid hydrolysis
Pomegranate peel	1.62 million	Cellulose 33%, Hemicellulose 13%, Lignin 8.5%	Ellagic acid	Ultrasound-assisted extraction
Discarded carrots	1,75,000	Cellulose 81%, Hemicellulose 9%, Lignin 2.5%, Pectin 7.5%	Hemicellulose–Pectin fractions purified	Hydrothermal
Sugar beet pulp	120 million	Hemicellulose 24%, Cellulose 44%, Lignin 2–4%, Pectin 20–30%	Pectin	Enzymatic treatment by cellulases, xylanases
Potato peelings	70–140 mil	Cellulose 55.25%, Hemicellulose 11.71, Lignin 14.24, Moisture 10.0%, Ash 8.8%	Phenolic acids	Extraction and fractionation by acidified water/ethanol-based solvents

properties. These should no longer be categorized as such but as raw materials to produce various value-added products [81] that have wide applications in different industries (Figure 5.1).

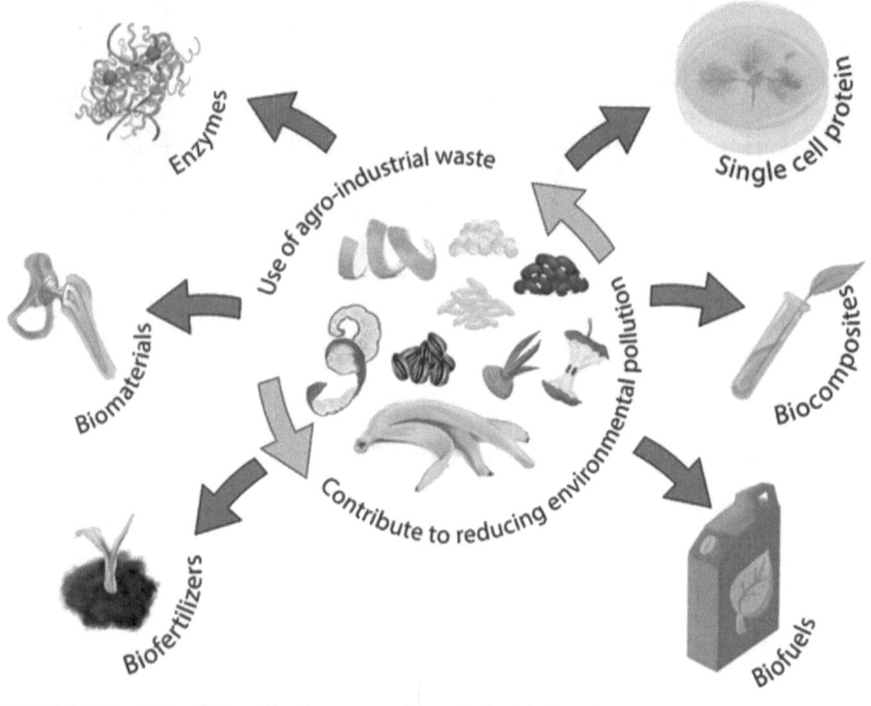

FIGURE 5.1 Potential application uses of agro-industrial wastes.

5.3.1 BIOFUELS

One of the great needs is the production of energy in a renewable way due to the scarcity of fossil fuels, which is why various strategies have been developed for the conversion of biomasses into fuels in recent years. Echaroj et al., [23] produced 33.4% biofuel through the optimization of alco-thermal liquefaction at 320°C, using as a catalyst the synthesis of modified graphene oxide, they found that the thermal characteristic of bio-oil is like gasoline, so this can be mixed with gasoline and used. In another research, Rani et al., [56] used wheat straw to obtain biogas through anaerobic digestion, performing a pretreatment based on KOH at room temperature to improve its biodegradability, where the highest cumulative biogas production was

558 mL with KOH at 2%, concluding that this pretreatment can be effective to improve biogas production from wheat straw and different sources of lignocellulosic biomasses.

5.3.2 BIOCOMPOUNDS

Recently, agro-industrial wastes have been used as raw materials to obtain products that provide health benefits. In recent years, interest in the health of the intestinal microbiota has increased, by means of xylan hydrolysis, xylooligosaccharides can be produced, which have prebiotic properties, and are not digestible but fermentable, and benefit the metabolism of probiotic bacteria. In this way, xylooligosaccharides transfer diverse health benefits, covering different systems, such as gastrointestinal, neurological, cardiovascular, inflammatory, etc. This makes them attractive in different pharmaceuticals, agriculture, and food industries. However, these productions costly, so research on new technologies for the complete recovery of xylan from lignocellulosic biomass is required as they are economic resources [28]. Seesuriyachan et al., [65] performed ultra-high-pressure pretreatment at 100 MPa for 10 min of corn cobs to improve enzyme accessibility for xylan to xylooligosaccharide conversion obtaining a maximum yield of 35.6 mg/g of substrate, in addition to containing antioxidant potential ABTS and FRAP, unlike untreated corn cobs. The pineapple industry is one of the agro-chains, which generates more waste, about 27 million tons are grown worldwide per year [40]where of the total fruit, about 50% represent waste, including peels, stems, leaves, and cores [70], which are potential sources of significant compounds, as they are rich in sugars, fibers, and micronutrients [59]. Sepúlveda et al., [67] used pineapple residues, mainly peels and cores, for the extraction of glycosides by autohydrolysis using water as a solvent for extraction at different temperatures, concentrations (w/v) and times, finding that the best conditions to produce glucose and fructose were 27.6 g/L and 33.8 g/L, respectively at 150°C for 30 minutes with a 1:10 w/v solid-liquid ratio, and the highest number of total polyphenols was 1.75 g/L and was obtained at 200C for 30 minutes with a 1:10 w/v ratio, which have diverse applications as additives food, prebiotics, and bio-preservatives. In 2013, 37.22 million tons of carrots were produced worldwide [77], of which approximately 30% is generated as waste due to various factors such as discarding, either by size, diameter, or shape, not counting the by-products generated in the food industry such as carrot peels, pulp or pomace; waste that is a cheap and sustainable source for the production of various value-added products, as it also contains various bioactive compounds [6]. The food and pharmaceutical industries have recently become interested in the development of bioactive products based on residues associated with the inhibition of free radicals [19]. For example, ellagic acid is a thermostable bioactive molecule, possessing therapeutic properties such as antioxidant, antiviral, cardioprotective, anti-inflammatory, anticancer, etc. Due to extraction costs, these compounds in pomegranate peel have not been highly exploited. Muñiz-Márquez et al., [49] performed ultrasound-assisted extraction of ellagic acid from pomegranate peel, evaluating factors of temperature, sonication time, solvent concentration and liquid-solid ratio, finding that the best extraction conditions were 93.6°C for 55.23 minutes in 75.23% aqueous ethanol with 3.27 mL/g liquid-solid ratio, obtaining a recovery of 19.47 mg/g of this valuable compound, which can be used in the food and pharmaceutical industry. Potatoes are one of the most commercial crops worldwide, the residues generated by potato peelings range from 15 to 40% depending on their consumption. Phenolic compounds are one of the main potato compounds, which have potential applications in the food industry for their antioxidant and antibacterial properties [80]. Sánchez et al., [62] investigated the extraction of phenolic acids from potato peels recovering quantitatively with acidified water/ethanol-based solvent more than 90% after two extractions, allowing the sustainable recovery of secondary metabolites from these residues, turning them into an alternative as food preservatives. Avocado peels and seeds are simply discarded by only using the pulp, which originates solid residues ranging between 21% and 30% [43] these residues contain polyphenols with antioxidant and antimicrobial power.

5.3.3 BIOMATERIALS

Various types of waste can be used as substrates for the production of biomaterials, but recently there has been a greater focus on agro-industrial waste, which has a high production rate worldwide [55]. Vijila et al., [78] reduced up to 87% fluorides in water with 0.5 mg/100 mL of an adsorbent made from rice husk residues with surface modification by HCl of 0.1 mol/dm^3 at 110°C with continuous agitation. Mentioning that the efficacy of this is associated with the increase in the concentration of the adsorbent and other parameters. According to the WHO, high fluoride levels in water cause dental and bone shortages in living beings, so it is necessary to eliminate them and thus make it safe for human use. This gives another opportunity for the utilization and research of these wastes for the creation of more biodegradable methods of different natural adsorbents that cover the removal of different pollutants from water. In addition to being consumed fresh, Mango is a fruit where its pulp is also used to obtain different products such as jams, juices, canned food, etc., in the food industry, but nevertheless, between 35% and 60% of the total fruit is discarded. From the seed alone, approximately more than 1 million tons of waste are generated that are not used and are rich in carbohydrates, proteins, lipids, vitamins, and functional properties [74] so they can be used. Henrique et al., [32], took advantage of mango seed waste, to give it a useful application and reduce its environmental impact, extracting, and characterizing cellulose nanocrystals by acid hydrolysis of 20 mL of 11.21 M H_2SO_4 at 40°C for 20 minutes per gram of cellulose, obtaining a yield of 28% with crystals of high crystallinity, good thermal stability and an aspect ratio of approximately 34.1 length/diameter, concluding that the nanocrystals obtained from mango seeds have great potential as reinforcing agents for the manufacture of nanocomposites; since another need is the development of biomaterials, renewable or biodegradable products. Several research has been directed to the study of nanocrystals from lignocellulosic substrates, as reinforcing agents of natural or synthetic polymer matrices. Cellulose nanocrystals have a wide variety of applications such as barrier films, antimicrobial films, transparent films, flexible screens, biomedical implants, pharmaceuticals, fibers, and textiles, etc. Ramos-Andrés et al., [54] performed the extraction of waste carrot pulp in a hydrothermal treatment in a continuous flow reactor at three different temperatures and pressurized to then be subjected to ultrafiltration and diafiltration with membranes for the purification of hemicelluloses and pectins, where they obtained purified solid fractions between 73.1% and 100% by weight of different molecular weights and low polydispersity, concluding that from these fractions biodegradable films can be obtained. Sugar beet is one of the crops with the highest production worldwide, and its pulp is one of the most abundant residues; worldwide, 120 million tons of wet matter are generated [12]. After sugar extraction, large amounts of residue are generated, consisting of approximately one-third cellulose, one-third hemicellulose, and one-third pectin, which is a good emulsifier. The extraction of pectin is commonly done by acid hydrolysis; however, this can reduce its molecular weight and affect its properties, so it must be extracted with more environmentally friendly technologies. Abou-Elseoud et al., [2] performed the extraction of pectin from sugar beet pulp by ultrasound-assisted enzymatic treatments showing a higher yield with the use of mixed enzyme mixtures than separately.

The sum of efforts made by scientists to take advantage of agro-industrial waste is abundant; however, industrial scaling is limited, and priority should be given to the use of environmentally friendly technologies in all processes and to close the waste cycle to avoid generating more pollutants to the environment.

5.3.4 AGRO-INDUSTRIAL WASTES: POTENTIAL SUBSTRATES FOR MICROBIAL BIOMASS PRODUCTION

To increasing the benefits of utilizing these wastes; implies using ecological and economic techniques and technologies [17]. Biological methods based on hydrolysis and saccharification by extracellular enzymes are environmentally friendly and do not develop inhibitors [48]. One of the biological methodologies is the use of microorganisms, which due to their excellent metabolism, can evolve, adapt, and thus, through controlled conditions (such as temperature, pH, inoculum, among others) carry out the biotransformation of high-value products [73]. Through fermentation technology, it is possible to obtain bioactive compounds and secondary metabolites that can be produced from microorganisms. Fermentation technology using agro-industrial waste as a substrate for the growth of microorganisms and to obtain high-value products allows the reuse of such waste, reducing costs and environmental pollution; however, it is necessary to study and research the capacity of microorganisms for the recovery of products, as well as the approach of experimental designs and the optimization of growth conditions, in order to obtain positive yields and high-quality compounds. There are two types of fermentation. Submerged fermentation and solid fermentation. In submerged fermentation, in which microbial growth occurs in a liquid medium, and in solid fermentation, in which the development of microorganisms occurs a solid medium in the absence of water particles [15]. The product produced or obtained depends largely on the composition of the substrate used; the use of agro-industrial wastes must be cost-effective, of high quality, non-renewable, multifunctional, etc. [81].

5.4 THE IMPORTANCE OF USING AGRO-INDUSTRIAL WASTE FOR SINGLE-CELL PROTEIN

Microorganisms can be considered allies to use food technologies and economic and ecological alternatives due to their attractive properties [47]. They have advantages compared to the production of conventional proteins such as vegetable and livestock proteins. Microorganisms can be cultivated in any season of the year and save water [24, 50]. Microbial protein is the biomass of unicellular microorganisms such as bacteria and multicellular microorganisms such as fungi and can be obtained from raw materials and wastes as carbon sources [24, 50]. Microbial protein is already a well-known concept, as it has been reported that yeasts were used to supply protein requirements since the first World War [75]. Single-cell proteins result in dried cells obtained from microorganisms that can be used as a protein supplement in human and animal food [9]. Today there is a need for alternative food production due to alarming global overpopulation, evident climate change, limited natural resources, as well as concern for human health [36]. Single-cell protein (SCP) production turns out to have an advantage in the face of these issues, as SCP production is more environmentally friendly, requires less water, less land area, and its effect on climate change is less than in the case of agriculturally derived proteins [14]. One of the main objectives of sustainable development is "zero hunger," and today malnutrition is one of the most severe problems in some countries around the world, specifically protein-energy malnutrition [10]. SCP production is crucial because it can replace in the future the deficiencies of conventional protein sources [13]. Although protein feed is mostly provided of animal origin, its production is costly and environmentally polluting since it requires 100 times more water than the same amount of production of other proteins such as those from vegetable sources [10] added to the fact that not everyone has the possibility of acquiring them, since there are estimates where the population increase will reach 8.1 billion people by 2025 and increase even more by 2050 to 9.6 billion, where approximately 3 billion people will belong to the middle class, this growth leads to a lower quality of life, poverty, and hunger, especially in developing countries where malnutrition is one of the most serious problems of public health [10, 57].

5.4.1 SINGLE-CELL PROTEIN PRODUCTION BY VARIOUS MICROORGANISMS

Therefore, alternative food sources that are of high nutritional value, low cost, and rapid synthesis are currently being sought [38], adding food security and sustainable food production. The production of SCP can be carried out by fermentation on a wide variety of substrates, such as industrial waste, where it can reach more profitable levels [16]. The use of microorganisms capable of growing on diverse carbon sources, which also exist in large quantities in the world and are not used, makes this alternative ecological, interesting, and low cost. Table 5.3 shows the use of different sources of agro-industrial wastes as substrates for obtaining unicellular protein through fermentation processes.

5.4.2 BACTERIAS

Bacteria are attractive because of their high protein content of 50 to 80% and their short growth time [50] they can grow on a wide variety of substrates and their protein is of higher quality than those of other microorganisms. An example is the bacterium *Methylophilus* spp., which is useful in animal feed [51]. However, they have disadvantages, such as their small size, low density, and high nucleic acid content; when ingested too much, they produce uric acid precipitation, causing kidney stones. Another disadvantage is that in relation to yeasts and fungi is the people belief that all bacteria are harmful [550]. For bacterial strains to be considered for SCP production they must meet requirements such as heat, oxygen, foam generation, growth rate, pH, and yield throughout fermentation, final product, etc. [13]. He et al., [30] used a wild-type strain *R. sphaeroides* Z08 to treat soybean wastewater and generate single-cell proteins by developing a cost-effective and nutrient-added process. Latex rubber sheet wastewater with fermented pineapple extract was efficiently treated under microaerobic light conditions using *R. palustris*, the biomass contained 65% protein and the unicellular protein contained higher protein, methionine, and threonine content than soybean meal (with a protein content of 37%) [37]. Soybean hulls obtained from soybean oil extraction are inexpensive food ingredients with a high-fiber content. Two *Bacillus subtilis* strains MR10 and TK8 were isolated from Tua-Nao, a traditionally fermented soybean seed in northern Thailand. The protein content after fermentation was 25.6% for MR10 and 26.6% for TK8 [79].

5.4.3 FUNGI

Fungi provide 30 to 50% of proteins, these are preferred for their chemical composition and amino acid profile [41]. In addition to providing the same, they can provide other nutrients such as some vitamins; mainly from the vitamin B complex such as riboflavin, niacin, thiamine, biotin, pantothenic acid, choline, pyridoxine, glutathione, p-amino benzoic acid, streptogenin, and folic acid, however, due to their high nucleic acid content they require prior processing to reduce them [58]. Aruna [7] and Khan Yousufi [35]

TABLE 5.3 Main Microorganisms Used to Produce Unicellular Protein from Agro-Industrial Wastes

Microorganism	Substrate	Culture Conditions	Yield
		Bacteria	
Rhodobacter sphaeroides	Soybean wastewater	Temperature: 12–25°C Light: Microaerobic DO: 0.5–1.0 mg L^{-1}	0.28 g g^{-1} Biomass 52% Protein
Rhodopseudomonas palustris	Pineapple extract	Inoculum: 2% v/v ph: 7 Light: Microaerobic COD: 2000 mg L Substrate concentration: 2% v/v fermented pineapple extract	65% Protein
Bacillus subtilis	Soybean hulls	Temperature: 37°C	25.6, 26.6% Protein
		Fungi	
Cladosporium cladosporioides, Aspergillus ochraceus, Aspergillus niger, Aspergillus flavus, Penicillium citrinum, Monascus ruber, Fusarium semitectum, Fusarium sp[1] and *Fusarium sp*[2]	Rice bran	Temperature: Ambient Inoculum: 5 × 10^6 cells/mL Moisture: 60–65%	9.59%, 10.25%, 10.63%, 10.46%, 10.19%, 10.25%, 10.03%, 10.19%, 10.17% Crude protein
Trichoderma viride	Pineapple peel	Temperature: 30°C Inoculum 2.7×10^6 Moisture: 60% Substrate concentration: 50 g of pineapple peels	14.89% Protein crude 12.02% True protein
Aspergillus oryzae Rhizopus oliogosporus	Fruit residue mixtures	Temperature: 28°C Substrate concentration: 1:1	57.3 mg/100 g, 61.2 mg/100 g Protein

TABLE 5.3 *(Continued)*

Microorganism	Substrate	Culture Conditions	Yield
Aspergillus niger	Potato starch	Temperature: 32.8°C Inoculum: 1.78% pH 6.67	24.86% True Protein
		Yeasts	
Candida tropicalis	Sugar cane bagasse	Temperature: 30°C Inoculum: 15 g L^{-1} pH 5.0 Agitation: 150 rpm Substrate concentration: SBBH 1:4	0.28 g g^{-1} Biomass 0.17 g g^{-1} Protein
Candida utilis	Orange peel	Temperature: 30°C Inoculum: 1×10^6 cel/mL Agitation: 150 rpm Substrate concentration: 10% w/v	15.71 g L^{-1} Biomass 6.22% Protein
Kluyveromyces marxianus	Mixtures from the food industry	Temperature: 30°C Inoculum: 1 g without agitation	33.7% Protein w/w dm 25.5% w/w dm Fat
Saccharomyces cerevisiae	Fruit and vegetable waste blends	–	39% Protein
Yarrowia lipolytica	Sugar cane molasses	Temperature: 28°C Inoculum: 400 mL pH: 6.0 Agitation: 300–600 rpm Aeration: 1–4 L/min Oxygenation: 30%	151.2 g L Protein

managed to increase the crude protein content of pineapple peel by solid fermentation using the addition of ammonium sulfate from 4.5% to 14.89%, obtaining protein-rich *T. viride* fungal biomass that can be used as animal feed. Liu et al., [42] through a double fermentation with a mutant strain of *A. niger* H3 and *B. licheniformis* using industrial waste from potato starch processing, optimized the fermentation conditions increasing the protein value of the dry residue, showing potential in large-scale industrial application. Rice bran is one of the crops found in large quantities. Valentino et al. [76] performed a solid-state fermentation for 9 species of endophytic fungi to determine their potential as single-cell protein producers, obtaining with the *A. niger* strain, enriching rice bran, so it could be a possible source of single-cell protein. Fruit juice residues are generated in large quantities [35] performed the combination of 36 residues of apple, papaya, orange, pomegranate rind, watermelon peel, pineapple, mango, guava, and banana with *A. oryzae* and *R. oligosporus* fungi, obtaining the highest production with the combination of pomegranate rind and guava peel for *A. oryzae* with 57.3 mg per 100 g of a substrate and for *R. oligosporus* 61.2 mg per 100 g of a substrate with the combination of pineapple peel and pomegranate peel, concluding that the combination of fruit wastes can be used as substrates for the production of unicellular proteins for food or as feed.

5.4.4 YEAST

Yeasts have a lower protein content, between 45% and 65%, compared to bacteria; however, these are the most accepted and used for the production of unicellular protein [66]. Humans have preferred these for their use in fermentation throughout history, for the production of beer, wine, and bread. In addition, it has advantages such as ease of harvesting, lower nucleic acid content compared to bacteria and larger size [50]. Additionally, its concentration of essential amino acids is higher compared to other microorganisms, especially lysine [38]. Magalhães et al., [44] having these advantages and a good taste, yeasts are suitable for use as food additives [25] used sugarcane bagasse, which is one of the most abundant agricultural residues worldwide, as a carbon source substrate by making a sugarcane bagasse hemicellulosic hydrolysate (SBBH) supplemented with yeast extract They carried out a fermentation that was monitored for a total of 96 hours, concluding that considering the high protein content in this substrate (higher than 40%), the *C. tropicalis* strain presents a great potential for protein production, since the percentage of protein production was higher than expected against other yeast and fungi strains. Carranza-Méndez et al., [14] used orange peel waste as a sole carbon source for the growth of *C. utilis*, optimizing, and evaluating different mineral salts by submerged fermentation, showing that orange peel is an agro-industrial waste with potential for the growth of *C. utilis*. Gervasi et al., [25] performed single-cell protein production through simple aerobic fermentation without pretreatment of fruit and vegetable mixtures and realize residue-free production. Also, Aggelopoulos et al., [3] performed a comparison of the fat and protein content of mixtures of orange peel waste, potato pulp, molasses, and cheese whey fermented with *S. cerevisiae, kéfir* and *K. marxianus.* Considered for enriching livestock feed. Yan et al., [82] used sugar cane molasses, used cooking oil and crude glycerol using *Y. lipolytica* for protein production. The cheapest and most effective residue for its production turned out to be sugar cane molasses. They performed an *in vivo* oral feeding test in fish, showing that the unicellular protein obtained is an excellent feed additive.

5.4.5 ALGAE

Algae are autotrophic organisms characterized by a wide genetic diversity. For their growth, in addition to inorganic nutrients, mainly nitrogen and phosphorus, they need water, carbon dioxide (CO_2) and light [60]. Microalgae can be a good source of SCP; however, they are mostly targeted for the potential production of omega-3 fatty acids [33]. Microalgae have the ability to produce cell biomass with a significant proportion of SCP (up to 70%), through the conversion of solar energy. Microalgae are unicellular microorganisms characterized by autotrophic growth, using light and CO_2 as energy and carbon suppliers [68]. Heterotrophic growth is characterized by the use of molasses, manure or other cheap organic materials such as agro-industrial waste as a carbon source. Putri et al., [53] performed the production of a unicellular protein with residues from tofu, cheese, and tempeh. They used the strain *Chlorella sp.* obtaining the highest total protein content (52.32%) with tofu waste. Wastewater has been reported as a culture medium. For the cyanobacterium *Arthrospira (Spirulina) platensis* with SCP yields of 48.59% or 56.17% on dry weight basis, depending on the culture medium. Based on recent algal SCP research, *Chlorella* sp. (up to 52%) and *Spirulina* (up to 50%) have been shown to be able to offer the highest protein content while providing healthy lipids and being considered environmentally friendly and very "green." It can be concluded that certain types of microalgae are successfully cultivated for animal and human consumption [53].

5.5 CHALLENGES AND FUTURE PROSPECTS

Considering that agro-industrial wastes produce massive biomass and are allowed to be converted into various value-added products, they can have a great positive effect both economically and environmentally. However, one of the main challenges is their conversion. Although various processes have been developed for the conversion of these wastes, current technologies are still vulnerable, costly, and the diverse composition of the feedstock limits their industrial scale-up. However, bioprocessing with the fermentation approach is very promising. During the development of this technology, by-products are released, so there is a need for effective measures, better understanding of their positive functions and manipulation of the process to avoid the emission of such by-products. Therefore, capital investment is needed for the production of these products and the development of satisfactory controlled bioprocesses. There is an urgent need for innovative, economically viable and environmentally friendly technology options. Another challenge is the lack of public knowledge about the treatment of these wastes. It is important to raise awareness in the agricultural and urban sectors through regulatory authorities. The union of the workers of this industry could manage the waste and establish strategies to reach a viable and stable path for the valorization of this waste.

On the other hand, food shortages and hunger in the future could benefit from the production of single-cell proteins. The single-cell protein must meet the nutritional requirements to be consumed by humans and animals. The quality of single-cell protein varies depending on the microorganism and substrate used. Therefore, it is necessary to investigate its nutritional quality such as amino acid composition, digestibility, and toxicity. Studying different microorganisms for its production helps to obtain better quality proteins, together with selecting adequate substrates. The use of agro-industrial wastes as raw material for its production is low cost and its process contributes to reducing environmental pollution. However, to ensure the success of their production, it is necessary to solve challenges in the technologies used. Their production requires scientific and industrial efforts. On the other hand, acceptance, and increasing consumer awareness of the use of non-conventional proteins such as those obtained from microorganisms. It is necessary to inform about their benefits and improve the perception of this type of alternative foods.

5.6 CONCLUSIONS

Currently, agro-industrial wastes have been used for the creation of different value-added products. Among the main applications is its use for the creation of biomaterials, biofuels, and biocomposites. Using agro-industrial wastes as raw materials reduces costs and minimizes their environmental impact. Interest in agro-industrial wastes is on the rise due to the concern for maintaining environmental balance and conservation. The use of these wastes for the production of value-added products allows researchers to extend their vision and at the same time reduce current environmental risks. The production of single-cell proteins can address some of the major future concerns, such as food shortages caused by global overpopulation and the unlimited use of natural resources. One of the great advantages is that it can be produced through low-cost substrates. Also, since microorganisms produce it, it is independent of the seasons and climatic conditions. The study and improvement of the quality of unicellular proteins can replace in the future proteins from conventional sources for human and animal consumption without side effects.

KEYWORDS

- **agro-industrial waste**
- **bacteria**
- **biocompounds**
- **biofuels**
- **biomaterials**
- **microbial biomass**
- **microorganisms**
- **single-cell protein**

REFERENCES

1. Abdel-Shafy, H. I., & Mansour, M. S. M., (2018). Solid waste issue: Sources, composition, disposal, recycling, and valorization. *Egyptian Journal of Petroleum, 27*(4), 1275–1290. https://doi.org/10.1016/J.EJPE.2018.07.003.

2. Abou-Elseoud, W. S., Hassan, E. A., & Hassan, M. L., (2021). Extraction of pectin from sugar beet pulp by enzymatic and ultrasound-assisted treatments. *Carbohydrate Polymer Technologies and Applications, 2*, 100042. https://doi.org/10.1016/J.CARPTA.2021.100042.
3. Aggelopoulos, T., Katsieris, K., Bekatorou, A., Pandey, A., Banat, I. M., & Koutinas, A. A., (2014). Solid state fermentation of food waste mixtures for single cell protein, aroma volatiles and fat production. *Food Chemistry, 145*, 710–716. https://doi.org/10.1016/J.FOODCHEM.2013.07.105.
4. Ahmad, A., Banat, F., & Taher, H., (2020). A review on the lactic acid fermentation from low-cost renewable materials: Recent developments and challenges. *Environmental Technology & Innovation, 20*, 101138. https://doi.org/10.1016/J.ETI.2020.101138.
5. Alzate, T. L. M., Jimenez, C. C., & Londoño, L. J., (2011b). Aprovechamiento de residuos agroindustriales para mejorar la calidad sensorial y nutricional de productos avícolas. *Producción + Limpia, 6*(1), 108–127. http://www.scielo.org.co/scielo.php?script=sci_arttext&pid=S1909-04552011000100010&lng=en&nrm=iso&tlng=es (accessed on 10 July 2023).
6. Amin, S., Jung, S., Kang, I., & Duval, A., (2021). Valorization of baby carrot processing waste. *Journal of Culinary Science and Technology*. https://doi.org/10.1080/15428052.2021.1879338.
7. Aruna, T. E., (2019). Production of value-added product from pineapple peels using solid state fermentation. *Innovative Food Science & Emerging Technologies, 57*, 102193. https://doi.org/10.1016/J.IFSET.2019.102193.
8. Behera, S. S., & Ray, R. C., (2016). Solid state fermentation for production of microbial cellulases: Recent advances and improvement strategies. *International Journal of Biological Macromolecules, 86*, 656–669. https://doi.org/10.1016/J.IJBIOMAC.2015.10.090.
9. Bertasini, D., Binati, R. L., Bolzonella, D., & Battista, F., (2022). Single cell proteins production from food processing effluents and digestate. *Chemosphere, 296*, 134076. https://doi.org/10.1016/J.CHEMOSPHERE.2022.134076.
10. Bessada, S. M. F., Barreira, J. C. M., & Oliveira, M. B. P. P., (2019). Pulses and food security: Dietary protein, digestibility, bioactive and functional properties. *Trends in Food Science & Technology, 93*, 53–68. https://doi.org/10.1016/J.TIFS.2019.08.022.
11. Bhatia, L., Johri, S., & Ahmad, R., (2012). *An Economic and Ecological Perspective of Ethanol Production from Renewable Agro Waste: A Review*. https://doi.org/10.1186/2191-0855-2-65.
12. Bonnin, E., Ralet, M. C., Thibault, J. F., & Schols, H. A., (2009). Enzymes for the valorization of fruit- and vegetable-based co-products. *Handbook of Waste Management and Co-Product Recovery in Food Processing, 2*, 257–285. https://doi.org/10.1533/9781845697051.3.257.
13. Bratosin, B. C., Darjan, S., Vodnar, D. C., & Durazzo, A., (2021). *sustainability Single Cell Protein: A Potential Substitute in Human and Animal Nutrition*. https://doi.org/10.3390/su13169284.
14. Carranza-Méndez, R. C., Chávez-González, M. L., Sepúlveda-Torre, L., Aguilar, C. N., Govea-Salas, M., & Ramos-González, R., (2022). Production of single cell protein from orange peel residues by *Candida utilis*. *Biocatalysis and Agricultural Biotechnology, 40*, 102298. https://doi.org/10.1016/J.BCAB.2022.102298.

15. Cassani, L., Marcovich, N. E., & Gomez-Zavaglia, A., (2022). Valorization of fruit and vegetables agro-wastes for the sustainable production of carotenoid-based colorants with enhanced bioavailability. *Food Research International, 152*, 110924. https://doi.org/10.1016/J.FOODRES.2021.110924.
16. Chacón, V. A., (2013). Perspectivas actuales de la proteína unicelular (SCP) en la agricultura y la industria. *Agronomía Mesoamericana, 15*(1), 93. https://doi.org/10.15517/am.v15i1.11939.
17. Chamorro, F., Carpena, M., Fraga-Corral, M., Echave, J., Riaz, R. M. S., Barba, F. J., Cao, H., et al., (2022). Valorization of kiwi agricultural waste and industry by-products by recovering bioactive compounds and applications as food additives: A circular economy model. *Food Chemistry, 370*, 131315. https://doi.org/10.1016/J.FOODCHEM.2021.131315.
18. Cury, R. K., Aguas, M. Y., Martinez, M. A., Olivero, V. R., & Chams, Ch. L., (2017). Residuos agroindustriales su impacto, manejo y aprovechamiento. *Revista Colombiana de Ciencia Animal–RECIA, 9*(S1), 122–132. https://doi.org/10.24188/recia.v9.ns.2017.530.
19. Da Silva, A. P. G., Sganzerla, W. G., Jacomino, A. P., Da Silva, E. P., Xiao, J., & Simal-Gandara, J., (2022). Chemical composition, bioactive compounds, and perspectives for the industrial formulation of health products from uvaia (*Eugenia pyriformis* Cambess–Myrtaceae): A comprehensive review. *Journal of Food Composition and Analysis, 109*, 104500. https://doi.org/10.1016/j.jfca.2022.104500.
20. Dahiya, S., Kumar, A. N., Shanthi, S. J., Chatterjee, S., Sarkar, O., & Mohan, S. V., (2018). Food waste biorefinery: Sustainable strategy for circular bioeconomy. *Bioresource Technology, 248*, 2–12. https://doi.org/10.1016/J.BIORTECH.2017.07.176.
21. Dey, T., Bhattacharjee, T., Nag, P., Ritika, Ghati, A., & Kuila, A., (2021). Valorization of agro-waste into value added products for sustainable development. *Bioresource Technology Reports, 16*, 100834. https://doi.org/10.1016/J.BITEB.2021.100834.
22. Duque-Acevedo, M., Belmonte-Ureña, L. J., Cortés-García, F. J., & Camacho-Ferre, F., (2020). Agricultural waste: Review of the evolution, approaches and perspectives on alternative uses. *Global Ecology and Conservation, 22*, e00902. https://doi.org/10.1016/J.GECCO.2020.E00902.
23. Echaroj, S., Pannucharoenwong, N., Ong, H. C., & Rattanadecho, P., (2021). Production of bio-fuel from alcohothermal liquefaction of rice straw over sulfated-graphene oxide. *Energy Reports, 7*, 744–752. https://doi.org/10.1016/J.EGYR.2021.07.081.
24. Fasolin, L. H., Pereira, R. N., Pinheiro, A. C., Martins, J. T., Andrade, C. C. P., Ramos, O. L., & Vicente, A. A., (2019). Emergent food proteins–towards sustainability, health and innovation. *Food Research International, 125*, 108586. https://doi.org/10.1016/J.FOODRES.2019.108586.
25. Gervasi, T., Pellizzeri, V., Calabrese, G., Di Bella, G., Cicero, N., & Dugo, G., (2018). Production of single cell protein (SCP) from food and agricultural waste by using *Saccharomyces cerevisiae. Natural Product Research, 32*(6), 648–653. https://doi.org/10.1080/14786419.2017.1332617.
26. Girelli, A. M., Astolfi, M. L., & Scuto, F. R., (2020). Agro-industrial wastes as potential carriers for enzyme immobilization: A review. *Chemosphere, 244*, 125368. https://doi.org/10.1016/J.CHEMOSPHERE.2019.125368.
27. Guimarães, I. C., Dos, R. K. C., Menezes, E. G. T., Rodrigues, A. C., Da Silva, T. F., De Oliveira, I. R. N., & Vilas, B. E. V. De. B., (2016). Cellulose microfibrillated suspension of carrots obtained by mechanical defibrillation and their application in edible starch

films. *Industrial Crops and Products, 89*, 285–294. https://doi.org/10.1016/J.INDCROP.2016.05.024.

28. Gupta, M., Bangotra, R., Sharma, S., Vaid, S., Kapoor, N., Dutt, H. C., & Bajaj, B. K., (2022). Bioprocess development for production of xylooligosaccharides prebiotics from sugarcane bagasse with high bioactivity potential. *Industrial Crops and Products, 178*, 114591. https://doi.org/10.1016/J.INDCROP.2022.114591.
29. Hassan, G., Shabbir, M. A., Ahmad, F., Pasha, I., Aslam, N., Ahmad, T., Rehman, A., et al., (2021). Cereal processing waste, an environmental impact and value addition perspectives: A comprehensive treatise. *Food Chemistry, 363*, 130352. https://doi.org/10.1016/J.FOODCHEM.2021.130352.
30. He, J., Zhang, G., & Lu, H., (2010). Treatment of soybean wastewater by a wild strain *Rhodobacter sphaeroides* and to produce protein under natural conditions. *Frontiers of Environmental Science and Engineering in China, 4*(3), 334–339. https://doi.org/10.1007/s11783-010-0239-5.
31. Hemalatha, M., & Brinda, L. A., (2021). Catalytic hydrolysis of fruit waste using magnetic carbon acid catalyst for bioethanol production. *Waste and Biomass Valorization, 12*(2), 971–983. https://doi.org/10.1007/s12649-020-01019-z.
32. Henrique, M. A., Silvério, H. A., Flauzino, N. W. P., & Pasquini, D., (2013). Valorization of an agro-industrial waste, mango seed, by the extraction and characterization of its cellulose nanocrystals. *Journal of Environmental Management, 121*, 202–209. https://doi.org/10.1016/J.JENVMAN.2013.02.054.
33. Jones, S. W., Karpol, A., Friedman, S., Maru, B. T., & Tracy, B. P., (2020). Recent advances in single cell protein use as a feed ingredient in aquaculture. *Current Opinion in Biotechnology, 61*, 189–197. https://doi.org/10.1016/J.COPBIO.2019.12.026.
34. Katare, V. D., Madurwar, M. V., & Raut, S., (2020). Agro-industrial waste as a cementitious binder for sustainable concrete: An overview. In: *Sustainable Waste Management: Policies and Case Studies* (pp. 683–702). Springer Singapore. https://doi.org/10.1007/978-981-13-7071-7_62.
35. Khan, Y. M., (2012). To determine protein content of single cell protein produced by using various combinations of fruit wastes and two standard food fungi. *International Journal of Advanced Biotechnology and Research, 3*(1), 533–536. http://www.bipublication.com (accessed on 10 July 2023).
36. Khoshnevisan, B., Tsapekos, P., Zhang, Y., Valverde-Pérez, B., & Angelidaki, I., (2019). Urban biowaste valorization by coupling anaerobic digestion and single cell protein production. *Bioresource Technology, 290*, 121743. https://doi.org/10.1016/J.BIORTECH.2019.121743.
37. Kornochalert, N., Kantachote, D., Chaiprapat, S., & Techkarnjanaruk, S., (2014). Use of *Rhodopseudomonas palustris* P1 stimulated growth by fermented pineapple extract to treat latex rubber sheet wastewater to obtain single cell protein. *Annals of Microbiology, 64*(3), 1021–1032. https://doi.org/10.1007/s13213-013-0739-1.
38. Kurcz, A., Błażejak, S., Kot, A. M., Bzducha-Wróbel, A., & Kieliszek, M., (2018). Application of industrial wastes for the production of microbial single-cell protein by fodder yeast candida utilis. *Waste and Biomass Valorization, 9*(1), 57–64. https://doi.org/10.1007/S12649-016-9782-Z.
39. Lenihan, P., Orozco, A., O'Neill, E., Ahmad, M. N. M., Rooney, D. W., & Walker, G. M., (2010). Dilute acid hydrolysis of lignocellulosic biomass. *Chemical Engineering Journal, 156*(2), 395–403. https://doi.org/10.1016/J.CEJ.2009.10.061.

40. Lim, Z. E., Thai, Q. B., Le, D. K., Luu, T. P., Nguyen, P. T. T., Do, N. H. N., Le, P. K., et al., (2020). Functionalized pineapple aerogels for ethylene gas adsorption and nickel (II) ion removal applications. *Journal of Environmental Chemical Engineering, 8*(6), 104524. https://doi.org/10.1016/J.JECE.2020.104524.
41. Linga, V. R., Ravindra, P., Ravinder, R., & Rao, L. V., (2003). Studies on *Aspergillus oryzae* mutants for the production of single cell proteins from deoiled rice bran yeast and lactic acid bacteria for probiotic applications. View project Biodiesel from *Jatropha curcas* using immobilized lipase View project. *Studies on Aspergillus oryzae Mutants for the Production of Single Cell Proteins from Deoiled Rice Bran.* https://www.researchgate.net/publication/232541260 (accessed on 10 July 2023).
42. Liu, B., Li, Y., Song, J., Zhang, L., Dong, J., & Yang, Q., (2014). Production of single-cell protein with two-step fermentation for treatment of potato starch processing waste. *Cellulose, 21*(5), 3637–3645. https://doi.org/10.1007/s10570-014-0400-6.
43. López-Cobo, A., Gómez-Caravaca, A. M., Pasini, F., Caboni, M. F., Segura-Carretero, A., & Fernández-Gutiérrez, A., (2016). HPLC-DAD-ESI-QTOF-MS and HPLC-FLD-MS as valuable tools for the determination of phenolic and other polar compounds in the edible part and by-products of avocado. *LWT, 73*, 505–513. https://doi.org/10.1016/J.LWT.2016.06.049.
44. Magalhães, C. E. B., Souza-Neto, M. S., Astolfi-Filho, S., & Matos, I. T. S. R., (2018). Candida tropicalis able to produce yeast single cell protein using sugarcane bagasse hemicellulosic hydrolysate as carbon source. *Biotechnology Research and Innovation, 2*(1), 19–21. https://doi.org/10.1016/J.BIORI.2018.08.002.
45. Martins, R. S. F., Gonçalves, F. G., Segundinho, P. G. De. A., Lelis, R. C. C., Paes, J. B., Lopez, Y. M., Chaves, I. L. S., & Oliveira, R. G. E. De., (2021). Investigation of agro-industrial lignocellulosic wastes in fabrication of particleboard for construction use. *Journal of Building Engineering, 43*, 102903. https://doi.org/10.1016/J.JOBE.2021. 102903.
46. Masaphy, S., Vanti, G. L., & Zabari, L., (2022). Laccase enhancement and antifungal toxicity reduction: Bidirectional influences between pomegranate peel extract and *Morchella conica* mycelium activity. *Bioresource Technology Reports, 17*, 100936. https://doi.org/10.1016/J.BITEB.2021.100936.
47. Matassa, S., Boon, N., Pikaar, I., & Verstraete, W., (n.d.). *Microbial Protein: Future Sustainable Food Supply Route with Low Environmental Footprint.* https://doi.org/10.1111/1751-7915.12369.
48. Mora-Sandí, A., Ramírez-González, A., Castillo-Henríquez, L., Lopretti-Correa, M., Roberto Vega-Baudrit, J., Almeida De, S. M., Barata, C. D. R., & Delattre, C., (2021). *Polymers Persea Americana Agro-Industrial Waste Biorefinery for Sustainable High-Value-Added Products.* https://doi.org/10.3390/polym13111727.
49. Muñiz-Márquez, D. B., Wong-Paz, J. E., Aguilar-Zárate, P., Sepúlveda, L., Buenrostro-Figueroa, J., Ascacio-Valdés, J. A., & Aguilar, C. N., (2021). Effect of ultrasound on the extraction of ellagic acid and hydrolysis of ellagitannins from pomegranate husk. *Environmental Technology & Innovation, 24*, 102063. https://doi.org/10.1016/J.ETI.2021.102063.
50. Nasseri, A. T., Rasoul-Amini, S., Morowvat, M. H., & Ghasemi, Y., (2011). Single cell protein: Production and process. *American Journal of Food Technology, 6*(2), 103–116. https://doi.org/10.3923/ajft.2011.103.116.
51. Øverland, M., Tauson, A. H., Shearer, K., & Skrede, A., (2010). Evaluation of methane-utilizing bacteria products as feed ingredients for monogastric animals. In: *Archives of Animal Nutrition* (Vol. 64, No. 3, pp. 171–189). https://doi.org/10.1080/17450391003691534.

52. Paritosh, K., Kushwaha, S. K., Yadav, M., Pareek, N., Chawade, A., & Vivekanand, V., (2017). *Food Waste to Energy: An Overview of Sustainable Approaches for Food Waste Management and Nutrient Recycling*. https://doi.org/10.1155/2017/2370927.
53. Putri, D., Ulhidayati, A., Musthofa, I. A., & Wardani, A. K., (2018). Single cell protein production of *Chlorella* sp. using food processing waste as a cultivation medium. *IOP Conference Series: Earth and Environmental Science, 131*, 012052. https://doi.org/10.1088/1755-1315/131/1/012052.
54. Ramos-Andrés, M., Aguilera-Torre, B., & García-Serna, J., (2021). Production of purified hemicellulose-pectin fractions of different molecular weight from discarded carrots by hydrothermal treatment followed by multistep ultrafiltration/diafiltration. *Journal of Cleaner Production, 321*, 128923. https://doi.org/10.1016/J.JCLEPRO.2021.128923.
55. Ranganathan, S., Dutta, S., Moses, J. A., & Anandharamakrishnan, C., (2020). Utilization of food waste streams for the production of biopolymers. In: *Heliyon* (Vol. 6, No. 9). Elsevier Ltd. https://doi.org/10.1016/j.heliyon.2020.e04891.
56. Rani, P., Bansal, M., & Pathak, V. V., (2022). Experimental and kinetic studies for improvement of biogas production from KOH pretreated wheat straw. *Current Research in Green and Sustainable Chemistry, 5*, 100283. https://doi.org/10.1016/J.CRGSC.2022.100283.
57. Reihani, S. F. S., & Khosravi-Darani, K., (2019). Influencing factors on single-cell protein production by submerged fermentation: A review. *Electronic Journal of Biotechnology, 37*, 34–40. https://doi.org/10.1016/J.EJBT.2018.11.005.
58. Ritala, A., Häkkinen, S. T., Toivari, M., & Wiebe, M. G., (2017). Single cell protein—State-of-the-art, industrial landscape and patents 2001–2016. *Front. Microbiol., 8*. https://doi.org/10.3389/fmicb.2017.02009.
59. Roda, A., & Lambri, M., (2019). Food uses of pineapple waste and by-products: A review. In: *International Journal of Food Science and Technology* (Vol. 54, No. 4, pp. 1009–1017). Blackwell Publishing Ltd. https://doi.org/10.1111/ijfs.14128.
60. Romero, M. L. D. L. Á., Guevara, M. Á., Gómez, B. J., Arredondo-Vega, B., Cortez, R., & Licet, B., (2017). Producción de pigmentos procedentes de *Arthrospira maxima* cultivada en fotobiorreactores. *Revista Colombiana de Biotecnología, 19*(1), 108–114. https://doi.org/10.15446/rev.colomb.biote.v19n1.59671.
61. Ruiz, R. L. G., Zamora, G. V. M., Pescuma, M., Van, N. C., Mozzi, F., & Sánchez, B. J. A., (2021). Fruits and fruit by-products as sources of bioactive compounds. Benefits and trends of lactic acid fermentation in the development of novel fruit-based functional beverages. *Food Research International, 140*, 109854. https://doi.org/10.1016/J.FOODRES.2020.109854.
62. Sánchez, M. A. F., Mudge, E., Gänzle, M. G., & Schieber, A., (2014). Extraction and fractionation of phenolic acids and glycoalkaloids from potato peels using acidified water/ethanol-based solvents. *Food Research International, 65*(PA), 27–34. https://doi.org/10.1016/J.FOODRES.2014.06.018.
63. Saravanan, A., Kumar, P. S., Mat, A. N. S., Jeevanantham, S., Karishma, S., Yaashikaa, P. R., Chew, K. W., & Show, P. L., (2021). A review on bioconversion processes for hydrogen production from agro-industrial residues. *International Journal of Hydrogen Energy*. https://doi.org/10.1016/J.IJHYDENE.2021.08.055.
64. Sarkar, N., Ghosh, S. K., Bannerjee, S., & Aikat, K., (2012). Bioethanol production from agricultural wastes: An overview. *Renewable Energy, 37*(1), 19–27. https://doi.org/10.1016/J.RENENE.2011.06.045.

65. Seesuriyachan, P., Kawee-ai, A., & Chaiyaso, T., (2017). Green and chemical-free process of enzymatic xylooligosaccharide production from corncob: Enhancement of the yields using a strategy of lignocellulosic destructuration by ultra-high-pressure pretreatment. *Bioresource Technology, 241*, 537–544. https://doi.org/10.1016/j.biortech.2017.05.193.
66. Sengupta, S., Mondal, A. K., Bhowal, J., & Bhattacharya, D. K., (2012). *Utilization of Fruit Wastes Producing Single Cell Protein.* Development of nondairy food product View project production of extruded food product view project utilization of fruit wastes in producing single cell protein. https://www.researchgate.net/publication/285112220 (accessed on 10 July 2023).
67. Sepúlveda, L., Romaní, A., Aguilar, C. N., & Teixeira, J., (2018). Valorization of pineapple waste for the extraction of bioactive compounds and glycosides using autohydrolysis. *Innovative Food Science & Emerging Technologies, 47*, 38–45. https://doi.org/10.1016/J.IFSET.2018.01.012.
68. Sharif, M., Zafar, M. H., Aqib, A. I., Saeed, M., Farag, M. R., & Alagawany, M., (2021). Single cell protein: Sources, mechanism of production, nutritional value and its uses in aquaculture nutrition. *Aquaculture, 531*, 735885. https://doi.org/10.1016/J.AQUACULTURE.2020.735885.
69. Sharma, A., Singh, G., & Arya, S. K., (2020). Biofuel from rice straw. *Journal of Cleaner Production, 277*, 124101. https://doi.org/10.1016/J.JCLEPRO.2020.124101.
70. Shivamathi, C. S., Gunaseelan, S., Soosai, M. R., Vignesh, N. S., Varalakshmi, P., Kumar, R. S., Karthikumar, S., et al., (2022). Process optimization and characterization of pectin derived from underexploited pineapple peel biowaste as a value-added product. *Food Hydrocolloids, 123*, 107141. https://doi.org/10.1016/J.FOODHYD.2021.107141.
71. Sidi-Yacoub, B., Oudghiri, F., Belkadi, M., & Rodríguez-Barroso, R., (2019). Characterization of lignocellulosic components in exhausted sugar beet pulp waste by TG/FTIR analysis. In: *Journal of Thermal Analysis and Calorimetry* (Vol. 138, No, 2, pp. 1801–1809). Springer Netherlands. https://doi.org/10.1007/s10973-019-08179-8.
72. Tchounwou, P. B., Kaniapan, S., Pasupuleti, J., Patma, N. K., Abubackar, H. N., Aminu, U. H., Oladosu, T. L., et al., (2022). *A Review of the Sustainable Utilization of Rice Residues for Bioenergy Conversion Using Different Valorization Techniques, Their Challenges, and Techno-Economic Assessment.* https://doi.org/10.3390/ijerph19063427.
73. Torres-León, C., Chávez-González, M. L., Hernández-Almanza, A., Martínez-Medina, G. A., Ramírez-Guzmán, N., Londoño-Hernández, L., & Aguilar, C. N., (2021). Recent advances on the microbiological and enzymatic processing for conversion of food wastes to valuable bioproducts. *Current Opinion in Food Science, 38*, 40–45. https://doi.org/10.1016/j.cofs.2020.11.002.
74. Torres-León, C., Rojas, R., Contreras-Esquivel, J. C., Serna-Cock, L., Belmares-Cerda, R. E., & Aguilar, C. N., (2016). Mango seed: Functional and nutritional properties. *Trends in Food Science & Technology, 55*, 109–117. https://doi.org/10.1016/J.TIFS.2016.06.009.
75. Ukaegbu-Obi, K., (2016). Single cell protein: A resort to global protein challenge and waste management. In: *J Microbiol Microb Technol* (Vol. 1, No. 1). https://www.researchgate.net/publication/318947586.
76. Valentino M. J. G., Ganado L. S., & Undan J. R. (2016). Single cell protein potential of endophytic fungi associated with bamboo using rice bran as substrate. *Advances in Applied Science Research, 7*(3), 68–72. https://www.primescholars.com/articles/single-cell-protein-potential-of-endophytic-fungi-associated-with-bamboo-using-rice-bran-as-substrate.pdf (accessed on 10 July 2023).

77. Varanasi, S., Henzel, L., Sharman, S., Batchelor, W., & Garnier, G., (2018). Producing nanofibers from carrots with a chemical-free process. *Carbohydrate Polymers, 184*, 307–314. https://doi.org/10.1016/J.CARBPOL.2017.12.056.
78. Vijila, B., Gladis, E. H. E., Jose, J. M. A., Sharmila, T. M., & Joseph, J., (2021). Removal of fluoride with rice husk derived adsorbent from agro-waste materials. *Materials Today: Proceedings, 45*, 2125–2129. https://doi.org/10.1016/J.MATPR.2020.09.729.
79. Wongputtisin, P., Khanongnuch, C., Kongbuntad, W., Niamsup, P., Lumyong, S., & Sarkar, P. K., (2014). *Use of Bacillus Subtilis Isolates from Tua-Nao Towards Nutritional Improvement of Soya Bean Hull for Monogastric Feed Application.* https://doi.org/10.1111/lam.12279.
80. Wu, D., (2016). Recycle technology for potato peel waste processing: A review. *Procedia Environmental Sciences, 31*, 103–107. https://doi.org/10.1016/J.PROENV.2016.02.014.
81. Yaashikaa, P. R., Senthil, K. P., & Varjani, S., (2022). Valorization of agro-industrial wastes for biorefinery process and circular bioeconomy: A critical review. *Bioresource Technology, 343*, 126126. https://doi.org/10.1016/J.BIORTECH.2021.126126.
82. Yan, J., Han, B., Gui, X., Wang, G., Xu, L., Yan, Y., Madzak, C., Pan, D., Wang, Y., Zha, G., & Jiao, L., (2018). Engineering *Yarrowia lipolytica* to simultaneously produce lipase and single cell protein from agro-industrial wastes for feed. *Scientific Reports, 8*(1), 758. https://doi.org/10.1038/s41598-018-19238-9.

CHAPTER 6

Microalgal and Fungal Biotechnology: A Path Forward in Assistance of the Sustainable Development Goals for the Management of Natural Resources

RUBY VARGHESE,[1,2] P. LOCHANA,[1] NAMITHA VIJAY,[2] and YOGESH BHARAT DALVI[2]

[1]*Department of Chemistry, School of Sciences, Jain Deemed to be University, Bangalore, Karnataka, India*

[2]*Pushpagiri Research Center, Pushpagiri Institute of Medical Sciences and Research Center, Tiruvalla, Pathanamthitta, Kerala, India*

ABSTRACT

Today the world is confronting humongous challenges like poverty, food scarcity, inequity, and climate changes. To administer these issues, the UN has proposed 17 goals to achieve an overarching paradigm that leads to sustainable development. Advancements in biotechnology offer solutions for stabilizing, securing, and tackling the urgent challenges currently faced by the world population. Emerging applications in the field of fungal and microalgal biotechnology have the potential to make a significant contribution to solving some of the challenges, by transitioning from the traditional oil-based economy to a bio-economy, reducing aquatic pollutants, enhancing the food supply to the human population by hindering emission of greenhouse gases, and sustainably produce constant sources of food, fuel, textiles, construction material, locomotive, and transportation industries, and more. The present chapter describes how Fungal and microalgal biotechnology can assist in achieving sustainable development goals framed by United Nations.

Bioresources and Bioprocess in Biotechnology for a Sustainable Future. Leonardo Sepúlveda Torre, Juan Carlos Contreras-Esquivel, Ann Rose Abraham, & A. K. Haghi (Eds.)

6.1 INTRODUCTION

Sustainable development goals (SDGs) framed by United Nations are a set of 17 contrived objectives for the well-being of our future generation by safeguarding our planet [48]. All these goals are interlinked, i.e., achieving one of the SDGs will influence the outcome of other SDGs, in such a way that any development in socio-economic-environmental sectors shall be well balanced [50]. Due to the poor interface between research fraternities and stakeholders; the implementation of SDGs in various countries is not properly executed [47]. In such an instance educating the biotechnological application to the non-academic stakeholders can prevent greater technological setbacks in achieving sustainability [30].

Biotechnology has multifaceted properties, which has created innumerable products and therapies to combat diseases, poverty, hunger, global warming, and employment with the sole goal to attain sustainability, and development through availing the emerging applications of nanotechnology, microbiology, pharmaceuticals, and sustainable farming (Figure 6.2). With the advanced low input biotechnology, various microorganisms or photosynthetic organisms are used for numerous sustainable activities such as bioremediation, biofuel production [5], renewable energy [40], sewage/wastewater treatment [25, 42], food, and medicine [2, 3, 9], and conversion of nonarable land to arable land [23, 27] (Figure 6.1).

Two of such incredible organisms are microalgae and fungi. Microalgae are photosynthetic organisms that utilize sunlight to convert CO_2 into organic matter. Though most of them are autotrophic, some can utilize carbon from an external source (heterotrophs). Due to their cosmopolitan nature, macromolecule-rich biomass with the capability to produce pigment, and secondary metabolites, they are extensively used in biotechnological applications such as biofuel, soil fertility, against plant pathogen, conversion of toxin effluents, and biorefinery as algal biochar [37, 48].

After the insect kingdom, the fungal kingdom plays a pivotal role in sustainability. As for a healthy ecosystem, the presence of fungi is essential, as they breakdown organic matters, and forms a symbiotic relationship with plants, with the use of the biotechnological application various products are developed such as Fungal foams which lead to sustainability (alternative to pernicious plastic material) [4], Pharmaceutical proteins [29], fungal recombinant enzymes for commercial purposes [16], and so on. This chapter shed light on the five applications of microalgae and fungi through biotechnology to attain sustainable development goals framed by United Nations (Figure 6.3).

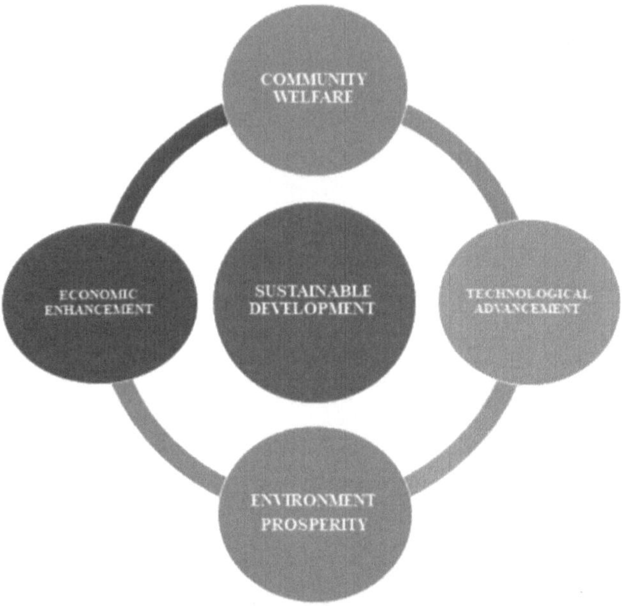

FIGURE 6.1 Four interlinked spheres of sustainable development.

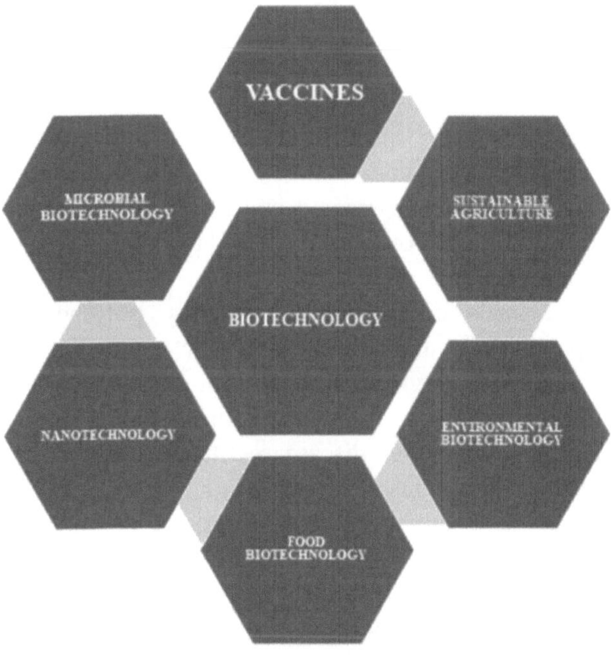

FIGURE 6.2 Biotechnological applications: thrust areas.

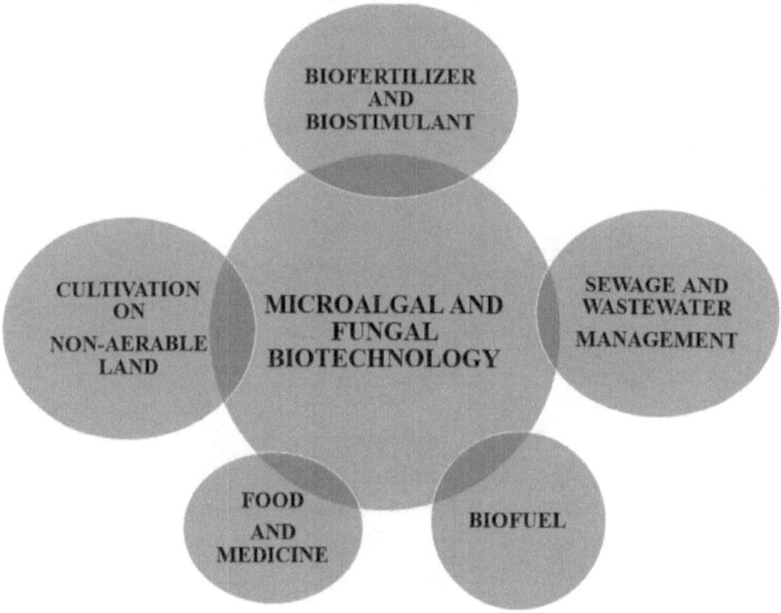

FIGURE 6.3 Five major applications of microalgal and fungal biotechnology in achieving sustainable development.

6.2 BIOFERTILIZER AND BIOSTIMULANT

Improved crop quality and quantity are necessary to enable global food security to feed the expanding human population sustainably during the current climatic changes. The agronomy and agro-industries have long viewed microalgae as a rich source of plant biostimulants and as a remarkable [22] investment prospect.

Microalgal extracts contain phytohormones, polysaccharides, carotenoids, and phycobilins that are essential for controlling plant growth, development, and defense system. In light of this, microalgal extracts can be utilized as sustainable sources for plant biostimulation. Microalgae increase plant development and confer disease resistance while posing minimum environmental costs, which results in high output in crops [21, 34]. Biostimulants from microalgae encourage root development, germination, flowering, fruit production, and good crop yield while providing resistance against various biotic and abiotic stress factors [22]. Microalgal biomass is a highly concentrated, stable, and transportable form of manure nutrients. The NPK (nitrogen, phosphorus, and potassium) concentration of the algal biomass is substantially higher than that of organic fertilizers available on the market. Even after using the microalgal biomass from wastewater treatment to make biofuel, it can still be used as manure for agriculture. The nation's economy and ecology will grow as a result of this zero-waste biorefinery strategy. For sustainable organic agriculture, these biofertilizers improve the soil's micro- and macronutrients and promote plant growth. This proves it to be an efficient biofertilizer [24].

As the applications of microalgae in agriculture are not yet explored completely, high-throughput phenotyping [11] and omic platforms, two recent biotechnological developments, are showing to be a promising answer in illuminating the underlying mechanisms of biostimulant activity and will afterward enable the development of innovative products with more research data.

Fungal biofertilizer inoculants increase crop output by promoting several biochemical processes that promote nutrient uptake, stimulate plant growth hormones, inhibit bacteria growth, and accelerate the breakdown of organic wastes. Fungi's innate ability to promote crop growth and lessen dependency on manmade chemical products accounts for much of its biocontrol potential [41]. The various enzymes produced by the fungi act as biostimulant and promotes plant growth and development [35].

Arbuscular mycorrhizal fungi (AMF), a class of soil fungi coexists with the majority of crops, having the capacity to release mineral nutrients for their plant hosts from the soil. AMF is known for promoting plant growth, protecting plants, and enhancing soil quality. They have gained widespread acceptability as organic farming manure in addition to being widely used in agricultural systems, particularly in areas where the pursuit of sustainable cultivation is prioritized [41]. To increase productivity without using excessive amounts of chemical fertilizers or pesticides, it is very advantageous to reap the benefits of relationships between fungi and the plant [53].

6.3 FOOD AND MEDICINE

Nutrient-rich microalgae have developed as a sustainable food source, offering a possible answer for world food security and reducing environmental problems brought on by the increase in land-based food production. A major advance in algal biotechnology has transformed algae into a potent "cell factory" for food production, spurring the rapid expansion of the algal bioeconomy in the food sector [26]. It is mainly focused on *Spirulina, Chlorella,* and *Dunalliella* for now. The market for human health food made from microalgae biomass is estimated to be worth roughly 2.5 billion US dollars in the future. Hence, the industry is booming with employment opportunities [39].

Similarly, food products with a fungal base are showing promising results in the food sector. These food products are very nutritious due to their high number of amino acids, fiber, low quantity of saturated fat, and high level of fungal protein digestibility. Due to their innate ability to secrete a variety of enzymes into the growth media, fungal cell factories are extensively used in the brewing, winemaking, and bread-making industries [33].

The use of microalgal bioactive compounds for the prevention and treatment of diseases like cancer, HIV/AIDS, malaria, diabetes, obesity, etc., is currently demonstrating encouraging results [28]. The diatom *Phaeodactylum tricornutum* has been successfully modified to produce and secrete human IgG antibodies against the surface protein of the hepatitis B virus by microalgal biotechnology. The major benefit is that the released antibodies are already in pure form, and labor-intensive and expensive purification stages are so made unnecessary [18]. The establishment of cell-cell fusion between HIV gp160 and CD4-expressing HeLa cells is inhibited by sulfated polysaccharides extracted from *Navicula directa,* known as Naviculum, which is essential in lowering HIV-1 infection [39].

Due to its increasing potential in medicine and pharmaceuticals, fungal biotechnology is a significant player in the global industry. With the discovery of the antibiotic penicillin, the use of fungus in medicine made significant progress. Following this, numerous bacterial infections that were once lethal are now curable. The widely used antifungal agent Griseofulvin is derived from fungi. *Griseofulvin* is used to treat dermatophytes extensively [54].

6.4 FUNGAL-ASSISTED ALGAL FLOCCULATION FOR THE BIODIESEL PRODUCTION

With the deprivation of natural resources and everchanging climatic conditions, there is an increasing demand for an alternative source of energy. Biodiesel made from biomass or vegetable oils is the best fuel alternative because biodiesel's raw materials can be sustainably supplied [46] and environmentally friendly [7, 51].

Biodiesel from microalgae and fungi is preferred as compared to conventional sources because of the following reasons:

- Small and marginal cultivation land is enough for microalgal and fungal biomass production [12]. The production of biomass will not interfere with the production of food derived from crops.
- The entire process of biodiesel production from either fungus or microalgae has been explored through technological advancements, like bio-flocculation. A most efficient and low-cost strategy for effective biomass production using both algal and fungal species like *Nannochloropsis oceanica*, *Chlorella* sp., *Tetraselmis suecica*, *Botryococcus* sp., *Chlamydomonas* sp., *Morticella elongate*, *Picochlorum* sp., *Saccharomyces* sp., *Aspergillus niger*, *Aspergillus nomius*, *Aspergillus* sp., and *Fumosorosea*. Both species when harvested together, more oil was produced which can be utilized as biofuel as compared to their conventional counterpart such as plants [49].
- About 60% of lipid (dry weight) can be accumulated by microalgae whereas fungi can garner < 20% (w/w) of lipid [45, 52].
- Disruption of the microalgal cell wall is an imperative step in the path of biofuel production. With the advances in technology, the production of cell wall-free algal protoplasts after the co-cultivation with fungal cells is a cost-effective oil extraction technique. Similarly, fungal cell walls are made up of chitin, whose degradation through pyrolysis represents an attractive method for the generation of biofuel [38].
- Due to the three important parameters such as growth rate, greater lipid accumulation, and the composition of essential fatty acids, fungi, and microalgae can be used as raw materials for the production of sustainable and renewable biofuels.

6.5 SEWAGE/WASTEWATER MANAGEMENT

The traditional way of water treatment is associated with the use of hazardous products and consequently their accumulation in the environment, to overcome this, sustainable, and eco-friendly treatment techniques are being developed. Numerous types of microalgae efficiently remove pathogens, heavy metals, pesticides, organic, and inorganic pollutants, and nitrogen and phosphorus from wastewater. The primary mechanism of removing pollutants involves storing them or utilizing them by microalgal cells [19].

Advantages of microalgal treatment over traditional treatment methods:

- The ability of microalgae to transform contaminants into useful biomass. This is important from an economic standpoint. Traditional wastewater treatment methods, in contrast, rely on expensive energy and chemicals for the same.
- Pathogens in the wastewater can be removed just by composting the microalgal biomass with the green waste of the wastewater [36].
- Microalgal bioremediation could be integrated with existing treatment methods or adopted as the single biological method for efficiently treating wastewater [6].
- Microalgae have higher biomass yield, with productivities estimated to be over 70 metric tons per hectare per year (ha1 yr1) of dry weight.
- Arable soil or fresh water is not necessary for the growth of micro-algae. It can be done by using wastewater [24].
- The co-cultivated microbes interact cooperatively, improving the overall uptake of nutrients.
- Microalgal systems are typically more resilient to oscillations in the environment [43].

The use of fungal biotechnology in wastewater treatment is a booming field of study. High-strength wastewater is treated using filamentous fungus. The wastewater organics are transformed by fungi into very useful biochemicals and fungal proteins, as well as highly dewaterable fungal biomass. The fungal biomass can be effectively separated from the combined liquid post-treatment [44].

One of several advantages of the fungal process is the enzyme-mediated activity that provides the solution to the treatment of waste streams containing hazardous or xenobiotic organic pollutants. The enzymes are produced during all phases of the fungal life cycle and are present even at low pollutant concentrations. In the liquid state bioconversion of the sludge treatment process, filamentous fungi *Aspergillus niger* and *Penicillium corylophilum* are utilized to entrap and immobilize solid sludge particles to produce bigger pellets or flocs, which improve the ultimate separation and biodegradation [1]. The biosolids' structure is altered by the fungal filamentous mycelia, which improves the bioseparation, dewaterability, and filterability [32]. Multiple useful biochemical compounds can be obtained from a single fungus species through concurrent fungal treatment. For instance, proteinase, lipase, lysozyme, and chitin are all produced by *Aspergillus oryzae*. This in turn leads to efficient water treatment and simultaneous byproduct recovery.

The biomass of the fungi *Aspergillus niger* shows 100% efficiency in removing heavy metals lead and cadmium from wastewater. Fungal wastewater treatment is a desirable and sustainable alternative to traditional practice as it uses inexpensive organic substrate as a feed to produce high-value fungal byproducts along with concurrent wastewater remediation [44].

6.6 CULTIVATION ON NON-ARABLE LAND

To maintain sustainable development, the utilization of chemical fertilizers needs to be replaced with organic fertilizers or comparative substitutes, which can be environmentally friendly and can help in perpetuating sustainability.

6.6.1 *FUNGAL BIODIVERSITY AND THEIR ROLE IN SOIL HEALTH*

Fungi improve soil fertility in the following ways:

- Degrading organic matter and converting it into biomass, CO_2, nitrogen, and organic acids [55].
- It also reduces nitrogen content in the soil by converting it into protein.
- It forms a synergistic relationship with plants by providing soil nutrients to the plant in exchange for sugar, for example, AMF [8]. Thus, the implementation of AMF can improve soil fertility, by reducing environmental impacts (reduction in the use of chemical fertilizers), and thereby imparting high crop yield [10].

The non-arable lands are sometimes occupied as landfills which eventually becomes a dumpster for the disposal of toxic substances. Mycoremediation is a method through which ubiquitously found fungi can convert or degrade toxic effluents into less harmful substances, as fungi are toxin resistant. Then arises a question of why fungi are better than bacteria or any other microorganisms when it comes to bioremediation? The answer is simple, fungi can explore their surroundings, uninhibitedly in the hunt for new substrates while the movement of the bacteria depends on an external factor such as fluid and direct contact [115, 20].

Fungi biodiversity can also improve the soil quality by acting as fungicides, for example, the addition of chitin increases the number of fungi and bacteria in the soil thereby degrading the pathogenic fungi cell wall [13]. Fungi also play a major role in biocontrol and as biostimulants in horticulture [31]. AMF together with plant growth promotes rhizobacteria (PGPR) and synergistically produces a favorable impact on horticultural plant growth and microbial diversity in soil [8, 14].

Maintaining soil fungal biodiversity, and understanding the relationship of fungus towards the ecosystem, interventions of fungal biodiversity in the form of disease control, and mycoremediation with help of the right tools, are essential parameters to compare functional structures between ecosystems and envisage responses to environmental changes would be a useful advance [17].

In the case of Microalgae, these photosynthetic organisms do not require arable lands, fresh water, and larger cultivation area like terrestrial plants. It can be grown in a short duration and with being optimized by modifying laboratory conditions. Hence they can be grown in brackish/waste/sewage water and still provides a high yield such as 6 g/l of photoautotrophic Distilled water biomass and 10 g/l of heterotrophic DW biomass.

6.7 CONCLUSION

Sustainable development fortifies the socio-economical, environmental, and human well-being by safeguarding the planet. However, the anticipated catastrophe of SDG attainment suggests the poor interface between researchers and stakeholders which causes insufficiency in existing knowledge on technology to balance socio-economical, environmental, and public health (Figure 6.1). Out of 17 SDGs, the present chapter showed the attainment of five goals through the application of fungal and microalgal biotechnology.

For SDG-6, Microalgal biomass, and with the symbiotic relation between AMF and plants, sustainable agriculture can be carried out without costing the nation's economy and being favorable to the ecosystem by not using chemical fertilizers still having higher yield. Similarly, Microalgae bioremediate the sewage water into useful biomass, while fungi can convert toxic effluents present in the water into less harmful compounds which are in support of SDG-6. Microalgae and fungus (certain species) are nutrient-rich sustainable food sources due to the presence of high content amino acid, fiber, and low saturated fat, which is in support of SDG-2. For SDG-12, bioactive compounds from fungus and microalgae have been extensively used in treating various diseases such as HIV, malaria, obesity, cancer, and dermatological disorders. Fungal-assisted algal flocculation is used for Biodiesel production and Microalgae individually convert the contaminants in the sewage/wastewater treatment to generate biomass which becomes the local source of biofuel which is in support of SDG-7. Microalgal biostimulants and fungal biostimulants can improve soil conditions by either growing on marginal lands or converting non-arable land to arable land by providing soil fertility with high product yield which is in support of SDG-15 (Figure 6.4).

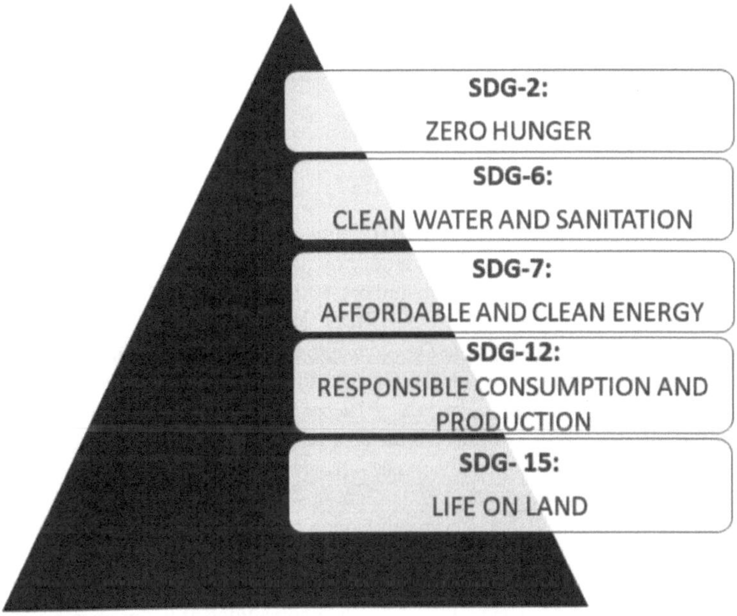

FIGURE 6.4 Fungal and microalgal biotechnology: In support of SDGs framed by UN.

Advancements in academics and technology with proper communication between scientists, and stakeholders need to be overcome to guarantee that fungal and microalgal biotechnology can promote the achievement of SDG, including optimization of production at every level and economical large-scale cultivation with considerable funding to accelerate microalgal and fungal technological innovation. Hence, the implementation of microalgal and fungal biotechnology at a large scale would be eventful, by ameliorating the environmental impacts with the provision of sustainable alternatives.

KEYWORDS

- **arable land**
- **biofertilizer**
- **biofuel**
- **biostimulant**
- **microalgal biotechnology**
- **sustainable development goals**
- **wastewater treatment**

REFERENCES

1. Alam, M. Z., & Fakhru'l-Razi, A., (2003). Enhanced settleability and dewaterability of fungal treated domestic wastewater sludge by liquid state bioconversion process. *Water Research, 37*(5), 1118–1124. https://doi.org/10.1016/S0043-1354(02)00452-9.
2. Abdel-Azeem, A. M., Yadav, A. N., Yadav, N., & Usmani, Z., (2021). *Industrially Important Fungi for Sustainable Development*. Springer International Publishing.
3. Abebe, G. M., (2020). The role of bacterial biofilm in antibiotic resistance and food contamination. *International Journal of Microbiology, 2020*.
4. Abhijith, R., Ashok, A., & Rejeesh, C. R., (2018). Sustainable packaging applications from mycelium to substitute polystyrene: A review. *Materials Today: Proceedings, 5*(1), 2139–2145.
5. Akinsemolu, A. A., (2018). The role of microorganisms in achieving the sustainable development goals. *Journal of Cleaner Production, 182*, 139–155.
6. Al-Jabri, H., Das, P., Khan, S., Thaher, M., & Abdulquadir, M., (2020). Treatment of wastewaters by microalgae and the potential applications of the produced biomass—A review. *Water, 13*(1), 27. https://doi.org/10.3390/W13010027.
7. Antolın, G., Tinaut, F. V., Briceno, Y., Castano, V., Perez, C., & Ramırez, A. I., (2002). Optimization of biodiesel production by sunflower oil transesterification. *Bioresource Technology, 83*(2), 111–114.
8. Azcón-Aguilar, C., & Barea, J. M., (1997). Applying mycorrhiza biotechnology to horticulture: Significance and potentials. *Sci. Hortic., 68*, 1–24. doi: 10.1016/S0304-4238(96)00954-5.
9. Yadav, A. N., Mishra, S., Kour, D., Yadav, N., & Kumar, A., (2020). *Agriculturally Important Fungi for Sustainable Agriculture*. Cham: Springer.
10. Bender, S. F., & Van, D. H. M. G., (2015). Soil biota enhance agricultural sustainability by improving crop yield, nutrient uptake and reducing nitrogen leaching losses. *Journal of Applied Ecology, 52*(1), 228–239.
11. Bulgari, R., Franzoni, G., & Ferrante, A., (2019). Biostimulants application in horticultural crops under abiotic stress conditions. *Agronomy, 9*(6), 306. https://doi.org/10.3390/AGRONOMY9060306.

12. Chisti, Y., (2007). Biodiesel from microalgae. *Biotechnology Advances, 25*(3), 294–306.
13. Cretoiu, M. S., Korthals, G. W., Visser, J. H. M., & Van, E. J. D., (2013). Chitin amendment increases soil suppressiveness toward plant pathogens and modulates the actinobacterial and oxalobacteraceal communities in an experimental agricultural field. *Appl. Environ. Microbiol.m 79*, 5291–5301. doi: 10.1128/AEM.01361-13.
14. De Coninck, B., Timmermans, P., Vos, C., Cammue, B. P. A., & Kazan, K., (2015). What lies beneath: Belowground defense strategies in plants. *Trends Plant Sci., 20*, 91–101. doi: 10.1016/j.tplants.2014.09.007.
15. De Wet, M. M. M., & Brink, H. G., (2021). Fungi in the bioremediation of toxic effluents. In: *Fungi Bio-Prospects in Sustainable Agriculture, Environment and Nano-Technology* (pp. 407-431). Academic Press.
16. Devi, R., Kaur, T., Guleria, G., Rana, K. L., Kour, D., Yadav, N., & Saxena, A. K., (2020). Fungal secondary metabolites and their biotechnological applications for human health. In: *New and Future Developments in Microbial Biotechnology and Bioengineering* (pp. 147-161). Elsevier.
17. Frąc, M., Hannula, S. E., Bełka, M., & Jędryczka, M., (2018). Fungal biodiversity and their role in soil health. *Frontiers in Microbiology, 9*, 707.
18. Hempel, F., & Maier, U. G., (2012). An engineered diatom acting like a plasma cell secreting human IgG antibodies with high efficiency. *Microbial Cell Factories, 11*(1), 1–6. https://doi.org/10.1186/1475-2859-11-126/FIGURES/3.
19. Hoffmann, J. P., (1998). Wastewater treatment with suspended and nonsuspended algae. *JournalofPhycology,34*(5),757–763.https://doi.org/10.1046/J.1529-8817.1998.340757.X.
20. Hurst, C. J., Crawford, R. L., Garland, J. L., & Lipson, D. A., (2007). *Manual of Environmental Microbiology*. American Society for Microbiology Press.
21. Kang, Y., Kim, M., Shim, C., Bae, S., & Jang, S., (2021). Potential of algae–bacteria synergistic effects on vegetable production. *Frontiers in Plant Science, 12*, 556. https://doi.org/10.3389/FPLS.2021.656662/BIBTEX.
22. Kapoore, R. V., Wood, E. E., & Llewellyn, C. A., (2021). Algae biostimulants: A critical look at microalgal biostimulants for sustainable agricultural practices. *Biotechnology Advances*, 49. https://doi.org/10.1016/J.BIOTECHADV.2021.107754.
23. Khalvati, M. A., (2022). Marginal lands as an important potential to produce biohydrogen. *International Journal of Hydrogen Energy*.
24. Khan, S. A., Sharma, G. K., Malla, F. A., Kumar, A., Rashmi, & Gupta, N., (2019). Microalgae based biofertilizers: A biorefinery approach to phycoremediate wastewater and harvest biodiesel and manure. *Journal of Cleaner Production, 211*, 1412–1419. https://doi.org/10.1016/J.JCLEPRO.2018.11.281.
25. Kour, D., Kaur, T., Devi, R., Yadav, A., Singh, M., Joshi, D., & Saxena, A. K., (2021). Beneficial microbiomes for bioremediation of diverse contaminated environments for environmental sustainability: Present status and future challenges. *Environmental Science and Pollution Research, 28*(20), 24917–24939.
26. Kusmayadi, A., Leong, Y. K., Yen, H. W., Huang, C. Y., & Chang, J. S., (2021). Microalgae as sustainable food and feed sources for animals and humans – Biotechnological and environmental aspects. *Chemosphere, 271*, 129800. https://doi.org/10.1016/J.CHEMOSPHERE.2021.129800.
27. Lafarga, T., Sánchez-Zurano, A., Morillas-España, A., & Acién-Fernández, F. G., (2021). Extremophile microalgae as feedstock for high-value carotenoids: A review. *International Journal of Food Science & Technology, 56*(10), 4934–4941.

28. Lauritano, C., Andersen, J. H., Hansen, E., Albrigtsen, M., Escalera, L., Esposito, F., Helland, K., et al., (2016). Bioactivity screening of microalgae for antioxidant, anti-inflammatory, anticancer, anti-diabetes, and antibacterial activities. *Frontiers in Marine Science, 3*, 1, 2. https://doi.org/10.3389/FMARS.2016.00068.
29. Li, Q. Z., Zheng, Y. Z., & Zhou, X. W., (2019). Fungal immunomodulatory proteins: Characteristic, potential antitumor activities and their molecular mechanisms. *Drug Discovery Today, 24*(1), 307–314.
30. Lindenfeld, L. A., Hall, D. M., McGreavy, B., Silka, L., & Hart, D., (2012). Creating a place for environmental communication research in sustainability science. *Environmental Communication: A Journal of Nature and Culture, 6*(1), 23–43.
31. López-Bucio, J., Pelagio-Flores, R., & Herrera-Estrell, A., (2015). Trichoderma as biostimulant: Exploiting the multilevel properties of a plant beneficial fungus. *Sci. Hortic. 196*, 109–123. doi: 10.1016/j.scienta.2015.08.043.
32. Mannan, S., Fakhru'l-Razi, A., & Alam, M. Z., (2005). Use of fungi to improve bioconversion of activated sludge. *Water Research, 39*(13), 2935–2943. https://doi.org/10.1016/J.WATRES.2005.04.074.
33. Meyer, V., Basenko, E. Y., Benz, J. P., Braus, G. H., Caddick, M. X., Csukai, M., De Vries, R. P., et al., (2020). Growing a circular economy with fungal biotechnology: A white paper. *Fungal Biology and Biotechnology, 7*(1), 1–23. https://doi.org/10.1186/S40694-020-00095-Z.
34. Michalak, I., & Chojnacka, K., (2015). Algae as production systems of bioactive compounds. *Engineering in Life Sciences, 15*(2), 160–176. https://doi.org/10.1002/ELSC.201400191.
35. Michellin, K. N. J., & Thatheyus, A. J., (2021). Fungi, fungal enzymes and their potential application as biostimulants. *Microbiome Stimulants for Crops*, 305–314. https://doi.org/10.1016/B978-0-12-822122-8.00024-8.
36. Molazadeh, M., Ahmadzadeh, H., Pourianfar, H. R., Lyon, S., & Rampelotto, P. H., (2019). The use of microalgae for coupling wastewater treatment with CO_2 biofixation. *Frontiers in Bioengineering and Biotechnology, 7*, 42. https://doi.org/10.3389/FBIOE.2019.00042/BIBTEX.
37. Mona, S., Malyan, S. K., Saini, N., Deepak, B., Pugazhendhi, A., & Kumar, S. S., (2021). Towards sustainable agriculture with carbon sequestration, and greenhouse gas mitigation using algal biochar. *Chemosphere, 275*, 129856.
38. Muradov, N., Taha, M., Miranda, A. F., Wrede, D., Kadali, K., Gujar, A., & Mouradov, A., (2015). Fungal-assisted algal flocculation: Application in wastewater treatment and biofuel production. *Biotechnology for Biofuels, 8*(1), 1–23.
39. Mutanda, T., Naidoo, D., Bwapwa, J. K., & Anandraj, A., (2020). Biotechnological applications of microalgal oleaginous compounds: Current trends on microalgal bioprocessing of products. *Frontiers in Energy Research, 8*. https://doi.org/10.3389/FENRG.2020.598803/FULL.
40. Obileke, K., Onyeaka, H., Meyer, E. L., & Nwokolo, N., (2021). Microbial fuel cells, a renewable energy technology for bio-electricity generation: A mini-review. *Electrochemistry Communications, 125*, 107003.
41. Odoh, C. K., Eze, C. N., Obi, C. J., Anyah, F., Egbe, K., Unah, U. V., Akpi, U. K., & Adobu, U. S., (2020). *Fungal Biofertilizers for Sustainable Agricultural Productivity* (pp. 199–225). https://doi.org/10.1007/978-3-030-45971-0_9.
42. Peter, A. P., Khoo, K. S., Chew, K. W., Ling, T. C., Ho, S. H., Chang, J. S., & Show, P. L., (2021). Microalgae for biofuels, wastewater treatment and environmental monitoring. *Environmental Chemistry Letters, 19*(4), 2891–2904.

43. Salama, E. S., Hwang, J. H., El-Dalatony, M. M., Kurade, M. B., Kabra, A. N., Abou-Shanab, R. A. I., Kim, K. H., et al., (2018). Enhancement of microalgal growth and biocomponent-based transformations for improved biofuel recovery: A review. *Bioresource Technology, 258*, 365–375. https://doi.org/10.1016/J.BIORTECH.2018.02.006.
44. Sankaran, S., Khanal, S. K., Jasti, N., Jin, B., Pometto, A. L., & Van, L. J. H., (2010). Use of filamentous fungi for wastewater treatment and production of high value fungal byproducts: A review. *Critical Reviews in Environmental Science and Technology, 40*(5), 400–449. https://doi.org/10.1080/10643380802278943.
45. Sergeeva, Y. E., Galanina, L. A., Andrianova, D. A., & Feofilova, E. P., (2008). Lipids of filamentous fungi as a material for producing biodiesel fuel. *Applied Biochemistry and Microbiology, 44*(5), 523–527.
46. Sheehan, J., Camobreco, V., Duffield, J., Graboski, M., & Shapouri, H., (1998). *An Overview of Biodiesel and Petroleum Diesel Life Cycles*. Distributed by the Office of Scientific and Technical Information, U.S. Department of Energy.
47. Suldovsky, B., McGreavy, B., & Lindenfeld, L., (2017). Science communication and stakeholder expertise: Insights from sustainability science. *Environmental Communication, 11*(5), 587–592.
48. Sutherland, D. L., McCauley, J., Labeeuw, L., Ray, P., Kuzhiumparambil, U., Hall, C., & Ralph, P. J., (2021). How microalgal biotechnology can assist with the UN sustainable development goals for natural resource management. *Current Research in Environmental Sustainability, 3*, 100050.
49. Ummalyma, S. B., Gnansounou, E., Sukumaran, R. K., Sindhu, R., Pandey, A., & Sahoo, D., (2017). Bioflocculation: An alternative strategy for harvesting of microalgae: An overview. *Bioresource Technology, 242*, 227–235.
50. UNGA (United Nations General Assembly) Transforming our world: The 2030 Agenda for Sustainable Development. A/RES/70/1-2015.
51. Vicente, G., Martınez, M., & Aracil, J., (2004). Integrated biodiesel production: A comparison of different homogeneous catalysts systems. *Bioresource Technology, 92*(3), 297–305.
52. Wrede, D., Taha, M., Miranda, A. F., Kadali, K., Stevenson, T., Ball, A. S., & Mouradov, A., (2014). Co-cultivation of fungal and microalgal cells as an efficient system for harvesting microalgal cells, lipid production and wastewater treatment. *PloS One, 9*(11), e113497.
53. Xavier, L. J. C., & Boyetchko, S. M., (2002). Arbuscular mycorrhizal fungi as biostimulants and bioprotectants of crops. *Applied Mycology and Biotechnology, 2*, 311–340. https://doi.org/10.1016/S1874-5334(02)80015-6.
54. Yadav, A. N., Kaur, T., Devi, R., Kour, D., Yadav, N., Abdel-Azeem, A. M., Yadav, A., & Ahluwalia, A. S., (2021). *Bioprospecting for Biomolecules from Industrially Important Fungi: Current Research and Future Prospects*, 767–791. https://doi.org/10.1007/978-3-030-85603-8_23.
55. Žifčáková, L., Vetrovský, T., Howe, A., & Baldrian, P., (2016). Microbial activity in forest soil reflects the changes in ecosystem properties between summer and winter. *Environ. Microbiol., 18*, 288–301. doi: 10.1111/1462-2920.13026.

CHAPTER 7

Biomass Production and Partial Characterization of Agavinase from *Aspergillus kawachii* via Submerged Fermentation: New Insights for a Sustainable Future

OSCAR FERNANDO VÁZQUEZ-VUELVAS,[1,2]
LAURA LETICIA VALDEZ-VELÁZQUEZ,[2]
ARMANDO PINEDA-CONTRERAS,[2] MARIO ALBERTO GAITAN-HINOJOSA,[2]
RODRIGO MACIAS-GARBETT,[1] EMILIO MENDEZ-MERINO,[3] and
JUAN CARLOS CONTRERAS-ESQUIVEL[1]

[1]Laboratory of Applied Glycobiotechnology, Academic Group of Food Science and Technology, School of Chemistry, Universidad Autonoma de Coahuila, Blvd. Venustiano Carranza Esq. Jose Cardenas Valdez s/n, Saltillo, Coahuila, Mexico

[2]Biochemical Engineering and Bioprocess Laboratory, Chemical Science Faculty, Universidad de Colima, Km 9 Carr. Colima-Coquimatlán s/n. Coquimatlán, Colima, Mexico

[3]Sigma Alimentos, San Pedro Garza Garcia, Nuevo Leon, Mexico

ABSTRACT

Aspergillus kawachii is a filamentous fungus widely used in the production of Japanese alcoholic and brewing beverages due to its enhanced saccharification capacity of structural polysaccharides such as cellulose, hemicellulose, and pectin. Agavin is a reserve polysaccharide in agave plants, which may be

Bioresources and Bioprocess in Biotechnology for a Sustainable Future. Leonardo Sepúlveda Torre, Juan Carlos Contreras-Esquivel, Ann Rose Abraham, & A. K. Haghi (Eds.)

used as a carbon source for fungal metabolites production. A submerged fermentation was carried out with *A. kawachii* with glucose, saccharose, and agavin as a carbon source to study the production of the enzyme agavinase. Biomass production, pH, and total sugars were measured during the cultivation to analyze the fungal growth and the feasibility of the secretion of agavinase. The enzyme activity to detect agavinase was carried out using agavin and fractions of crude culture broths from the three substrates. A chromatography of molecular size exclusion was applied to desalt and separate the enzyme extract from the rest of the components to identify agavin activity. Agavin activity was not detected in glucose extracts; low values were shown in the sucrose extract and were better for the agavin extracts. Agavinase is demonstrated to be produced inductively by *Aspergillus kawachii* in agavin-rich media, and the glucose as substrate evidenced the absence of agavinase, and then it is not a constitutive enzyme for the fungus.

7.1 INTRODUCTION

Aspergillus kawachii is a filamentous fungus widely used to elaborate *shōchū*, a traditional Japanese spirit produced from the saccharification of starchy materials such as rice, barley, and sweet potato [22]. Other structural polysaccharides have been employed as a carbon source to produce several metabolites for industrial purposes. Hemicellulose [10] and pectin [6, 7] are examples of bioprocessed substrates by *A. kawachii* to obtain hydrolysates, which demonstrate its fungal degrading and other lytic capabilities. Moreover, some Korean beverages, such as miso, use *A. kawachii* to carry out glycolysis of cereals to provide monosaccharides for alcoholic fermentation.

Agave tequilana Weber Var. Azul is an important bioresource for the Mexican economy, as this *Agavaceae* plant represents the raw material to produce several traditional fermented and distilled spirits such as pulque or tequila [2]. The main carbohydrate in the agave pines is fructan, a polysaccharide of fructose, which, as with inulin or levan, agave fructans possess structural particularities [12]; this is the reason for naming them agavins. Fructans are also present in several plants as Jerusalem artichoke (*Heliantus tuberosus*), the taproot of chicory (*Cichorium intybus*), the tubers of dahlia (*Dahlia variabilis*), the bulbs of tulip (*Tulipa gesneriana*) and onion (*Allium* cepa) [18]. As reserve carbohydrate of plants, fructans have acquired industrial relevance as a functional food; in fact, *A. tequilana* is used as a commodity to produce agavins, a powder food with prebiotic properties, which because of its value as a dietary fiber and sweetener, it is used to synthesize fructooligosaccharides and other food economically valuable supplements [1, 21].

From the biotechnological point of view, the production of hydrolysates of fructans is crucial to elaborate fructose syrups, which are conventionally manufactured by heating methods. However, the use of hydrolytic enzymes to saccharify the fructan biopolymer has been usually reported time ago. The use of inulinases, mainly produced by filamentous fungi such as *Aspergillus awamori*, *Aspergillus niger*, or *Aspergillus orizae* employing inulins or levans, has been often reported [19]. Since the capacity of *Aspergillus* sp. to contribute to inulinase production, *Aspergillus kawachii* was also experimented with and documented to produce an extracellular enzyme inulinase from yacon tubercle inulin culture (*Smallanthus sonchifolius* Poepp. & Endl) as a substrate in a liquid [4]. Nonetheless, several reports document the production of inulinases from *Aspergillus spp.*; there is a knowledge gap on the production of an inducible fructanase enzyme employing agavins as a substrate with *Aspergillus kawachii.* This report describes the conditions to culture *A. kawachii* for extracellular production of an enzymatic extract, a kinetic study of fungal growth using several carbon sources, the evaluation of the exo-agavinase activity of the crude extract, and an initial step of fractionation through chromatography of size exclusion.

7.2 EXPERIMENTAL METHODS AND MATERIALS

7.2.1 MATERIALS

Glucose and sucrose were acquired from Sigma Aldrich (Mexico), agave fructan was procured from CP Ingredientes (Guadalajara, Mexico). Materials such as potato dextrose agar (PDA) (Difco), tryptone (Bacto), and Tween 80 (Hycel) were provided by different suppliers. Mineral salts used such as K_2HPO_4, KCl, $MgSO_4 \cdot 7H_2O$, and $FeSO_4 \cdot 7H_2O$ were analytical grade salts. Filtering membranes of 0.45 μm were provided by Millipore.

7.2.2 FUNGAL MICROORGANISM

Aspergillus kawachii NRBC4308 was supplied by the Biological Resource Center NITE (Tokyo, Japan) and activated from preservation in 80 mL PDA Erlenmeyer flasks during 6–8 days at 30°C [7]. Afterward, spore harvesting was carried out by adding 20 mL of Tween 80 at 0.02% (v/v) to the PDA fungal culture, and the spore collection was done by rotating a sterile magnetic bar over the surface. The spore suspension was counted by utilizing a microscope with a Neubauer chamber.

7.2.3 FUNGAL FERMENTATION

The culture broth was prepared using a previously reported protocol [6] with some modifications of composition and carbon source (g/L): K_2HPO_4 1 g; KCl 0.5 g; $MgSO_4 \cdot 7H_20$ 0.5 g; $FeSO_4 \cdot 7H_2O$ 0.01 g; tryptone 5 g, and carbon source 10 g (glucose, sucrose, and agavin). The broth was adjusted to pH 4.00 with H_3PO_4 1 M and sterilized in 500 mL conical flasks with 100 mL of culture solution for 15 min at 121°C and then inoculated with 10^6 spores per flask. The liquid fermentation was carried out in a New Brunswick shaker incubator with orbital agitation at 200 RPM and 30°C for 84 h. Three replicants were withdrawn every 12 h to analyze pH and biomass production. The culture broth was filtered with generic wide pore size filter paper to separate the fungal mycelium for determining the biomass production by a gravimetric method. The mycelium-free culture broth was analyzed for pH and immediately passed through 0.45 μm membranes (45 mm diameter). The filtered extract was stored at – 20°C for additional assays.

7.2.4 BIOMASS PRODUCTION

Biomass production was determined by a gravimetric method to study the fungal growth kinetics throughout the fermentation. The mycelium was separated by filtration, washed, and rinsed with 200 mL of deionized water employing a piece of wide pore-size filter paper weighed previously. The biomass was dried at 60°C for 48 h. The final weights were determined, and the calculated results were expressed in g/L.

7.2.5 TOTAL SUGAR DETERMINATION

To correlate the fungal consumption of substrates with its growth, the sucrose, and agavin culture broths were assayed for total sugars by the colorimetric phenol sulfuric acid method [13], glucose culture broth by the glucose oxidase-peroxidase method [17] using sucrose, agavin, and glucose as standards, respectively.

7.2.6 AGAVINASE ACTIVITY

Agavinase activity was estimated by the reducing sugar released from agavin in a hydrolytic enzyme reaction. A reaction test tube was prepared to contain 240 mL of agavin substrate solution (1500 ppm) in 50 mM acetate buffer, pH 4.5, and 10 mL of enzyme extract were added. A control substrate test tube was prepared with 240 mL of the same agavin acetate buffer, and 10 mL of acetate buffer was added. A control enzyme test tube was mixed with 240 mL of buffer acetate and 10 mL of enzyme extract. The test tubes were incubated at 40°C for 60 min. Afterward, the Somogyi-Nelson method was carried out to quantify the amount of reducing sugars in the samples using a calibration curve of fructose as standard [15, 20]. One unit of enzymatic activity (U/mL) is defined as the amount of enzyme required to liberate 1 µmol of reducing sugars per minute per milliliter under the assay conditions. The calculation formula to quantify the exo agavinase activity is the following:

$$\text{Agavinase activity}\left(\text{U}/\text{mL}\right) = M * V_s * \frac{\text{DF}}{\text{V}_\text{e} * t} \quad (1)$$

where; M is the concentration of reducing sugar equivalent fructose in the sample (µmol/mL); V_s is the total volume of enzymatic reaction mixture (mL); V_e is the volume of the enzyme (mL); DF is the dilution factor if the enzyme was diluted; and $t_{(min)}$ is the incubation time for reaction in minutes (min).

7.2.7 AGAVINASE ENZYME KINETICS

The analysis was carried out by an enzymatic reaction test tube with 2400 mL of agavin substrate solution (1500 ppm) in 50 mM acetate buffer, pH 4.5, and 100 mL of enzyme extract were added. A control substrate test tube was prepared with 2400 mL of the same agavin acetate buffer, and 100 mL of acetate buffer was added. A control enzyme test tube was mixed with 2400 mL of buffer acetate and 100 mL of enzyme extract. The test tubes were incubated at 40°C for 240 min, and a duplicate sample was taken from the three test tubes every 48 minutes to quantify the amount of reducing sugars.

7.2.8 PROTEIN ASSAY

Protein concentration was measured by the Lowry method [11]. The calibration curve was made with bovine serum albumin as standard.

7.2.9 GEL PERMEATION CHROMATOGRAPHY (GPC) OF ENZYMATIC EXTRACTS

Culture broths proceeding from the three carbon sources were submitted to a desalting procedure to separate enzymatic, sugar, and mineral components by a size exclusion chromatography (SEC) employing a Fast Protein Liquid Chromatography (FPLC) equipment AKTA Prime Plus (Amersham Biosciences, Uppsala, Sweden). GPC is a chromatographic technique based on separating dissolved solids by size, with the particularity of allowing to permeate first larger molecules and retain smaller molecules. The latter can be further eluted with the continuous flow of the mobile phase. The columns to separate the crude extracts were HiTrap Desalting (Amersham Biosciences, Uppsala, Sweden) 5 mL with Sephadex G-25 resin as a stationary phase and HiPrep 2610 Desalting 60 mL. First, the crude extracts from culture broths proceeding from the fungal growth study were lyophilized and rehydrated with a minimum of distilled water. They were then filtrated by GPC with the HiTrap column, using an extract sample of 0.5 mL eluted with 1 mL/min with distilled water, collecting 1 mL fractions. The crude extracts from endpoint fermentations were also lyophilized and rehydrated, and then filtered through a HiPrep column, which was used with an extract sample of 15 mL and eluted with 5 mL/min with distilled water and collecting 5 mL fractions.

7.2.10 MONODISPERSE AGAVIN SOLUTION

A 150 g/L of agavin was prepared and filtered by SEC to separate high molecular weight polymer chains from small ones to evaluate the agavinase activity with a substrate with initial diminished reducing power. The SEC procedure with the agavin solution was fractioned with a HiPrep column; 15 fractions were analyzed for total sugars and reducing sugars. The fractions with a high concentration of total sugars and a low concentration of reducing sugars were pooled to prepare a standard agavin solution to estimate agavinase *exo* activity.

7.3 RESULTS AND DISCUSSION

7.3.1 BIOMASS PRODUCTION

The fungal growth behavior with the different substrates was analyzed to identify the best conditions concerning the fermentation time. The substrates glucose, sucrose, and agavin represent a different metabolic challenge to the fungus *A. kawachii* due to the structure of the substrates presenting an upward trend in structural complexity to express and secretion hydrolytic enzymes to the broth. The behavior of the different substrates' trends is evaluated to define a constitutive or inductive nature of the agavinase enzyme.

The fungal growth with each of the three substrates is shown in Figure 7.1. The lag and logarithmic phases of the three cultures are similar, demonstrating a biomass production of upper values of 4 g/L at 72 and 84 h of fermentation. The mycelial weight corresponding to the fermentation with glucose, sucrose, and agavin was 4.16 g/L, 4.35 g/L, and 4.37 g/L, respectively. These biomass production results resemble the reported in the literature for *A. kawachii* using glucose as substrate and for *A. japonicus* when grown in a sucrose culture media during a production study of an analogous fructofuranosidase [3]. Although the growths are similar, the stationary phase is clearly shown by the culture with glucose and agavin, not the same for sucrose culture biomass. However, the total sugar concentration for sucrose and glucose (Figure 7.2) exhibits an almost total consumption of both at 84 h (0.02 g/L and 0.18 g/L, respectively). The residue of agavin at the same fermentation time was observed with 1.19 g/L. This performance coincides with the complexity of the structure of agavins, which is characterized by the presence of long-branched fructose chains, contrary to the simple glucose and fructose sugars [12].

7.3.2 pH ANALYSIS

The pH of the three fermentations practically showed the same behavior. The pH adjustment of culture media to 4 was set to favor the segregation and activity of the hydrolytic acid-stable enzymes of *A. kawachii*. The cultures did not exhibit relevant pH changes during the adaptation phase; from hour 36 to 60, the pH decreased to a value of 3.2, as shown in Figure 7.3. At 60 h of fermentation, the fungal cultures reach a predominant citric acid production, reflected through low pH values due to the uncomplex availability of carbon sources [16]. Nonetheless, the fungal growing biomass production (Figure 7.1) coincided with the augmentation of the pH values of the culture broth for glucose culture. This condition of incremented pH values is defined by a phagocytosis stage associated to the stationery phase, where the decrement of the fungal population predominates, and ammoniacal metabolic residues result in increments of pH values [8]. Finally, the culture broth for glucose and saccharose culture show a pH of 5.5 and 4.6, respectively, differing from the pH of 3.7 for agavin broth, whose exponential stage maintains at 84 h due to the complexity of the polysaccharide.

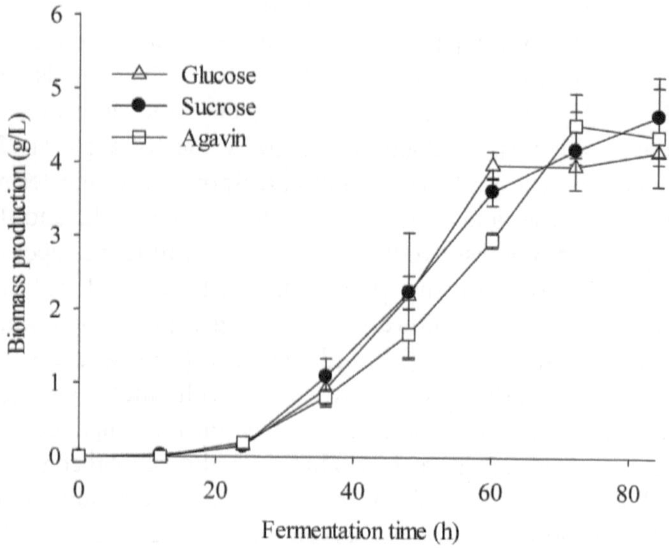

FIGURE 7.1 Kinetic profile of *A. kawachii* using different substrates to evaluate biomass production.

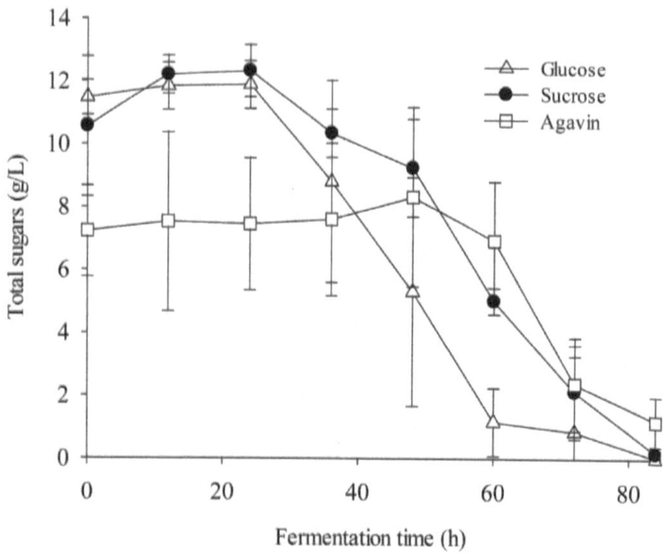

FIGURE 7.2 Consumption profile of the carbon sources in presence of *A. kawachii*.

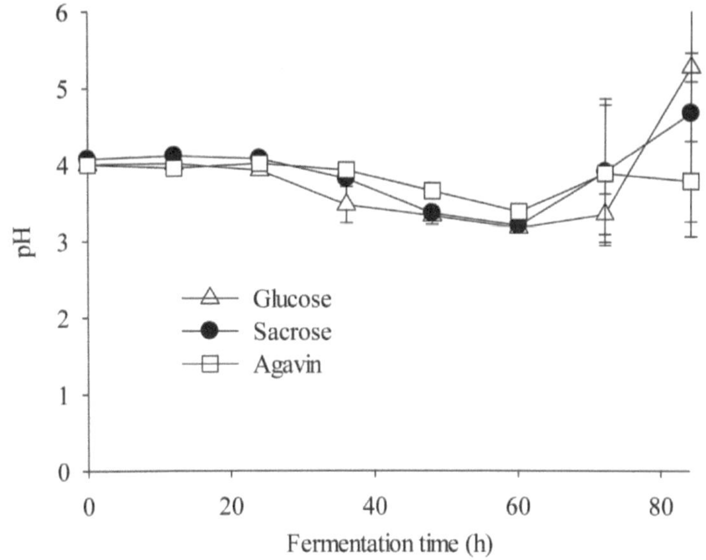

FIGURE 7.3 pH profile of the culture broths with different substrates in presence of *A. kawachii.*

7.3.3 TOTAL SUGAR ANALYSIS

The concentration of the carbon source was measured to know the consumption profile of substrates by *A. kawachii* in the culture broth. This consumption correlates with other bioprocesses' parameters, such as biomass production and pH, as described and discussed above. The total sugar profile showed differences among the three carbon sources, as shown in Figure 7.2. When observing the graph, the trend of consumption agrees with biomass production (Figure 7.1), but the beginning of exponential decrement of total sugars correlates with exponential biomass increments, even the behavior of the pH changes. As distinguished from the intake of glucose and sucrose by *A. kawachii*, whose exponential decrement in the broth started at 24 h, agavin represented a more complex substrate, starting the exponential decrement until 60 h, contrarily to the behavior of total sugars from glucose culture, which at that time, the consumption reached 88% and almost 99% at 84 h. These results were similar to those reported previously for the fermentation with glucose, a carbon source to induce polygalacturonase synthesis for the herein used microorganisms [6]. *A. kawachii* required 84 h to show an 88% concentration of total sugars in the agavin culture, showing the metabolic difficulty that agavin represents for the fungus. For agavin culture, as shown in the pH profiles (Figure 7.3), an acidified medium is the permissive condition to activate the hydrolytic agavinase enzymes to become carbohydrates bioavailable to *A. kawachii*. Finally, sucrose culture resembled glucose culture for the fungus; substrate consumption reached 99% at 84 h.

7.3.4 AGAVINASE ENZYMATIC ACTIVITY

The culture broths analyzed for detection of agavinase were selected at times of fermentation of 60 h. The extracts were lyophilized, rehydrated, and fractioned by GPC. The 10 fractions were analyzed for agavinase activity using monodispersed agavin as the substrate of the enzymatic reaction, and the results are shown in Figure 7.4. As can be observed, fractions from GPC of glucose extract did not exhibit agavinase activity. The corresponding sections from the saccharose extract resulted in positive agavinase presence on fraction 1. The fractions from the agavin extract clearly show agavinase activity mainly on fractions 2, 3, and 4, with 2.13 mU/mL, 4.53 mU/mL, and 2.1 mU/mL, respectively. Fraction 6 showed a value of 1 mU/mL. The agavinase activity absence from glucose culture indicates that agavinase enzyme is not constitutively produced by *A. kawachii* in this condition. Likewise, agavinase activity showed by GPC fraction number 1 from the sucrose extract may be associated with sucrase activity due to sucrose molecule is part of an end of the polymeric chains of agavins [14], in agreement with the mechanism of enzymatic activity of fructosyltransferases, whose initial substrate is sucrose [5]. Thus, extracts from sucrose and agavin were tested for an enzyme kinetic performance to evaluate the catalytic activity during 240 min; results are shown in Figure 7.5.

The agavinase kinetic performance was evaluated using extracts from sucrose and agavin cultures (Figure 7.4). Evaluated extract samples proceeded from 60, 72, and 84 h fermentation times. The agavinase activity was tested using an acetate buffered solution of monodispersed agavinase (high content of carbon chains, low content of reducing sugars). The 84-h sucrose extract showed agavinase activity at the end of the enzyme kinetic test value of 5.91 mU/mL, showing the presence and the constitutive production of agavinase enzyme. Around 60 and 72 h sucrose extracts exhibited negligible agavinase activity (data not shown). On the other hand, the agavin extract showed an enzyme activity of 27.6 mU/mL and 38.2 mU/mL at the end of the enzyme kinetic test, respectively for extracts for agavin cultures of 84 h and 60 h.

The catalytic results shown by agavin, and sucrose extracts demonstrate that *A. kawachii* produces, not only constitutively but also inductively, the agavinase enzyme.

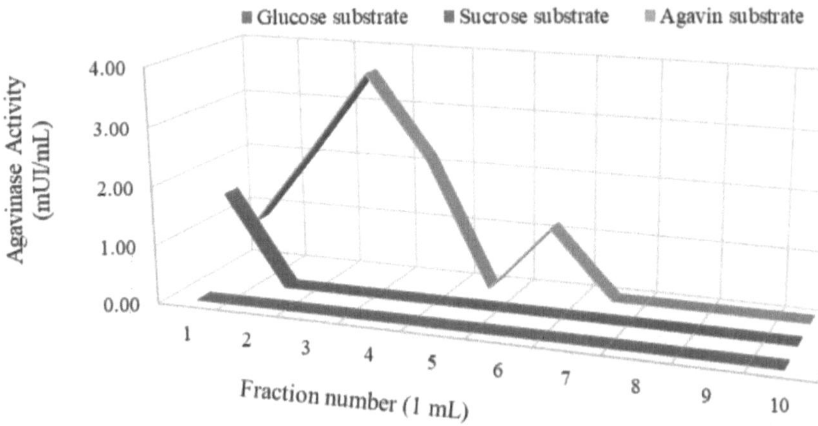

FIGURE 7.4 Agavinase enzymatic activity shown by the GPC fractions of different culture extracts.

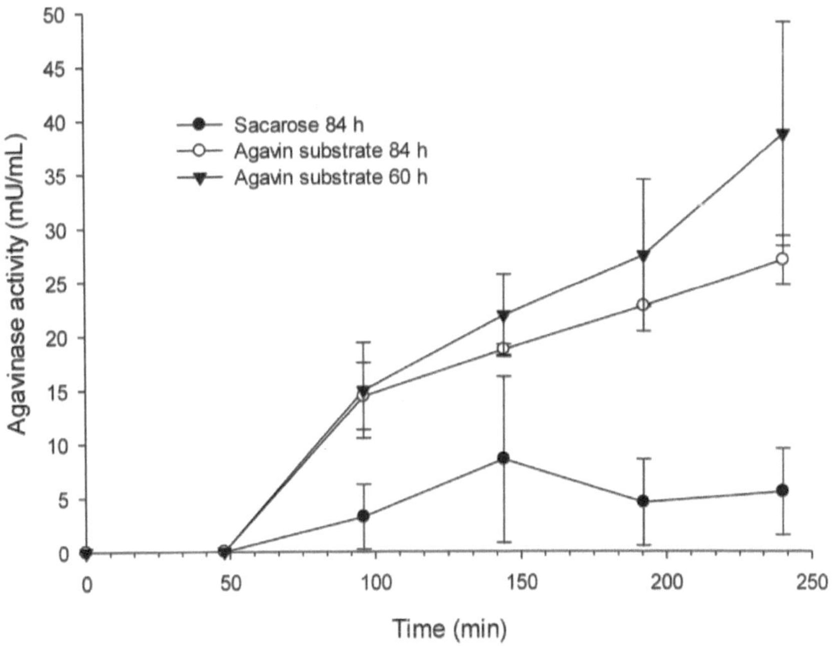

FIGURE 7.5 Enzymatic kinetics displayed by the most agavinase-active GPC pooled fractions.

Since the agavinase activity shown in Figure 7.5 is slightly low, an endpoint fermentation to 60 h of agavin as a substrate was carried out, and culture broth was lyophilized and fractionated by GPC chromatography at a preparative level. Around 15 fractions of 5 mL were obtained, and agavinase activity, total sugars, and protein were also assayed to show a characterization profile of the extract employed as agavinase; Figure 7.6 displays the results. As can be observed, protein content is present in all fractions, except for the first three, whose content represents the volume of the mobile phase to equilibrate column at the beginning of the GPC and are free of material from the sample. Fractions 4, 5, 6, and 7 showed agavinase activity with the acetate buffered monodisperse agavin solution in a 240 min endpoint enzymatic reaction, with values of 1.98 U/mL, 3.13 U/mL, 3.13 U/mL, and 1.33 U/mL, respectively, and whose specific activities are 0.15 U/mg, 0.025 U/mg, 0.01 U/mg and 0.003 U/mg of protein (data not shown). The total sugar content of the extract, whose content was diluted along the 15 fractions, exhibits a low concentration in the catalytically and analytically relevant fractions 5, 6, and 7, with values around 60 mg/L of the analyte. Such finding is fundamental because long-chain carbohydrates of fructan are bonded fructosidically through the anomeric carbon of the pentose, the molecular entity that manifests the reducing power of a saccharide. The reducing power is not expressed if the anomeric carbon is engaged in a fructosydic bond. Then, the reducing sugars from agavin extracts allow the detection of more fructose released from monodispersed agavin during the enzymatic reaction to yield better agavinase activity values.

7.4 CONCLUSIONS

The filamentous fungus *Aspergillus kawachii* grew satisfactorily with the different substrates to evaluate the production of agavinase enzyme. The growing kinetic of the fugal biomass for the three substrates indicated a successful adaption. Glucose and sucrose were completely metabolized at 84 h, while the agavin substrate represented a more complex carbon source for the microorganism, leading to the expression and secretion of metabolites with agavin depolymerizing skills. The evaluation of the agavinase activity using a system of semi-purification by diluting crude enzymatic extracts by GPC chromatography allowed the detection of catalytic capacity through the release of fructose units in an enzyme kinetic test with agavin as substrate. The enzymatic extracts with agavinase activity were from sucrose and agavin cultures but not from the glucose culture broth. These results primarily demonstrate that *Aspergillus kawachii* does not secrete, with a constitutive nature, an agavinase enzyme. Regarding the sucrose extract, the agavinase enzyme was not detected because a possible action of an invertase activity was reflected by a low value of reducing sugars in the enzyme activity test. The production of a larger volume of the culture broth of *A. kawachii* in agavin substrate, with the suitable GPC chromatography, was adequately applied to detect the presence of the agavinase enzyme. These latter results demonstrated the inductive nature of the agavinase enzyme expressed and secreted by *Aspergillus kawachii*, and then, the polysaccharides coming from *Agave tequilana* represent a potential bioprocessing carbohydrate source for food industry purposes.

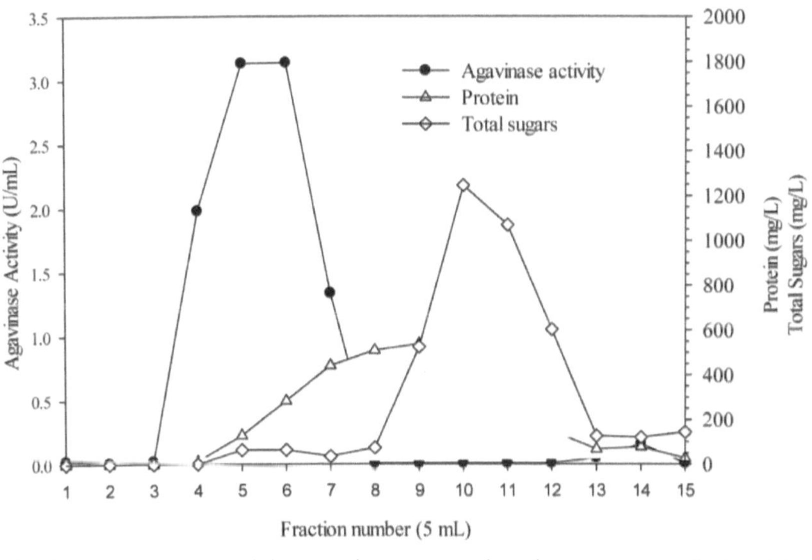

FIGURE 7.6 Agavinase activity, protein content and total sugar content displayed by the different purified extract fractions.

ACKNOWLEDGMENTS AND FINANCIAL SUPPORT

This study was partially supported by a grant "Fortalecimiento de Cuerpos Académicos En Formación (CAEF)" SEP PRODEP México 2021 to "UCOL-CA-117 Bioquímica aplicada," and also partially supported by "Apoyos a la Ciencia de Frontera: Fortalecimiento y Mantenimiento de Infraestructura de Investigación de Uso Común y Capacitación Técnica 2021" grants 316686, 317007, and 316608.

KEYWORDS

- ***Agave tequilana***
- **agavinase**
- **agavins**
- ***Aspergillus kawachii***
- **biomass**
- **gel permeation chromatography**
- **liquid fermentation**
- **protein assay**

REFERENCES

1. Arrizon, J., Morel, S., Gschaedler, A., & Monsan, P., (2012). Fructanase and fructosyltransferase activity of non-*Saccharomyces* yeasts isolated from fermenting musts of Mezcal. *Bioresource Technology, 110*(0), 560–565. https://doi.org/http://dx. doi.org/10.1016/j.biortech.2012.01.112.
2. Ávila-Fernández, Á., Galicia-Lagunas, N., Rodríguez-Alegría, M. E., Olvera, C., & López-Munguía, A., (2011). Production of functional oligosaccharides through limited acid hydrolysis of agave fructans. *Food Chemistry, 129*(2), 380–386. https://doi.org/http://dx.doi.org/10.1016/j.foodchem.2011.04.088.
3. Chen, W. C., & Liu, C. H., (1996). Production of β-fructofuranosidase by *Aspergillus japonicus*. *Enzyme and Microbial Technology, 18*(2), 153–160. https://doi.org/http://dx.doi.org/10.1016/0141-0229(95)00099-2.
4. Chesini, M., Neila, L. P., Fratebianchi De La, P. D., Rojas, N. L., Contreras-Esquivel, J. C., Cavalitto, S. F., Ghiringhelli, P. D., & Hours, R. A., (2013). *Aspergillus kawachii* produces an inulinase in cultures with yacon (*Smallanthus sonchifolius*) as substrate. *Electronic Journal of Biotechnology, 16*(3). http://www.ejbiotechnology.info/index.php/ejbiotechnology/article/view/v16n3-13 (accessed on 10 July 2023).
5. Chesinia, M., Wagner, E., Baruque, D. J., Vita, C. E., Cavalitto, S. F., Ghiringhelli, P. D., & Rojas, N. L., (2018). Protein expression and purification. *Protein Expression and Purification, 147*, 29–37.
6. Contreras-Esquivel, J. C., Hours, R. A., Voget, C. E., & Mignone, C. F., (1999). *Aspergillus kawachii* produces an acidic pectin releasing enzyme activity. *Journal of Bioscience and Bioengineering, 88*(1), 48–52. https://doi.org/10.1016/s1389-1723(99)80174-1.
7. Contreras-Esquivel, J. C., & Voget, C. E., (2004). Purification and partial characterization of an acidic polygalacturonase from *Aspergillus kawachii*. *Journal of Biotechnology, 110*, 21–28. https://doi.org/10.1016/j.jbiotec.2004.01.010.
8. Iwashita, K., (2002). Recent studies of protein secretion by filamentous fungi. *Journal of Bioscience and Bioengineering, 94*(6), 530–535. https://doi.org/http://dx.doi.org/10.1016/S1389-1723(02)80191-8.

9. Ah-Jin, K., Jung-Nam, C., Jiyoung, K., Hyang, Y. K., Sait-Byul, P., Soo-Hwan, Y., Ji-Ho, C., et al., (2012). Metabolite profiling and bioactivity of rice koji fermented by aspergillus strains. *Journal of Microbiology and Biotechnology, 22*(1), 100–106.
10. Koseki, T., Okuda, M., Sudoh, S., Kizaki, Y., Iwano, K., Aramaki, I., & Matsuzawa, H., (2003). Role of two α-l-arabinofuranosidases in arabinoxylan degradation and characteristics of the encoding genes from shochu koji molds, *Aspergillus kawachii* and *Aspergillus awamori*. *Journal of Bioscience and Bioengineering, 96*(3), 232–241. https://doi.org/10.1016/s1389-1723(03)80187-1.
11. Lowry, O., Rosenbrough, N., Lewis, F. A., & Randall, R. J., (1951). Protein measurement with the Folin fenol reagent. *Journal of Biological Chemistry, 193*(75), 265–275.
12. Mancilla-Margalli, N. A., & López, M. G., (2006). Water-soluble carbohydrates and fructan structure patterns from agave and dasylirion species. *Journal of Agricultural and Food Chemistry, 54*(20), 7832–7839. https://doi.org/10.1021/jf060354v.
13. Michel, D., Gilles, K. A., Hamilton, J. K., Rebers, P. A., & Smith, F., (1956). Colorimetric method for determination of sugars and related substances. *Analytical Chemistry, 28*(3), 350–356.
14. Muñoz-Gutiérrez, I., Rodríguez-Alegría, M. E., & López-Munguía, A., (2009). Kinetic behavior and specificity of b-fructosidases in the hydrolysis of plant and microbial fructans. *Process Biochemistry, 44*, 891–898.
15. Nelson, N., (1944). A photometric adaptation of the Somogyi method for de determination of glucose. *Journal of Biological Chemistry, 153*, 375–380.
16. Papagianni, M., (2004). Fungal morphology and metabolite production in submerged mycelial processes. *Biotechnology Advances, 22*, 189–259. https://doi.org/10.1016/j.biotechadv.2003.09.005.
17. Ramos-de-la-Peña, A. M., Renard, C. M. G. C., Wicker, L., Montañez, J., De La Luz Reyes-Vega, M., Voget, C., & Contreras-Esquivel, J. C., (2012). Enzymatic liquefaction of jicama (*Pachyrhizus erosus*) tuberous roots and characterization of the cell walls after processing. *LWT – Food Science and Technology, 49*(2), 257–262. https://doi.org/http://dx.doi.org/10.1016/j.lwt.2012.07.030.
18. Ritsema, T., & Smeekens, S., (2003). Fructans: Beneficial for plants and humans. *Current Opinion in Plant Biology, 6*(3), 223–230. https://doi.org/https://doi.org/10.1016/S1369-5266(03)00034-7.
19. Singh, R. S., Singh, T., Hassan, M., & Kennedy, J. F., (2020). Updates on inulinases: Structural aspects and biotechnological applications. *International Journal of Biological Macromolecules, 164*, 193–210. https://doi.org/https://doi.org/10.1016/j.ijbiomac.2020.07.078.
20. Somogyi, M., (1951). Notes on sugar determination. *Journal of Biological Chemistry, 195*, 19–23.
21. Vazquez-Vuelvas, O. F., Chávez-Camacho, F. A., Meza-Velazquez, J. A., Mendez-Merino, E., Ríos-Licea, M. M., & Contreras-Esquivel, J. C., (2020). A comparative FTIR study for supplemented agavin as functional food. *Food Hydrocolloids, 103*, 105642.
22. Yoshizaki, Y., Takamine, K., Shimada, S., Uchihori, K., Okutsu, K., Tamaki, H., Ito, K., et al., (2011). The formation of β-damascenone in sweet potato shochu. *Journal of the Institute of Brewing, 117*(2), 217–223.

CHAPTER 8

Modern Agrotechnologies of Sustainable Agroculture

GENRIETTA E. MERZLAYA,[1] MICHAEL O. SMIRNOV,[1]
SERGEI I. NOVOSELOV,[2] and ALEXEY M. KOMELIN[2]

[1]*Pryanishnikov All-Russian Scientific Research Institute of Agrochemistry, Moscow, Russia*

[2]*Mari State University, Yoshkar-Ola, Russia*

ABSTRACT

The humus state of the treated soils is dynamic and depends on soil and climatic conditions and agrotechnical measures. The research made it possible to assess the magnitude and direction of changes in the humus regime of sod-podzolic soil under the influence of agrotechnologies and natural factors. As a result of the research, it was found that agrotechnologies using combined tillage in crop rotation, the application of perennial grasses, sideral crops, the use of straw and organic fertilizers optimized the humus state of the soil, which ultimately contributed to the sustainability of agricultural production.

8.1 INTRODUCTION

Humus is the most important indicator of soil fertility. Agrochemical, physical, physicochemical properties of the soil, its microbiological activity depend on its content [1, 2]. The amount of humus in the soil is determined by the ratio of its formation and mineralization processes. In virgin and fallow soils, the process of humus formation prevails over mineralization. In arable soils, with a lack of organic matter intake, on the contrary, mineralization

Bioresources and Bioprocess in Biotechnology for a Sustainable Future. Leonardo Sepúlveda Torre, Juan Carlos Contreras-Esquivel, Ann Rose Abraham, & A. K. Haghi (Eds.)

processes prevail over the processes of humus formation. This leads to a decrease in its content and deterioration of soil properties [3]. To stabilize and predict the humus content, it is necessary to know the laws of the influence of agrotechnical techniques on the humus state of soils. This would provide new opportunities for sustainable agriculture. In order to clarify these patterns, this study was conducted.

8.2 METHODS

Our research was carried out on sod-podzolic soils of the Mari El Republic of the Russian Federation. The problem of humus content is particularly acute just for sod-podzolic soils. They are characterized by low humus content, high acidity, poor physical properties and as a result, low productivity. The humus content in sod-podzolic soils of the Mari El Republic does not exceed 2.0%. The ratio of carbon of humic acids to carbon of fulvic acids is in the range of 0.5–0.7, in the illuvial horizons the ratio is 0.3–0.5. These soils have a humate-fulvate type of humus profile. Fulvic acids in them significantly predominate over humic acids. This type of soil is characterized by very low nitrogen availability. Common to sod-podzolic soils is the accumulation of nitrogen in the form of non-hydrolysable and difficulty hydrolyze organic compounds [4, 5, 7].

The methods adopted in agrochemistry [8] were used to determine nutrient elements and other agrochemical characteristics. It should be noted that the development and improvement of methods for monitoring and forecasting humus content in agricultural soils is an important scientific and practical task. When assessing the humus state of soils, the most reliable are the data obtained directly during the agrochemical analysis of the soil. When developing crop rotations and justifying agrotechnical techniques and measures, computational methods for assessing the humus state are used. At the same time, calculations are carried out either according to the removal of soil nitrogen by the harvest, or according to the normative indicators of the intake and mineralization of organic matter established according to field experimental data. Calculation methods are less accurate, since the processes of mineralization and humification are complex, depend on many factors, and the coefficients used in the calculations are very conditional. When calculating the balance of humus through nitrogen, indicators of a heterogeneous nature are compared. The humus content is estimated in the arable layer, and nitrogen consumption by plants also occurs from deeper soil layers. When using mineral and organic fertilizers, it is only possible to estimate approximately what proportion of nitrogen is consumed from the soil, and which from fertilizers. The intervals of nitrogen utilization coefficients are so large that one can only conditionally assume how much of it was absorbed by plants from fertilizers, and which from the soil. The data on the contribution of biological nitrogen are also ambiguous. All this introduces elements of conditionality into the calculations. Based on the conducted research, we propose a photo-biochemical approach to calculating the humus balance in crop rotation. The humus balance is calculated as the difference between its formation and mineralization.

8.3 DISCUSSION

The conducted studies have shown that the humus state of sod-podzolic soil was influenced by the method of tillage, the type of crop rotation and the use of organic fertilizers. The use of combined treatment in crop rotation in comparison with annual plowing reduced the rate of humus mineralization, and the use of organic fertilizer increased its content in the soil (Table 8.1). At the end of the second rotation of the crop rotation, the humus content in the unfertilized soil decreased with combined treatment from 1.85% to 1.75%, and with the use of annual plowing to 1.72%. The use of mineral fertilizers in two rotations of crop rotation in the amount of $N_{680}P_{365}K_{690}$ did not affect the change in the humus content in the soil. At the end of the second rotation of the crop rotation, it was at the level of unfertilized soil and amounted to 1.71% with the use of annual plowing, and 1.75% with combined treatment. n of peat-manure compost (PMC) at a dose of 80 t/ha had a positive effect on the humus content in the soil. When using dump plowing, its content in the soil was 1.96%, and with combined treatment – 2.08%.

TABLE 8.1 Influence of Tillage and Fertilizers on Humus Content (%; End of the Second Rotation of Crop Rotation, 1997)

Variant	Before Experience Laying Out (1990)	Soil Treatment	
		Plowing	Combined
Without fertilizers	1.85	1.72	1.75
$N_{680}P_{365}K_{690}$	1.85	1.71	1.75
PMK 80т/га ($N_{600}P_{456}K_{408}$)	1.85	1.96	2.08

The effect of crop rotation and fertilizers on the humus content in the soil was studied in the experiment (Table 8.2). At the end of the rotation, the

TABLE 8.2 Humus Content in Soil at the End of Rotation of Seven-Field Crop Rotations (%; 2008)

Variant	**Crop Rotation**			
	Bare Fallow, Winter Rye, Potatoes, Barley, Clover, Winter Wheat, Barley	**Bare Fallow + Mowing Siderate, Winter Rye, Potatoes, Barley, Clover, Winter Wheat, Barley**	**Full Fallow, Winter Rye, Potatoes, Barley, Clover, Winter Wheat, Barley**	**Green Fallow, Winter Rye, Potatoes, Barley, Clover, Winter Wheat, Barley**
Without fertilizers	1.76	1.76	1.80	1.82
$N_{275}P_{100}K_{285}$	1.74	1.75	1.77	1.79
Manure ($N_{275}P_{100}K_{285}$)	1.96	1.97	2.00	2.02
NPK + manure ($N_{550}P_{200}K_{570}$)	2.01	2.05	2.10	2.10
Humus content in the soil before the experiment (2002)	1.97	1.97	1.97	1.97

amount of humus in the arable layer of unfertilized soil in the crop rotation with bare fallow decreased.

A similar humus content was in the soil of the crop rotation with bare fallow when applying mowing siderate. In the unfertilized soil of the crop rotation with full fallow, its content was 1.80%, and with green fallow 1.82%. The effect of mineral fertilizers in average annual doses of $N_{39}P_{14}K_{41}$ on the humus content in the soil of all crop rotations was insignificant. In a seven-field crop rotation with clover, the application of 60 t/ha of manure (the saturation of the crop rotation with manure was 8 t/ha) ensured that the humus content in the soil was maintained at the initial level. Its content in the soil of the crop rotation with bare fallow was 1.96%, the crop rotation with full fallow was 2.00% and the crop rotation with green fallow was 2.02%. The combined use of mineral fertilizers and manure ensured an increase in the humus content in the soil of crop rotations with full and green fallow to 2.1%, and in the soil of crop rotations with bare fallow, the humus content was maintained at the initial level (Table 8.2).

During the development of the land the influence of agrotechnical techniques on the humus state of the soil affected as follows. Before laying the experiment, the soil of the sod fallow contained 1.94% humus and 8.8% of the total organic matter (Table 8.3).

The cultivation of agricultural crops changed both the content of total organic matter and humus in the arable soil layer. By the end of the second rotation of the four-field crop rotation (steam, winter rye, potatoes, barley), the organic matter content decreased to 5.1–6.9%, and humus to 1.78–1.87%. It should be noted that the straw of winter rye and barley remained in the field and was used as an organic fertilizer. The greatest decrease in total organic matter in the arable soil layer was during combined tillage in crop rotation, and humus – during moldboard plowing. A greater decrease in the total organic matter in the arable soil layer during combined processing was due to a lower intake of root residues of cultivated crops, and the greatest decrease in humus during plowing was associated with more favorable conditions for its mineralization. The maximum decrease in the humus content in the soil both against the background of combined and moldboard systems of basic tillage occurred in crop rotations with bare fallow. The use of green fallow in crop rotation had a positive effect on the content of organic matter and humus, but at the same time increased the deficiency of phosphorus and potassium in the soil (Table 8.3).

Studies of the fractional and group composition of humic substances showed that extracts No. 1 and No. 2 had the lowest content of humic acids

TABLE 8.3 Influence of Types of Fallows, the System of Basic Tillage in Crop Rotation and Fertilizers on the Content of Organic Matter and Humus in the Arable Soil Layer (2010–2017)

Type of Fallow in Crop Rotation	The System of Tillage in Crop Rotation	Fertilizer	Total Organic Matter (%)		Humus (%)	
			2010	2017	2010	2017
Bare	Moldboard plowing	Without fertilizers	8.8	6.4	1.94	1.78
		NPK	8.8	7.0	1.94	1.83
	Combined	Without fertilizers	8.8	5.3	1.94	1.82
		NPK	8.8	5.7	1.94	1.84
Full	Moldboard plowing	Without fertilizers	8.8	6.7	1.94	1.80
		NPK	8.8	7.1	1.94	1.81
	Combined	Without fertilizers	8.8	5.1	1.94	1.83
		NPK	8.8	5.8	1.94	1.83
Green	Moldboard plowing	Without fertilizers	8.8	6.9	1.94	1.87
		NPK	8.8	7.2	1.94	1.88
	Combined	Without fertilizers	8.8	5.8	1.94	1.87
		NPK	8.8	6.5	1.94	1.89

(ha) and the highest content of fulvic acids (fa) in the soil of crop rotation with bare fallow, and the best ratio C_{ha}/C_{fa} was in the soil with green fallow (Table 8.4).

TABLE 8.4 The Effect of the Type of Fallow on the Fractional and Group Composition of Humus (%)

C_{ha}			C_{fa}			C_{ha}/C_{fa}		
b.f.	**f.f.**	**g.f.**	**b.f.**	**f.f.**	**g.f.**	**b.f.**	**f.f.**	**g.f.**
Extract No. 1 0.1 n NaOH								
0.081	0.085	0.093	0.139	0.129	0.137	0.58	0.66	0.68
Extract No. 2 0.1 n NaOH After Decalcification								
0.199	0.201	0.206	0.224	0.209	0.212	0.89	0.96	0.97
Extract No. 3 0.02 n NaOH When Heated in a Water Bath								
0.165	0.172	0.184	0.164	0.170	0.172	1.01	1.01	1.07

Note: b.f.: Soil from bare fallow; f.f.: Soil from full fallow; and g.f.: Soil from green fallow.

The results of these experiments showed that the more intense the mechanical impact on the soil, the more humus mineralization occurs in it. On the one hand, this is due to the activation of microbiological processes, and on the other, as studies have shown, to the destabilization of humus in the soil due to photochemical destruction.

Studies have revealed that the effect of solar energy on the soil leads to an increase in the amount of easily hydrolysable nitrogen and labile humus substances (Table 8.5). An increase in their content was observed only in soil layers 0–2 and 2–5 cm. In the soil closed from sunlight, the content of easily hydrolysable nitrogen in these layers was 83.2 mg/kg, and in the soil exposed to sunlight, respectively, 87.7 and 84.6 mg/kg. Due to the photochemical factor in the 0–2 cm soil layer, the content of easily hydrolysable nitrogen increased by 4.5 mg/kg, and in the 2–5 cm layer by 1.4 mg/kg.

The effect of solar energy on the soil led to an increase in the content of labile humus substances. In the soil, closed from sunlight, during the fallow period, the content of mobile humus substances in the 0–2 cm layer increased from 0.215 to 0.225%, and in the natural state to 0.238%. In the underlying layers 2–5 and 5–10 cm, their content in the soil, respectively, was 0.216 and 0.231% and 0.228 and 0.234%. In the lower part of the arable layer at a depth of 10–20 cm, the content of labile humus substances in the soil has not changed. On average, over five years of research, over 20 days of open soil vaporization, the content of easily hydrolysable nitrogen increased from 81 mg/kg to 114 mg/kg, and mobile humic substances from 0.187% to 0.216%. These changes were less significant in the soil covered from the sun with a white cloth. The content of easily hydrolysable nitrogen was 96 mg/kg, and labile humic substances were 0.199%. Consequently, due to the photochemical factor, the content of easily hydrolysable nitrogen increased by 18 mg/kg, and labile humus substances – by 0.017% (Table 8.6).

TABLE 8.5 The Effect of Light Energy on the Content of Easily Hydrolysable (eh) Nitrogen in the Soil (mg/kg), and the Content of Labile Humus Substances (LHS) in the Soil (0.1 n NaOH; %), on Average for 2012–2014

Soil Layer (cm)	Original Content		After 20 Days of Fallowing			
			Open Ground		Protected Ground	
	N_{eh}	LHS	N_{eh}	LHS	N_{eh}	LHS
0–2	81.0	0.215	83.2	0.225	87.7	0.238
2–5	81.1	0.215	83.2	0.216	84.6	0.231
5–10	81.0	0.215	81.8	0.228	80.0	0.234
10–20	80.5	0.215	80.9	0.216	79.5	0.218

Studies conducted with soil irradiation in a laboratory installation revealed that the photochemical effect on the soil changed both the amount of labile humic substances and the ratio of humic and fulvic acids (Table 8.7). In the soil irradiated for 100 hours, the content of labile humic substances increased from 0.181% to 0.198%. At the same time, the amount of humic acids decreased by 0.060%, and the amount of fulvic acids increased by 0.077%. The ratio C_{ha}/C_{fa} decreased from 0.75 to 0.10.

The development and improvement of methods for monitoring and forecasting humus content in agricultural soils is an important scientific and practical task. The formation of humus occurs during the humification of organic matter of plants, soil animals, organic fertilizers, waste of organic origin. Plant residues and organic fertilizers differ significantly in their ability to form humus substances. The maximum amount of humus is formed when plowing plant residues and straw, and the minimum when embedding siderates. Humification processes are considered depending on the method of embedding organic fertilizers and plant residues in the soil. Deep plowing of organic matter provides anaerobic conditions conducive to regenerative processes and humus formation. Surface sealing creates aerobic conditions that activate mineralization processes (Table 8.8). Intensive mineralization of humus begins with plowing virgin and fallow lands and involving them in agricultural use. Dehumidification processes largely depend on the intensity of the impact on the soil. The more often there is a mechanical impact on the soil, the more intensive is the mineralization of humus. The mechanism of this process can be represented as follows. Mineralization of soil humus occurs due to photochemical destruction of humus under the influence of sunlight and a large group of microorganisms that use humus substances as a source of nutrition and energy. The effect of solar energy on the soil leads to the destruction of complex in composition and valuable in properties of humic acids and the formation of more mobile labile humic substances that are easily subjected to microbiological mineralization. The mineralization process depends on the total humus reserves, climatic conditions, the intensity of photochemical exposure, the type of crop rotation and the applied tillage system. Minimal mineralization of humus substances of the soil goes under perennial grasses. At the same time, soil fauna plays a significant role in mineralization processes. Shrews, moles, earthworms bring significant amounts to the surface of soil, which is further exposed to sunlight. Given that their activities are of a regular nature, their scale is very significant.

TABLE 8.6 The Effect of Light Exposure on the Content of Labile Humus Substances and Easily Hydrolysable Nitrogen in the Fallow Soil (in the Soil Layer 0–2 cm, Microfield Experience)

Year	Variant	N_{eh} (According to Cornfield) (mg/kg)		Labile Humus Substances (0.1 n NaOH) (C%)	
		Before Experience Laying Out	After of 20 Days of Fallowing	Before Experience Laying Out	After of 20 Days of Fallowing
2010	Protected ground	65	91	0.172	0.186
	Open ground	65	131	0.172	0.208
2011	Protected ground	91	109	0.181	0.199
	Open ground	91	120	0.181	0.218
2012	Protected ground	79	84	0.198	0.211
	Open ground	79	87	0.198	0.224
2013	Protected ground	87	90	0.154	0.161
	Open ground	87	98	0.154	0.176
2014	Protected ground	83	83	0.232	0.239
	Open ground	83	88	0.232	0.252
On average	Protected ground	81	96	0.187	0.199
	Open ground	81	114	0.187	0.216

TABLE 8.7 Group Composition of Labile Humus Substances of the Soil (%; Extract 0.1 n NaOH, Laboratory Experiment 1)

Variant	C_{tot}	C_{ha}	C_{fa}	C_{ha}/C_{fa}
Without irradiation	0.181	0.078	0.103	0.75
With irradiation	0.198	0.018	0.180	0.10

TABLE 8.8 Coefficients of Humification of Plant Residues and Organic Fertilizers (2000)

Plant Residues of Agricultural Crops and Types of Fertilizers	The Method of Embedding Organic Fertilizers and Plant Residues	
	Moldboard Plowing	Disking
Cereals, rapeseed, flax, annual, and perennial herbs	0.18	0.14
Potatoes, root vegetables, vegetables	0.06	0.05
Grain straw	0.18	0.13
Sideral fertilizer	0.04	0.03
Litter manure	0.06	0.05
Peat-manure compost 1:1	0.07	0.06

The mineralization of organic fertilizers and plant residues slows down during deep embedding and accelerates during surface embedding (Table 8.9).

TABLE 8.9 Humus Mineralization Coefficients (%) of Gross Reserves in the Soil (2000)

Cultivated Crops	Processing Method Soils in Crop Rotation	
	Moldboard Plowing	Surface
Cereals, grain legumes	1.8	1.4
Tilled crops	2.8	2.8
Annual herbs	1.6	1.5
Perennial herbs	0.15	0.15
Bare fallow	3.3	3.3

Photochemical irradiation is a surface phenomenon, and its effect on the soil depends on a number of factors: the zoning of soils, the intensity of tillage, the structure of crop rotations, applied agricultural technology. All this contributes its own features to the processes of mineralization and humification of organic matter and requires clarification in specific soil and climatic conditions.

8.4 CONCLUSIONS

1. The stability of the fertility of sod-podzolic soil is influenced by the method of tillage, the type of crop rotation and the use of organic fertilizers. The use of bare fallow and annual plowing in crop rotation activates the rate of mineralization of soil humus, and the application of organic fertilizer increases its content.
2. The combined use of green fallow and straw in crop rotation has a positive effect on the content of organic matter and its stabilization in the soil.
3. In crop rotation with clover, with a manure saturation of 8 t/ha per year, the humus content in the soil is maintained at the initial level.
4. Under the influence of the photochemical factor, the amount of easily hydrolysable nitrogen and labile humic substances in the soil increases, the content of humic acids decreases and the content of fulvic acids increases.
5. In order to reduce the photochemical effect of solar radiation on the soil and for the stability of the humus content, it is advisable to use combined tillage in crop rotation, replace bare fallows with full and green ones, cultivate perennial grasses, crop, and intermediate siderates, create a mulching layer from plant residues in the form of straw, stubble, etc. Thus, the use of technologies with the use of the agricultural practices described above contributes to increasing the sustainability of agriculture.

KEYWORDS

- **agricultural practices**
- **crop rotations**
- **fulvic acids**
- **labile humus substance**
- **light energy**
- **organic and mineral fertilizers**
- **soil humus**
- **sustainability**
- **tillage**

REFERENCES

1. Ivanova, E. I., Kazarinova, G. I., & Novoselov, S. I., (1979). Group and fractional composition of humus of sod-podzolic soils of the Mari ASSR. *Geography and Soil Fertility–Saransk* (pp. 132–138). In Russian.
2. Lykov, A. M., (1985). *Humus and Soil Fertility* (p. 192). Moscow: «Moskovskyi Rabochyi» («Moscow Worker» in Russian. In Russian.
3. Merzlaya, G. E., (2021). The study of the stability of agrocenoses with prolonged use of fertilizers on sod-podzolic soil. *Soil Science*, (3), 355–362. In Russian.
4. Novoselov, S. I., & Zavalin, A. A., (2013). *The Role of Photochemical Factor in the Destruction of Humus*, (1), 59–64. In Russian.
5. Semenov, V. M., Ivannikova, L. A., & Tulina, A. S., (2009). Stabilization of organic matter in soil. *Agrochemistry*, (10), 77–96. In Russian.
6. *Program and Methods of Research of Humus State of Soils of Long-Term Experiments of Geonet, Reference Sites and Polygons of Agroecological Monitoring*, (2008). (p. 306), M.: Pryanishnikov All-Russian Scientific Research Institute of Agrochemistry. In Russian.
7. Shevtsova, L. K., (1998). Transformation of humus of sod-podzolic soils in experiments with long-term use of fertilizers. *Soil Science*, (7), 825–831. In Russian.

PART II
Biotechnology Process Development

CHAPTER 9

Degradable Polymers for the Development of Effectiveness of Slow-Release Fungicide Formulations for Suppressing Potato Pathogens

EVGENIY G. KISELEV,[1,2] SVETLANA V. PRUDNIKOVA,[1] NADEZHDA V. STRELTSOVA,[1] TATIANA G. VOLOVA,[1,2] and SABU THOMAS[1,3]

[1]*School of Fundamental Biology and Biotechnology, Siberian Federal University, Krasnoyarsk, Russia*

[2]*Federal Research Center "Krasnoyarsk Science Center SB RAS," Institute of Biophysics SB RAS, Krasnoyarsk, Russia*

[3]*International and Interuniversity Center for Nanoscience and Nanotechnology, Mahatma Gandhi University, Kottayam, Kerala, India*

ABSTRACT

The purpose of the present study was to evaluate biological activity of slow-release fungicide formulations for suppressing potato pathogens. Fungicides embedded in degradable poly-3-hydroxybutyrate/wood flour matrix for pre-emergence soil application were studied. The slow-release fungicide formulations (azoxystrobin (AZ), azoxystrobin + mefenoxam, and difenoconazole (DIF)) were compared with their commercial analogs (Quadris, Uniform, and Score) in the *in vitro* experiments, using disk diffusion test, and in the *in vivo* experiments with potato cv. Krasnoyarskiy ranniy. In cultures of plant pathogens, the embedded fungicides showed an inhibitory effect comparable to their commercial analogs, most effectively controlling the growth of

Bioresources and Bioprocess in Biotechnology for a Sustainable Future. Leonardo Sepúlveda Torre, Juan Carlos Contreras-Esquivel, Ann Rose Abraham, & A. K. Haghi (Eds.)

colonies of *Phytophthora infestans*, *Alternaria longipes*, *Rhizoctonia solani*, and *Fusarium solani* (2.0–2.3 times relative to the negative control). In the *in vivo* experiments, with the same phytosanitary condition of seed tubers and the initial soil, potatoes were damaged by pathogens in different ways. In laboratory experiment, only *Rhizoctonia* species developed; in the groups with embedded fungicides, germination occurred earlier, potatoes grew better, and the area of damage of the plants was reduced to 10% or less; the weight of microtubers from one clone was 25–35% greater compared to the control; the yield increase ranged between 60% and 71.3%. In the field experiment, the embedded fungicides inhibited the development of *Phytophthora* and *Alternaria* in the rhizosphere during the entire growing season and reduced the area of pathogen damage to plants to 10–15%, which was less by a factor of two than in the groups of plants treated with commercial fungicides. The higher biological activity of the embedded fungicides ensured the maximum number of tubers undamaged by pathogens and the total yield of 22–23 t/ha, which exceeded the yields in the groups with commercial fungicides (18.4–20.8 t/ha). The slow-release fungicide formulations prepared by embedding fungicides in P(3HB)/wood flour degradable matrix are effective in protecting potatoes from pathogens and in increasing potato yields and have an advantage over their commercial analogs.

9.1 INTRODUCTION

The potato (*Solanum tuberosum L.*) is one of the most valuable farm crops cultivated around the world, whose tubers are used for food, for technical purposes, and for animal feed. The significance of the potato is growing as productivity of cereal crops is decreasing everywhere. Statistical data reported by FAO show that in 2019 world potato production reached 400 million tons harvested from 19.25 million hectares [2], and annual potato production is expected to increase by 1.6% over the coming years. Under optimal conditions, potato yield may potentially reach 60–100 t/ha, but the real yields are considerably lower because of the high incidence of diseases during the growing period and because of tuber infection that develops during storage. The losses may range between 15–20% and 80–100% [85]. Recent years have seen the alarming instability and decrease in potato yields caused by pests and diseases that severely damage this crop. The most numerous and widespread potato diseases are infections caused by microorganisms. The diverse common fungal, bacterial, and viral potato diseases include late blight, early blight, fusarium infection, black scurf, various mosaics, leaf roll, etc., which cause 10 to 60% yield losses [4, 55, 60]. The incidence of damaging diseases caused by soil microorganisms such as *Fusarium spp., Phytophthora* spp., *Verticillium* spp., *Sclerotinia* spp., *Rhizoctonia* spp., and *Pythium* spp. increases every year [45]. Thus, there is a need for developing effective means and techniques for protecting potato crops and increasing yields.

The only way to achieve high potato yields and reduce losses caused by pathogens is to implement a combination of protective measures such as the use of healthy seeds and disease-resistant potato varieties and the introduction of good farming practices, involving the use of effective pesticides, fertilizers, growth regulators, etc. [5, 6, 10, 33, 50].

One of the principal methods used today to control pathogens of potato is chemical protection of the plants: diverse pesticides are used to treat seed tubers, growing potato plants, and harvested potato tubers before they are placed in storage. In addition to farming practices and selection and application of pesticides, new findings are expected from the research aimed at developing novel formulations for protecting potatoes from diseases. The cause for concern is that many of the currently applied pesticides do not fully comply with safety requirements. Chemical fungicides are nonspecific, kill non-target organisms, are often toxic to animals and humans, and cause formation of resistant plant pathogen populations, which results in more frequent treatments and higher application rates of fungicides. In addition, the use of chemicals induces the development of pesticide poisoning, which unfavorably affects potato yields. The wide use of chemical pesticides pollutes the environment and causes environmental concerns, as most of them accumulate in biological objects, contaminate soil and water, kill beneficial organisms, and destroy the balance in natural ecosystems [11, 37]. Pesticides are the most dangerous creation of modern civilization, causing damage to beneficial biota and posing a threat to human health. Pesticides are reprotoxic [36] and teratogenic [47, 88], have epigenetic activity [25, 65], affect humoral immunity [31], and increase the long-term risks of cancer incidence [12].

Because of these factors, along with the high cost of chemical protection of potatoes, the necessity arises to develop new and environmentally safe means and ways to reduce the pesticide impact on both potato crops and natural ecosystems and the entire environment.

One of the modern approaches to plant pathogen control investigated by researchers is to enhance potato resistance to pathogens using methods of genetics and selection. Increasing genetic diversity of the stock material is a strategy in potato selection aimed at breeding pathogen resistant varieties [95]. However, because of the high virulence of pathogens, this resistance disappears in 10 years. Therefore, chemical methods still prevail in potato disease management.

In the EU countries, where green farming practices are becoming increasingly prevalent, the number and amounts of chemical pesticides used in agriculture are decreased, and the farmers tend to use microbiological formulations, the so-called effective microorganisms, to enhance soil fertility and suppress plant pathogens [46, 53, 62]. Biological pesticides, however, have disadvantages that affect their marketability: most of them are unable to eliminate the pest or pathogen population, only reducing its damage; biological pesticides usually need more time to take effect compared to chemicals; biological efficacy depends on the environmental conditions (temperature, moisture content); biological pesticides should be applied repeatedly. The large labor input to the production of biological pesticides and the high cost limits their use, and the market size for biological pesticides is still relatively small [42].

The newest line of research is development of new-generation pesticide formulations with controlled release of the active ingredient by embedding the active ingredients into biodegradable materials, which are degraded in soil-by-soil microflora to form harmless products, enabling sustained delivery of the pesticides. The advantages of such formulations are their long action and reduction in the number of treatments of potato crops; longer activity of unstable pesticides; lower toxicity for biota and reduced accumulation in the food chain; conversion of the liquid forms to the solid ones, making them safe for use and simpler to transport [9, 39].

Increasing attention has been directed to the development of microencapsulated pesticides used as aqueous dispersions and emulsions of microgranules to spray the vegetative parts of the plants or to treat seeds. Such forms enable closer contact with plants and have longer function. Development of such pesticides is aimed at increasing their stability in the environment, enhancing their effect on plant pathogens, and decreasing pesticide toxicity to the target crops. Considerable research effort has been recently devoted to the construction and investigation of fungicide formulations with the active ingredient encapsulated in various materials [26, 74]. Recent years have seen the publication of studies on micro- and nano-sized fungicide formulations to protect potato and other vegetable crops from pathogenic fungi including *Fusarium* and *Phytophthora*. A number of relatively recent studies reported encapsulation of the azoxystrobin (AZ) fungicide in silica, chitosan, and poly(lactide/glycolide) microparticles to produce slow-release formulations [89, 87, 92]. Similar research was performed with other fungicides such as the highly effective pyraclostrobin, which was encapsulated in polylactide/chitosan microparticles [3, 86], carbendazim, loaded into nanospheres of polylactic acid and poly(lactide/glycolide) [58], etc. Pesticides in the form of microparticle or microcapsule suspensions or emulsions have a number of disadvantages: complex composition, multistage, and complicated fabrication, and potential hazard to beneficial biota. The efficacy of such pesticide forms has been insufficiently studied, and their application may entail risks such as potential toxicity. Thus, possible environmental and biological consequences of using microencapsulated pesticides need to be thoroughly investigated.

One of the currently developed approaches to plant pathogen control is the use of controlled-release fungicide-loaded mulch films. The well-known technology of application of organic fertilizers followed by plastic mulching is considered as a promising approach to controlling soil-transmitted diseases of crops [43]. It is important to use mulch films made of biodegradable materials, which will be degraded in soil to produce harmless compounds and will not do any damage to the environment [57], such as carboxymethyl cellulose, chitosan, polylactic acid. The active ingredients released to soil as the films are degraded control the development of pathogenic fungi [7, 22, 32]. Although there are still rather few studies in this field, they show that biodegradable mulch films loaded with fungicides may be an effective tool to suppress soil pathogenic fungi causing plant diseases.

The data on the newest and most promising approach–fabrication and pre-emergence application of slow-release targeted pesticide formulations–are highly limited. Only a few studies report production of embedded fungicides for soil application. Thiram–a contact preventative fungicide–was entrapped in agar and alginate matrices, and its release kinetics to water and soil was described [61]. In another study, thiram was incorporated in nonwoven polylactic acid fibers [56]. A fungicide formulation based on zinc and manganese ions was prepared in the form of nanoparticles coated by polyethylene glycol; the formulation was tested against mildew and Pythiaceae fungi [76]. Controlled-release nano-formulations of mancozeb based on polyethylene glycol were described in a study by Majumder et al. [35]. Duellman et al., [14] reported loading bioactive eugenol preparations into chitosan-based matrix and producing formulations that were used to treat potatoes, effectively controlling potato blight. The Quadris fungicide formulation (with AZ as the active ingredient) based on polyacrylate gels was used as a pre-emergence fungicide and effectively controlled potato diseases (common scab–*Streptomyces scabies*, black dot–*Colletotrichum coccoides*, Fusarium disease–*Fusarium* sp., rubbery rot–*Geotothrichum candidum*, bacterial rot–*Pectobacterium* sp.); the formulation was developed and produced by researchers of Moscow State University (Russia) [63, 64].

The principal aspect to constructing slow-release pre-emergence pesticide formulations is the availability of appropriate materials that must have the following properties: controlled degradation in soil, environmental safety, and processability into granules, microcapsules, films, etc., by available methods, compatible with pesticide production technologies.

Polyhydroxyalkanoates (PHAs)–degradable microbial polymers–are among the most promising materials for fabricating formulations to protect crops from weeds and pathogens. These polymers can be processed into slowly degraded products using various methods, and these products can function in soil for extended periods of time [8, 24, 68, 71, 77]. Although research of PHAs as material for embedding pesticides has been started rather recently, results have been obtained that suggest high potential of these biopolymers for developing new-generation formulations for plant protection.

The slow-release formulations of pesticides embedded in PHA matrices described in the literature include fungicides Sumilex and Ronilan [59]; herbicides ametryn, atrazine, metsulfuron-methyl, 2-methyl-4-chlorophenoxyacetic acid [1, 18, 19, 30, 34]; insecticide malathion [67]. Jiangsu Changqing Agrochemical (China) developed and started the production of the highly effective fungicide-fenoxanil–loaded into poly-3-hydroxybutyrate microcapsules, which showed high parameters of controlled release and reduced environmental toxicity of the fungicide [94]. Recently, PHAs have been investigated as material for manufacturing mulch films loaded with pesticides and intended for both suppressing weeds and preventing plant root diseases, as noted above [7, 22].

The authors of the present study have recently carried out integrated research that included development of formulations of herbicides and fungicides embedded into the matrix of poly-3-hydroxybutyrate, P(3HB)–the most widely available and slowly degraded PHA. The research showed that the formulations effectively controlled weeds and plant pathogens. The results were summarized in the book "New generation formulations of agrochemicals: Current trends and future priorities" [80–82]. In order to reduce the cost of such formulations and make them easily accessible, pesticides were embedded not only into the very expensive pure P(3HB) but also into the polymer filled with inexpensive natural materials (clay, peat, wood flour). The efficacy of formulations was confirmed in vegetable (tomato, table beet) and cereal (wheat, barley) crops infested by weeds and infected by root rot pathogens [23, 51, 72, 78–82].

The present work, for the first time, reports results of studying the efficacy of slow-release formulations of fungicides embedded in the degradable P(3HB)/birch wood flour matrix in controlling potato pathogens.

9.2 MATERIALS AND METHODS

9.2.1 FUNGICIDES

Fungicides with different modes of action [54] were produced by Xi'an Taicheng Chem Co., Ltd, China:

- AZ (Methyl (2E)-2-(2-{[6-(2-cyanophenoxy)pyrimidin-4-yl]oxy} phenyl)-3-methoxyacrylate) (a strobilurin fungicide); ≥98.0%.
- DIF (3-chloro-4-((2RS,4RS;2RS,4SR)-4-methyl-2-(1H-1,2,4-triazol-1-ylmethyl)-1,3-dioxolan-2-yl)phenyl 4-chlorophenyl ether (a triazole fungicide); ≥95.0%.
- Mefenoxam [metalaxyl R-enantiomer] [methyl N-(2,6-dimethylphenyl)-N-(methoxyacetyl)alaninate] (a phenylamide fungicide); ≥91.0%.

In positive control groups, commercial fungicides were used:

1. **Quadris:** A suspension concentrate (with AZ as the active ingredient, 250 g/L). Certificate of product registration: 041-02-211-1 (25.12.2023) Registrant: OOO "Syngenta" (RF). Application rate 0.3 L/ha; solution application rate 80–200 L/ha. Hazard category 2.
2. **Score:** An emulsion concentrate (with DIF as the active ingredient, 250 g/L). Certificate of product registration: 041–02–171–1 (08.12.2023). Registrant: OOO "Syngenta" (RF). Application rate 0.3–0.5 L/ha; solution application rate 200–400 L/ha. Hazard category 3 (soil persistence hazard category: 2).
3. **Uniform:** An emulsion concentrate (with AZ, 321.7 g/L, and mefenoxam, 123.7 g/L, as the active ingredients). Certificate of product registration: 041-02-499-1 (29.12.2024); 041-02-499-1/113 (29.12.2024). Registrant: OOO "Syngenta" (RF). Application rate 1.3–1.5 L/ha; solution application rate 80–120 L/ha. Hazard category 3.

9.2.2 MATERIALS AND PRODUCTION OF SLOW-RELEASE FUNGICIDE FORMULATIONS

Fungicides were loaded into the blend of degradable microbial polymer–poly-3-hydroxybutyrate, P(3HB), and birch wood flour. Wood flour was produced by grinding wood of birch (*Betula pendula* Roth) using an MD 250–85 woodworking machine ("StankoPremyer" Russia). Then it was dried at 60°C for 120 h until it reached constant weight, and 0.5 mm mesh was used to separate the particle size fraction; degree of crystallinity 26%; onset of thermal decomposition 220°C. The process of polymer/wood flour blending was described in detail elsewhere [72].

The powder blends containing P(3HB), birch wood flour, and fungicide in preset ratios were processed by spheronization technique using a granulator (Formag, Germany) to produce fungicide formulations for protecting potato from pathogens. The patented technology (Patent RF 2733295C1) was described in detail elsewhere [79]. The granules contained the components in the following ratios: P(3HB):wood flour:fungicide=60:30:10 (when one fungicide was used–AZ or DIF) and 60:30:5:5 (wt.%) (when two fungicides were used–AZ + mefenoxam). The granules were 3.0 mm in diameter and 9 ± 1 mg in weight. Pesticide concentration was controlled by varying the number of granules buried in soil.

9.2.3 IN VITRO TESTING OF BIOLOGICAL ACTIVITY OF EMBEDDED FUNGICIDES

The fungistatic activity of experimental fungicide formulations (granules loaded with AZ, azoxystrobin + mefenoxam, or DIF) was compared to the fungistatic activity of the commercial formulations Quadris, Uniform, and Score in the tests with plant pathogens isolated from soil and tubers of potato cv. Krasnoyarskiy ranniy by diffusion method on wort agar. The tests were conducted in cultures of potato pathogens that had been isolated and identified previously. The pathogens were grown in lawn plates on potato-dextrose agar (PDA, HiMedia, India) at a temperature of 25°C for 5–7 d. Agar blocks with the colonies showing the best growth were cut out under aseptic conditions. Then, in a Petri dish with sterile wort agar medium, a block with mycelium (5 mm in diameter) and one of the tested fungicide forms were placed onto the agar on the opposite sides of the dish. In a Petri dish with malt extract agar, a block with culture grid and one of the tested fungicide forms were placed onto the agar on the opposite sides of the dish. The dishes were incubated for 7–10 d in a temperature-controlled cabinet at 25°C. To assess the effectiveness of the experimental granules with fungicides, diameters of the colonies in the treatments were compared to those in the positive and negative controls. Each procedure was performed in triplicate, and results were photographed.

9.2.4 POTATO CULTIVAR

Potato cv. "Krasnoyarskiy ranniy" was used in the study. This potato variety is included in the Russian State Register of Selection Achievements and approved for cultivation in the Central Black Earth, West-Siberian, and East-Siberian regions. Originators of the variety are Krasnoyarsk State Agrarian University (Russia) and A. G. Lorkh Russian Potato Research Center (Russia). This is an early maturing table variety. The potato variety has been acclimated to the continental subarctic climate of Siberia, with cold winter and short summer. It is a tasty variety and stores well. This potato variety is moderately susceptible to viral infections, early blight, and black scurf; its tubers are moderately susceptible, and its leaves and stems are highly susceptible to late blight. The average yield is 30–47 t/ha or 10–12 t/ha if harvested early; this variety has medium-starch (13–16%) tubers. The weight of the commercial-grade tuber is 90–140 g.

9.2.5 GROWING OF POTATOES IN LABORATORY

Potatoes were grown in an environmental chamber (Fitotron-LiA-2, Russia) in 2-L plastic pots filled with pre-characterized soil; one tuber was planted per pot. Diurnal cycles of temperature, light, and humidity were maintained in the six-step mode: night–early morning–late morning–afternoon–early evening–night. The temperature was varied between 10°C at night and 18C in the daytime during the first seven weeks of the experiment and between 14°C and 22°C in the following weeks. Light intensity was varied between 0 and 300 μmol/m^2/s in 100 μmol/m^2/s increments. Soil moisture content was maintained at a level no less than 50%. Two forms of fungicides were investigated–experimental formulations (buried in soil simultaneously with tuber planting): AZ granules, AZ + mefenoxam (AZ + MEF) granules, and DIF granules. Their commercial analogs were used as positive control: Quadris (with AZ as the active ingredient) and Uniform (AZ + mefenoxam)–both used as solutions to treat soil simultaneously with tuber planting, and

Score (DIF)–used as solution to spray stems and leaves at Day 22 of the experiment. Each experiment was performed in triplicate. In each treatment, potatoes were grown in 6 pots.

9.2.6 GROWING POTATOES IN THE FIELD

Potatoes were grown on microplots of the field located at the village of Drokino, the Yemelyanovskii District, Krasnoyarskii Krai (Siberia, Russia) (5605′03.5″N 92°46′04.2″E), following the recommendations by the RF Ministry of Agriculture [70]. Potatoes were planted manually; treatments and controls were arranged randomly. The negative control included intact plants. In the positive control, commercial fungicides Quadris and Uniform were used. In the treatments, experimental granules containing AZ and AZ + mefenoxam were used. The area of a microplot for each treatment/control group was 10 m^2; the area of the record plot was 5 m^2; 35 tubers were planted per plot; each experiment was conducted in quadruplicate. The total area of the experimental plot was 200 m^2.

Before planting, the quality of seed tubers was checked in accordance with GOST 33996-2016 Interstate Standard "Seed Potatoes" using an integrated sample (300 tubers of potato cv. Krasnoyarskiy ranniy). This method is used to detect and count the tubers with external and internal signs of diseases, injuries, and defects. Tubers affected by common scab, corky scab, and powdery scab are revealed. Tubers with more than 33.3% infected surface (more than 10% for black scurf) are considered as infected. Tubers affected by silver scab are considered to be infected only in the case of turgor loss or damaged eyes. Tubers affected by soft and dry rots are counted separately; tubers are cut to reveal black rot and ring rot.

Tubers were checked before planting and after harvesting, after the integrated sample was collected. Detection of infected tubers was based on external signs. Tubers with a maximum transverse diameter of 28–60 mm were classified in the standard category.

The planting was done manually on July 03, as the spring of 2021 was cold and too long. The fungicides were applied to soil at the time of potato planting. In the treatments, granules containing AZ and AZ + mefenoxam were buried in soil. In the positive control, Quadris and Uniform solutions were applied to the soil at the same concentrations of fungicides as those in the treatments using a hand-operated sprayer; the application rates were 3.0 and 1.4 L/ha, respectively–the rates accepted in Russia. To suppress weeds, on July 15, the plot was treated with the Escudo herbicide (0.015 kg/ha) mixed with the Adieu adjuvant (200 ml/ha); the consumption rates of the solutions were 3 L/ha and 200 L/ha.

9.2.7 AGROCHEMICAL CHARACTERIZATION OF SOIL

The agrochemical examination of the soil was carried out at the end of May and before planting potatoes, using conventional methods, according to the Russian state standards; humus, acidity, and major mineral components were determined as described elsewhere [83]. The agro-transformed garden soil used in the experiment was leached chernozem, which was almost black in color, smearing, and weakly structured. These properties are the basis for the high fertility of the soil and prolonged crop cultivation on it. The 0–20 cm topsoil layer was characterized by the high humus content (17.4%) and high nitrate nitrogen N-NO_3–120.0 mg/g; it was high in available phosphorus and potassium P_2O_5–151.2 mg/100 g; K_2O–80 mg/100 g soil); the pH of soil solution was 6.6 (close to neutral). No fertilizers were applied to soil during the field season. During the growing season of the crops, moisture content and temperature of the 0–20-cm soil layer were determined by thermogravimetric analysis using a TR–46908c electronic moisture meter and a temperature sensor on three sample plots within the study field. Soil temperature and moisture content were measured every 10–12 d.

9.2.8 A MICROBIOLOGICAL STUDY

The structure of the soil microbial community was analyzed by conventional microbiological methods [40]. Counting the total number of microfungi in soil samples was carried out by plating soil suspension on potato-dextrose agar with benzylpenicillin (1,000,000 U·L^{-1} of medium) to suppress the growth of bacteria. All plantings were done in triplicate from dilutions to 10^4. The plates were incubated at 25°C for 7–10 days. Microscopic examination of colonies was performed using an AxioStar microscope (Carl Zeiss). Microfungi were identified by cultural and morphological features according to the identification guides [69]. Counting the total number of bacteria was performed on meat peptone agar from dilutions to 10^6; the plates were incubated in the temperature-controlled cabinet at 30°C for 5–7 d.

To reveal plant pathogens in soil and potato tubers, phytosanitary analysis was performed by plating the samples on wort agar. To isolate plant pathogens, approximately 1-cm pieces were cut from the affected potato parts and placed on the nutrient medium in Petri dishes. Cultivation was conducted for 5–7 d at a temperature of 28°C, and then the isolates were passaged three times on wort agar to isolate pure cultures. Identification of the isolated pathogens was based on symptoms of the damaged tubers [93] and molecular-genetic features of the fungus isolates. The nucleotide sequences of the 28S rRNA gene fragments of the isolated fungi were compared with the sequences in the GenBank using the BLAST program for searching highly homologous sequences of the web resource.

9.2.9 MONITORING POTATO PLANT DEVELOPMENT DURING THE GROWING SEASON OF 2021

During the growing season, plants were regularly examined and photographed. The parameters monitored to study the effects of different modes of fungicide delivery on plant growth and development included the dates of developmental phases (shoot emergence and complete development of the shoots, budding, flowering, and physiological maturity) and biometric parameters (height and number of stems per plant and level of development of the roots).

Estimation of plant pathogen infection was based on the severity of the damage to plants and the prevalence of disease. The severity of damage to potato plants was rated on a 1–9 scale [93]: 9 indicating that no damage symptoms were observed; 8–1 to 10% of the leaf surface was damaged; there were a few spots on individual plants (up to 10 leaves were infected, in total approximately 50 spots per plants); 7–10 to 25% of the leaf surface was damaged (symptoms of infection were seen on almost all leaves of most plants, but the plants retained their normal shape, green was the prevailing color, the plants could smell of *Phytophthora* rot); 5–25 to 50% of the leaf surface was damaged (almost all plants were infected, but the prevailing color was still green although brown spots occupied a large part of the leaf surface area); 3–more than 50% of the leaf surface area was damaged (it was unclear whether green or brown color of the leaves prevailed, but the stems of most of the plants remained green); 1–all leaves were damaged; the stems were dying or dead.

In addition to the severity of damage, which characterized the surface area of stems and leaves (or tubers) infected by the plant pathogen, another parameter for assessing fungicide effectiveness was the prevalence of the disease in the agroecosystem. This parameter is the ratio of the number of infected plants (tubers) to the total number of plants (tubers) grown on the plot.

The prevalence of the disease of vegetative biomass (tubers) was determined taking into account the area of the damaged surface of leaves or tubers, using the following formula:

$$P = n/N \times 100$$

where; 'P' is disease prevalence; 'n' is the number of infected plants (tubers); 'N' is the total number of plants (tubers) in the sample.

The yield structure parameters were the number of the tubers per plant; the ratio of the saleable tubers (over 40 g in weight) to non-saleable tubers; the average weight of the saleable tuber; total yield. The potatoes that had been stored for two weeks after harvesting were analyzed to determine dry matter (GOST 28561-90), starch and moisture (GOST 7194-81), reducing sugars–by the potassium permanganate method, and nitrates–using a SOEK NUK-019-2 nitrate meter.

9.2.10 *STATISTICAL ANALYSIS*

Statistical analysis of the results was performed using the standard software package of Microsoft Excel and StatSoft STATISTICA 13. Arithmetic means and standard deviations were found [13].

9.3 RESULTS AND DISCUSSION

Biological activity of fungicides loaded into the degradable P(3HB)/birch wood flour blend was studied in the laboratory *in vitro* experiments in cultures of pathogens causing potato diseases, in pot experiments with potato plants cultivated in the growth chamber, and in the experiments with field-grown potato plants; the activity of the experimental fungicide formulations was compared to the activity of their commercial analogs.

Three types of fungicide formulations prepared as granules loaded with AZ, AZ + mefenoxam, and DIF were compared to their commercial analogs Quadris, Uniform, and Score. All active ingredients and fungicide formulations are approved for use in Russia [66] and are widely used in other countries to treat tubers before planting and to spray potato plants. Early blight of potato caused by *Alternaria solani* was more effectively controlled by spraying the plants with AZ and DIF compared to the treatment with AZ alone, especially when the infection percent was very high [44]. Spraying of seed tubers with AZ, as well as with chlorothalonil and fludioxonil, was found to be effective against *Fusarium graminearum*, which causes dry rot disease, leading to tuber damage, dry matter reduction in tubers, and formation of toxins in rotten tubers [73]. Another study [38] demonstrated that AZ exhibited different biological activity towards *Rhizoctonia* isolates obtained from infected potato plants grown in South Africa; the response of *Rhizoctonia* isolates to fungicide treatment varied from high sensitivity (EC50: <0.09 μg a.i./ml^{-1}) to insensitivity, with EC_{50} values exceeding 5 μg a.i./ml^{-1}. The data on fungicide sensitivity of *Rhizoctonia* isolates are critical for developing effective control strategies, facilitating monitoring of fungicide insensitive isolates in pathogen population. Field studies showed that spraying of potato plants with metalaxyl (a stereoisomer of mefenoxam–the active ingredient of uniform) along with other fungicides (dimethomorph, fenamidone + mancozeb, mancozeb) reduced late blight severity and increased tuber yields in the plains of Nepal [21]. The Ridomil fungicide (Syngenta) containing mancozeb + mefenoxam effectively protected potato varieties such as Belete, Gudene, and Jalene from *Phytophthora infestans*–the agent of late blight; the authors reported optimal frequencies of spraying for each of the potato varieties [20]. The effectiveness of metalaxyl as potato foliar spray to control *Pythium ultimum* was evaluated in another study [75]; in field trials, metalaxyl was found to be a more effective fungicide against *Pythium ultimum* than hymexazol and propamocarb-HCl when applied by foliar spray and in-furrow treatment. These data, which were published in the last 2–3 years, are indicative of the significance of the fungicides used in the present study to construct controlled-release formulations.

9.3.1 IN VITRO TESTING OF BIOLOGICAL ACTIVITY OF EMBEDDED FUNGICIDES

Biological activity of embedded fungicides was tested *in vitro* in cultures of the most commonly occurring and harmful potato pathogens: *Alternaria longipes, Fusarium solani, Phytophthora infestans*. These pathogens were isolated from the samples of the field soil and tubers of potato cv. Krasnoyarskiy ranniy and identified using conventional culture-based and molecular-genetic methods [52].

Three types of experimental fungicide formulations fabricated as granules were tested in cultures of 10 strains of phytopathogenic fungi: *Alternaria alternata, Alternaria longipes, Boeremia exigua, Phytophthora infestans, Rhizoctonia solani, Fusarium oxysporum, Fusarium redolens, Fusarium solani, Fusarium equiseti*, and *Fusarium vanettenii*. Fungicidal activity of the granules was evaluated using diffusion method on wort agar, by determining the degree of inhibition of mycelial growth (the average diameter of the fungus colony) caused by fungicides. The granules loaded with DIF, and the granules loaded with two fungicides–AZ + mefenoxam–were constructed and investigated for the first time. The fungicide formulation of AZ loaded into the blend of poly-3-hydroxybutyrate and a natural material (wood flour, peat, or clay) was fabricated and described in a previous study [79]; biological activity of the embedded AZ was evaluated and confirmed *in vitro*, in the culture of *Fusarium verticillioides*, a plant pathogen causing root rot of cereal crops.

All experimental fungicide formulations produced inhibitory effect on phytopathogenic microscopic fungi that was comparable to the effect of their commercial analogs (positive control). The embedded fungicides caused a 1.2- to 2.3-fold reduction in the average diameters of the colonies in different treatments relative to the negative controls. Granules with DIF had the strongest inhibitory effect followed by the formulation with two fungicides–AZ + mefenoxam. The amounts of the fungicides diffused to agar from the degradable granules placed onto the surface of the agar medium were sufficient to inhibit pathogen mycelium growth. The degree of inhibition varied across the taxa, including species. For example, among *Alternaria* species, *A. alternata* was less affected by both forms of fungicides (free and embedded ones) than *A. longipes*. A similar trend was noted for *Fusarium* species (Figure 9.1).

The species that showed the highest sensitivity to the experimental fungicide formulations included *P. infestans, A. longipes, B. exigua, R. solani*, and *F. solani*. Figure 9.1 shows some of the photographs of colonies of these phytopathogenic fungi growing in Petri dishes on wort agar in the presence of embedded fungicides and their commercial analogs versus negative control. The average diameter of the *A. longipes* colony in negative control (without fungicides) reached 7.6 ± 0.4 cm; in the presence of the embedded AZ and DIF, the diameter of the colony was reduced by a factor of 1.8. The commercial formulations controlled fungal growth equally effectively, reducing the diameter of the colony by a factor of 1.6–1.7.

In the negative control, the *F. solani* colony covered almost the entire surface of the agar medium in the Petri dish, and its diameter was 8.8 ± 0.2 cm (Figure 9.1). All experimental fungicide formulations were equally effective in inhibiting the growth of this pathogen: the diameter of the colonies affected by fungicides were decreased by a factor of 1.7–1.8. Commercial formulations Quadris and Uniform showed comparable biological activity, and Score controlled the growth of the colonies more effectively, reducing their diameter by a factor of 2.3 relative to the control.

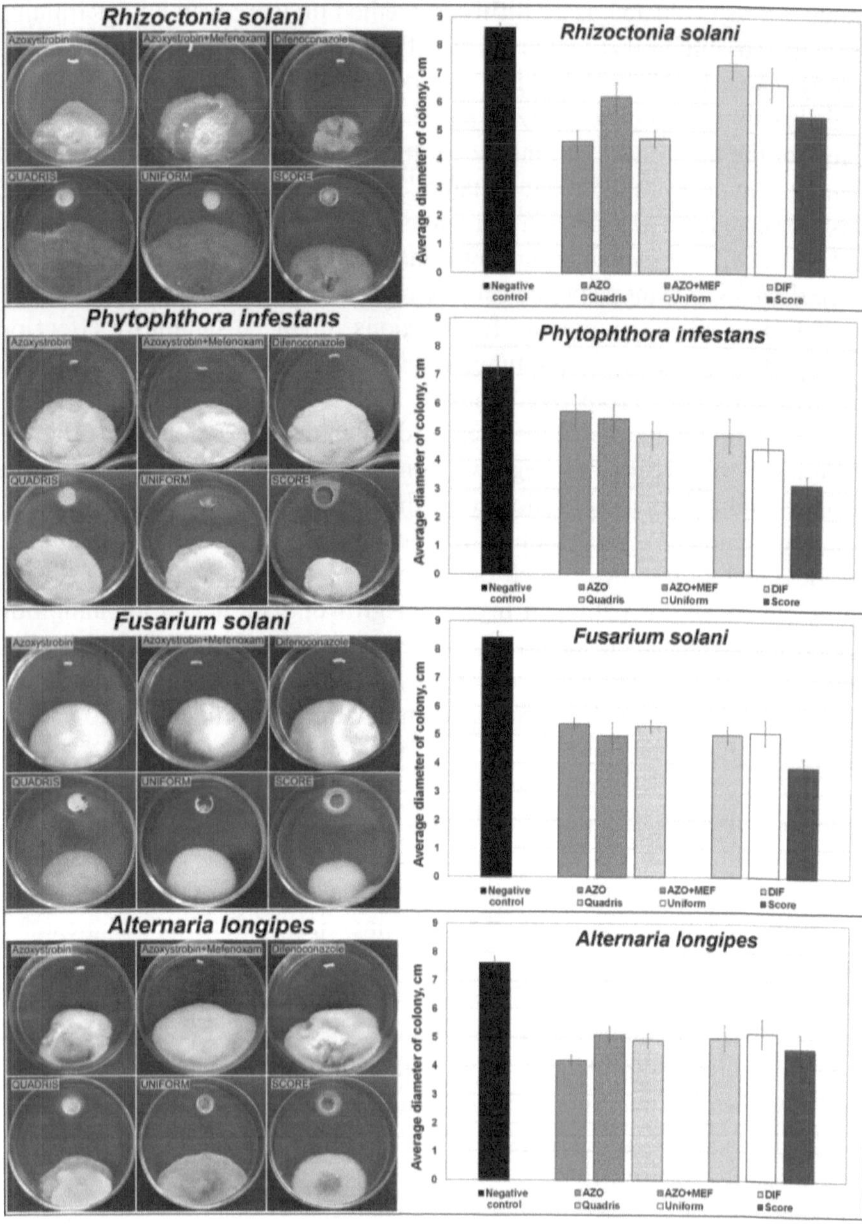

FIGURE 9.1 Inhibition of mycelial growth of phytopathogenic fungi *A. longipes*, *F. solani*, *R. solani*, and *P. infestans* by various forms of fungicides: treatment (granular embedded fungicides) and positive control (commercial fungicide formulations) versus negative control (colony growth with no fungicides in the medium).

Similar results were obtained for the phytopathogenic fungus *P. infestans*. In the negative control, with no fungicides, the diameter of the colony was 7.1 0.4 cm; all three experimental fungicide formulations were equally effective in inhibiting mycelial growth, and the diameter of the colony was reduced by a factor of 1.3. Of the commercial analogs, the Score fungicide, which contained DIF, produced the strongest inhibitory effect: the diameter of the colony was decreased by a factor of 2.2. The plant pathogen *R. solani* was the most sensitive to the embedded DIF and its commercial analog–Score. DIF reduced the diameter of the colonies relative to the control by a factor of 2.3 when the embedded fungicide was used and by a factor of 1.5 in the presence of Score. Commercial formulations Quadris (containing AZ) and Uniform (containing AZ + mefenoxam) were somewhat less effective in inhibiting *R. solani* growth, in contrast to their effect in cultures of other plant pathogens studied here. The experimental granules containing AZ alone and AZ + mefenoxam also had a considerable inhibitory effect on *R. solani* mycelial growth, decreasing the diameter of the colonies by a factor of 1.4 relative to the negative control.

Thus, experimental formulations with fungicides embedded in degradable matrix were effective against plant pathogens including the most harmful potato pathogens–*P. infestans*, *R. solani*, and *F. solani*, and their effect was comparable to that of their commercial analogs. Therefore, an *in vivo* study was performed to evaluate the effectiveness of embedded pesticides in protecting laboratory-grown potatoes.

9.3.2 A COMPARATIVE STUDY OF THE EFFECT OF EMBEDDED FUNGICIDES AND THEIR COMMERCIAL ANALOGS ON GROWTH AND DISEASE SUSCEPTIBILITY OF POTATOES CULTIVATED IN LABORATORY

The effect of the fungicide delivery mode on growth and disease susceptibility of potato cv. Krasnoyarskiy ranniy was studied in laboratory pot experiments in the environmental chamber. Mechanically undamaged tubers 5–6 cm in diameter were selected for planting. The experiment was started on July 13, 2021. Potatoes were grown in an environmental chamber (Fitotron-LiA-2, Russia) in 2-L plastic pots filled with agro-transformed pre-characterized soil. One tuber was planted per pot. Diurnal cycles of temperature, light, and humidity were maintained in the chamber. Two forms of fungicides were tested: the experimental embedded fungicides and their commercial analogs. The experimental granules carrying AZ, azoxystrobin + mefenoxam (AZ + MEF), and DIF were buried in soil together with tubers. The commercial formulations–Quadris (with AZ as the active ingredient) and Uniform (azoxystrobin + mefenoxam)–were applied to soil as solutions at the time of potato planting; Score (DIF) was used to spray potato plants at Day 22 of the experiment. Each experiment was performed in triplicate.

Phytosanitary analysis of potato tubers and microbiological analysis of the soil were conducted to detect possible sources of infection. Analysis of seed tubers showed that 1% of the tubers were infected with black scurf (*Rhizoctonia solani*) and 2% with common scab (*Streptomyces scabies*). Microbiological analysis of the soil before the experiment showed that the total abundance of organotrophic bacteria in soil samples was 5.4 ± 2.7 million CFUs per g soil. The dominant bacteria were *Bacillus* species (64.5%) including *B. pumilus, B. licheniformis*, and *B. muralis*, followed by *Streptomyces* (12.2%); the Gram-negative *Pseudomonas* constituted 7.7%. The abundance of fungi in the soil was 17.3 ± 3.5 x 1,000 CFUs/g soil. The dominant fungi were *Penicillium* (46.6%) and *Trichoderma* (22.3%) species. The soil also contained potato pathogens: *Alternaria alternate* (2.5%) and *Fusarium oxysporum* (7.8%).

Soil application of embedded fungicides had a favorable effect on potato plant germination, growth, and development. In the treatment group, where fungicide granules were buried in soil at the time of planting, emerging crops were observed 13 days after potato planting. In the negative control (intact plants) and in the positive control, seedlings emerged later, 18 days after planting. A similar positive effect of pre-planting treatment of the soil with AZ and mefenoxam (commercial formulation Uniform) on potato germination rate was described by other authors [28]. Soil application of uniform at a rate of 1.5 L/ha before planting potatoes significantly increased potato germination rate, led to earlier seedling emergence, and facilitated potato growth and development in the field.

In the current study, one month after planting (August 13, 2021), the plants were photographed and analyzed to characterize their vegetative organs and roots and to estimate the degree of damage by pathogens (Figures 9.2 and 9.3). The severity of damage to potato plants was rated on a 1–9 scale [93].

Analysis of the damage caused by pathogens to the plants at those time points detected only black scurf although the initial soil contained insignificant amounts of other pathogens (*Alternaria alternata* and *Fusarium oxysporum*). One month after planting, the severity of black scurf in all groups was given a score of 8 points, indicating that no more than 10% of the aboveground biomass was damaged by the disease. In all groups of plants except those treated with Score, at the base of the stems, there were focal necrotic areas (Figure 9.2), which developed into oval or oblong brownish spots (1 to 9) sunk in tissue but having distinct boundaries. No other damage was observed;

leaves were uninjured, showing no signs of pathogen development. The low damage to the plants was likely associated with the conditions of the experiment, in which plants were grown at stable temperature and humidity and with no contact with the outer environment. The only disease that affected plants was black scurf, which usually develops in humid conditions. Other pathogens present in the initial soil did not develop during laboratory experiment, and the fungicides effectively controlled them.

Intact plants (negative control) had significantly fewer stems and they were significantly shorter (Figure 9.3) compared to the treatment plants. The potato plants in the treatment and positive control had comparable numbers of stems and similar root length. Embedded fungicides facilitated growth of the aboveground parts of the plants. In the treatment groups, the height of the plants was greater by a factor of almost two compared with both negative and positive controls. The exception was the group treated with Uniform, in which the height of the plants was comparable to the height of the plants in the treatment group where two embedded fungicides (azoxystrobin + mefenoxam) were applied.

Soil application of pesticides should be based on microbiological studies. Microorganisms metabolize pesticides and other pollutants and serve as the major factor in transforming xenobiotics and removing them from the biosphere. An increase in pesticide concentrations leads to cumulative toxicity, resulting in suppression of microbial parameters of soil, reduction in diversity indices, and selection of resistant microorganisms [41, 84, 90, 91]. The comparison of the embedded and free fungicides applied to soil shows that, on the one hand, pesticides directly affect the structure of soil microbial community and metabolic potential of microorganisms and on the other hand, soil application of fungicides embedded in polymer matrix provides various soil bacteria and fungi with a supplementary growth substrate–polymer, stimulating their development.

By the end of the experiment (October 26), the abundance of the fungi had decreased in all groups of plants, including the negative control (Figure 9.4). The positive control plants treated with Quadris and Uniform did not differ from the negative control while their embedded analogs (azoxystrobin and azoxystrobin + mefenoxam) continued inhibiting the development of fungi, reducing their abundance by a factor of 1.8 and 2.3, respectively. The commercial and embedded DIF exhibited the strongest fungicidal effect, reducing the abundance of fungi by a factor of 4.6 and 3.1, respectively. However, taking into account the exposure of bacterial flora to stress, it could be assumed that Score destabilized the microbial community in the laboratory soil system while its embedded analog was less toxic.

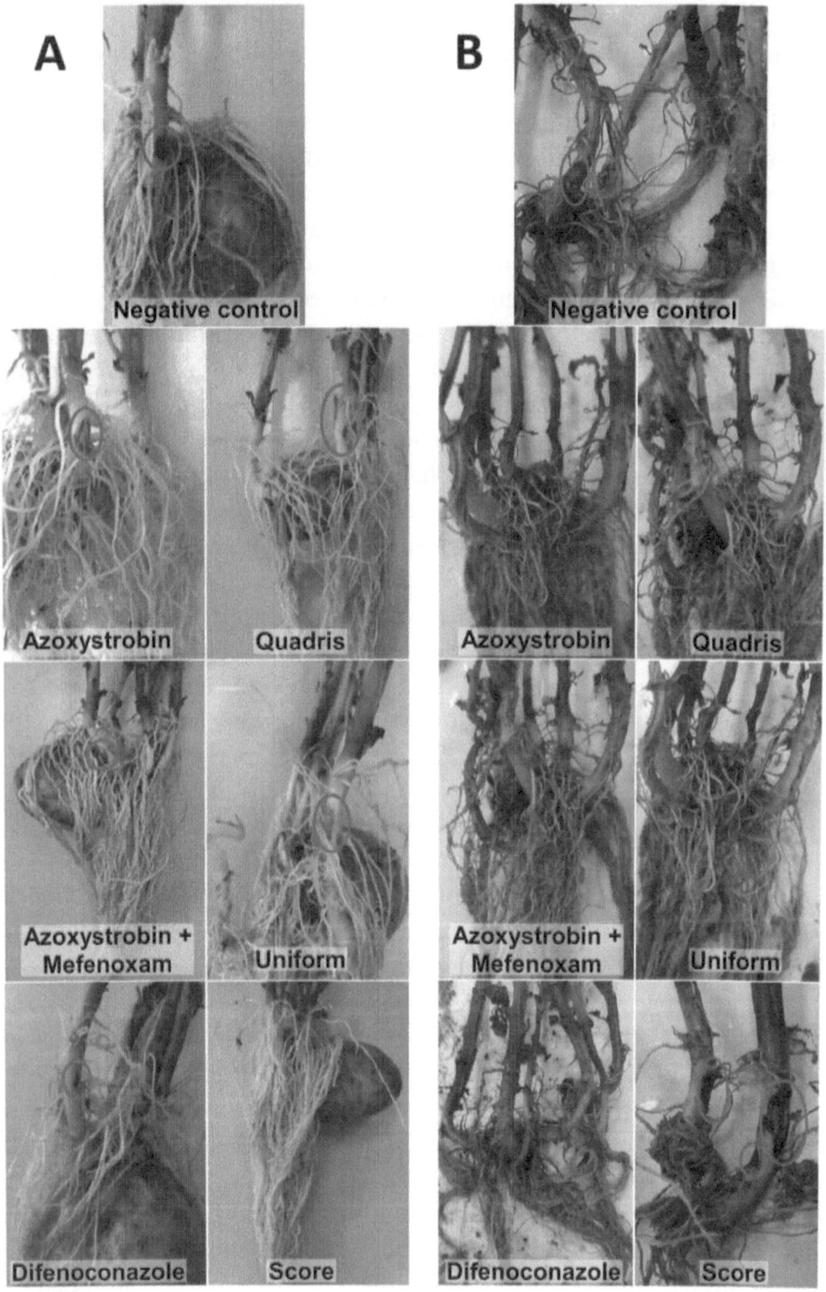

FIGURE 9.2 Photographs of potato plants of the variety Krasnoyarskiy ranniy grown in laboratory with different forms of fungicides on August 13, 2021 (A) one month after planting and October 26, 2021 (B) before harvesting; red color shows black scurf disease.

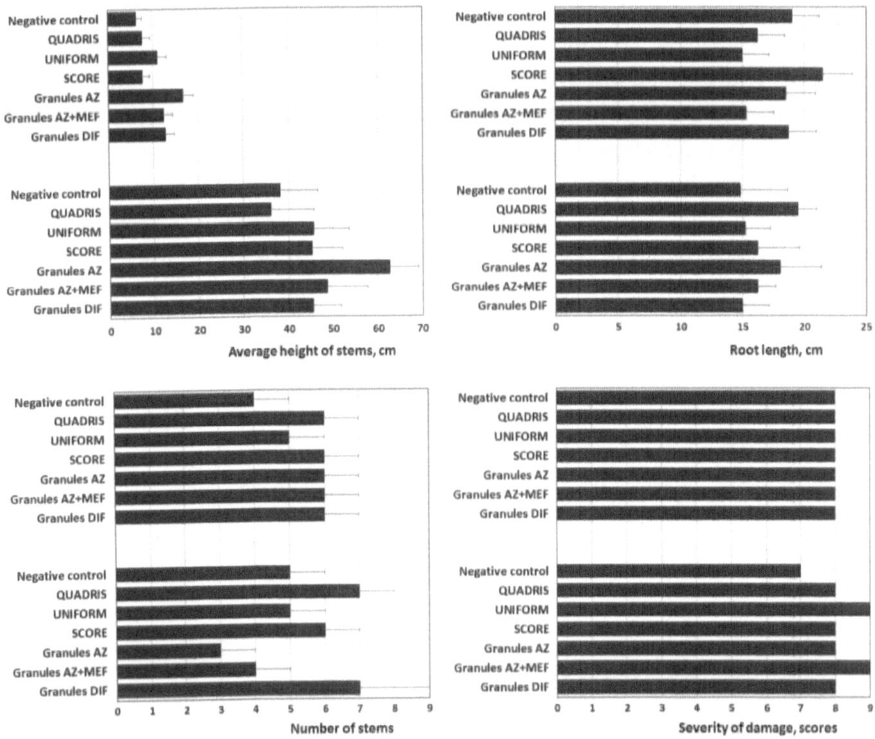

FIGURE 9.3 Biometrics of potato cv. Krasnoyarskiy ranniy grown in laboratory with different forms of fungicides on August 13, 2021 and October 26, 2021.

Identification of phytopathogenic fungi showed that *A. alternata* was isolated only from the initial soil but was not detected in the potato plant rhizosphere. That could be attributed to the severe competition between soil microorganisms in pot experiments. The *Fusarium* species that were isolated from the rhizosphere of intact plants were *F. oxysporum* and *F. solani*, whose total abundance increased during the experiment from 11.1 to 24.6%. One month after soil application of embedded fungicides, *F. oxysporum* (constituting 1%) was only isolated from the rhizosphere of the plants in the treatment with AZ granules. At the end of the experiment, all embedded fungicides suppressed the growth of *Fusarium* species while in the groups treated with Quadris and Uniform, *Fusarium* species constituted 5.3 and 3.1%, respectively (Figure 9.4).

These results are consistent with the data reported in the study of the effect of the pesticide delivery method on the microbial community of field soil [51]. The authors of that study found that embedded pesticides had a considerably weaker negative effect on soil microflora, including functional groups of microorganisms and primary degraders of poly-3-hydroxybutyrate, compared to free pesticides; moreover, sometimes, they produced a stimulating effect. Thus, prolonged-action pesticide formulations, which release pesticides gradually, preserve degradation potential of microflora–the main factor preventing accumulation of xenobiotics in the biosphere. The study of the effects of various pesticides demonstrated that the epoxiconazole, AZ, and tebuconazole fungicides embedded in the P(3HB)/natural material (peat, clay, or wood flour) blends, like their free forms, effectively controlled the growth of fungi and hence, reduced the abundance of phytopathogenic fungi: *Alternaria* and *Pythium* by 100%, *Fusarium* by 76.2%, and *Verticillium* by 66.7%.

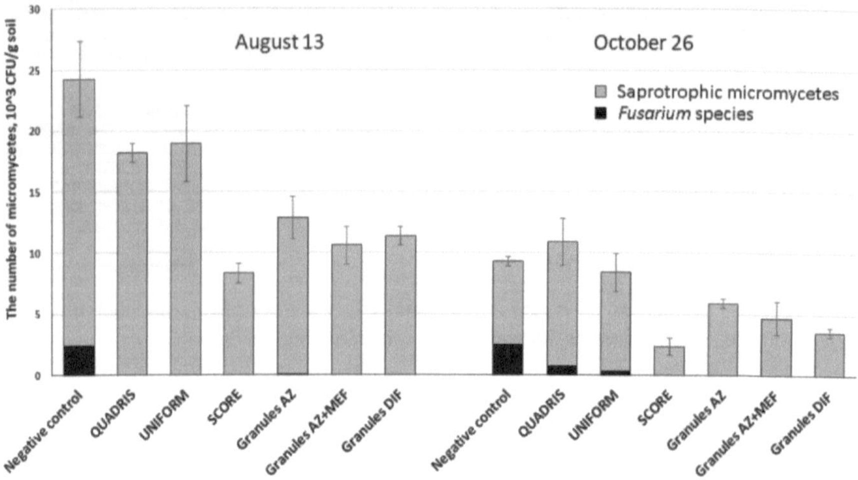

FIGURE 9.4 Total abundance of fungi (including plant pathogens of the genus *Fusarium*) in the potato rhizosphere in the presence of different fungicide formulations.

Around 60 days after the stage of fully sprouted seed potatoes (October 26, 2021), the experiment was completed, and results are shown in photographs of the roots of plants, where symptoms of black scurf are indicated (Figure 9.2). Biometrics of potato plants are shown in Figure 9.3. In the groups with embedded fungicides, the height of the plant aboveground parts was greater than in the negative control and comparable with the height of the plants in positive controls with Uniform and Score. Potato plants grown on the soil with embedded AZ had the tallest stems, 62.7 ± 6.4 cm, and they were 1.6 and 1.7 times taller than the stems in the group of intact plants and the group where Quadris was applied, respectively. The smaller number of the stems in that treatment could be compensated by the elongation of the aboveground parts of the plants. As the space for root development in the pots was limited, the length of the roots did not differ significantly between treatments and controls.

The severity of black scurf was the highest (7 points) in the group of intact plants, where up to 15–20% of the plant surface was affected; the plants, however, were of normal shape, and the green color prevailed. Plants in the treatments and positive control were less affected by black scurf (8 points): no more than 10% of their surface was damaged. In the groups with embedded AZ + mefenoxam and in the positive control with soil application of Uniform, no damage was found on the plant stems–9 points (no damage). No symptoms of fungal infection were detected on the leaves. Partial drying of some leaves and red color were caused by the spectrum of the light used in the growth chamber. Yet, drying of leaves was more pronounced in the group of plants sprayed with Score. No signs of bacterial or fungal infection were detected on roots of all plants in all experimental groups. The highest parameters of plant development were obtained in the groups with embedded AZ and AZ + mefenoxam and in the group treated with Uniform, which contains these two fungicides. In a study by Kuznetsova et al. [29], soil was treated with Uniform before planting potatoes, and the fungicides both effectively controlled *Pythium* and facilitated growth and development of potato plants, enabling the greatest increase in the yield of potatoes (mid-season potato cv. Lorkh) grown in pots in the greenhouse.

In the groups with embedded AZ and DIF, tuber formation parameters (the number of tubers per plant and their size) were higher than in the groups where the corresponding commercial formulations (Quadris and Score) were used. The yields were also higher. The yield of one plant grown in the environmental chamber varied from one to four microtubers with widely ranging average diameters. The mass of the tubers produced by one plant was the greatest in the groups with embedded AZ (12.8 ± 1.8 g) and AZ + mefenoxam (13.7 ± 2.3 g)–1.6 and 1.7 times greater than in the groups with the commercial analogs of these fungicides. The maximum increase in the potato yield achieved in the treatment groups relative to the positive controls was in the groups with embedded AZ (60%) and AZ + mefenoxam (71.3%).

Microbiological analysis of microtubers showed that in the negative control, most of the tubers were insignificantly damaged by black scurf (*R. solani*) and common scab (*S. scabies*). The damaged surface constituted less than 1% of the area. In all treatments and positive controls, there were fewer damaged tubers than in the negative control. In the groups with embedded

AZ + mefenoxam and their commercial analog, no surface indications of damage by plant pathogens were observed.

The embedded AZ and AZ + mefenoxam effectively protected potato plants from plant pathogens. Thus, laboratory experiment with embedded fungicides and their commercial analogs demonstrated the advantages of the experimental prolonged-action pesticide formulations, which had favorable effects on plant germination, growth, and development, facilitated tuber formation, and enhanced the quality of potatoes.

9.3.3 A COMPARATIVE STUDY OF THE EFFECT OF EMBEDDED FUNGICIDES AND THEIR COMMERCIAL ANALOGS ON GROWTH AND DISEASE SUSCEPTIBILITY OF POTATOES CULTIVATED IN THE FIELD

Micro-plot experiments aimed at testing the efficacy of embedded fungicides in controlling potato pathogens were conducted in Krasnoyarskii Krai (Siberia, Russia). The region is a considerable distance from the ocean and thus, has a continental subarctic climate. Mean annual air temperature ranges between 0.5 and 1.3°C, sometimes dropping to–2°C. The period of biological activity varies between 90 and 115 days. The heat summation is 1550–1800°C; the ground freezes to a depth of 1.5–3.0 m.

In 2021, May was cold, and warm weather set in late. June was also rather cold, and precipitation fell unevenly in time. The amount of precipitation in the first 20 days of June was low, and between June 21 and 30, the rainfall amount reached 100 mm. In July, the air temperature rose abruptly, and there was no rain in the first 10 days and insignificant rainfall amount for the rest of the month–50–60% of the average monthly precipitation. In August, the air temperature was 3°C higher than the long-term monthly average; the rainfall amount was 43 mm–below the average monthly precipitation. Thus, the 2021 growing season was not favorable enough for potato growth and development and for the activity of soil microorganisms–the main factor of biodegradation of the experimental fungicide formulations and fungicide delivery to plants.

Two forms of fungicides were used in the study: embedded fungicides (buried in soil during potato planting)–AZ granules and AZ + mefenoxam granules and free fungicides as commercial formulations Quadris and Uniform (both used as solutions to treat soil during planting of potatoes–early potato cv. Krasnoyarskiy ranniy.

Field germination rate of potato cv. Krasnoyarskiy ranniy was high, about 96–100%, in all groups of plants, as the cultivar was adapted to the soil and climate in the Krasnoyarsk region, where it was produced. Despite the low air and soil temperatures in the early stages of plant development followed by dry and hot weather, potato plant developmental phases (germination, budding, flowering, physiological maturity) began at dates usual for this potato variety and regardless of the form of fungicides applied to soil. Intact plants (negative control), plants growing in the presence of commercial fungicides (positive control), and plants in the treatment groups, where experimental fungicide formulations were used, entered each developmental phase simultaneously.

The mode of fungicide delivery influenced crop biometrics (Figure 9.5). In the period of shoot emergence, the number of stems in intact plants was significantly smaller than in the plants grown with embedded fungicides and plants grown with commercial fungicides (positive control); the difference was 1.1–2.3 stems (12.3–14.7%). The same trend was observed in the early stage of physiological maturity. Intact plants were less developed and 9–11 cm (12.6%) shorter than the plants in the positive control. The difference from the plants grown with embedded fungicides was even more considerable, reaching 14–15 cm (13.5%). When the plants entered the phase of physiological maturity, the height difference was observed between both controls and the treatment plants, which were grown with embedded AZ and AZ + mefenoxam. In both treatments, the plants were the most well-developed and intensely green in color; in the flowering phase, there were more flowers on them. These results are in good agreement with the findings obtained in laboratory pot experiments in the current study, in which fungicides (AZ alone and AZ + mefenoxam), especially embedded ones, facilitated growth and development of potato plants. They are also consistent with the literature data showing the favorable effect of the Uniform commercial formulation, containing AZ and mefenoxam, on the germination rate, growth, and development of potato plants grown in pots in greenhouses and in the field [28, 29].

The mode of fungicide delivery influenced the structure of soil microbial community, the presence of pathogenic fungi causing early blight and late blight, and the severity of these potato diseases. This is the radical difference between growing potatoes in the field, under complex and ever-changing environmental conditions, and laboratory experiments described above, in which the soil microbial composition and the quality of seed potatoes were the same, but the only pathogen detected on potatoes was that of black scurf.

The structure of soil microbial community of potato rhizosphere in the field differed substantially from the structure of laboratory soil microecosystems. The revealed differences were largely associated with the weather, chiefly soil temperature and moisture content, which influenced development and activity of microflora. In addition, a study comparing biological activities of the free and embedded fungicides towards plant pathogens should take into account that the activity of embedded fungicides is affected greatly by the rate with which active ingredients are released from the degradable polymer matrix. The release rate, in turn, depends on the activity of the soil microbial community as the main factor determining the degradation kinetics of the polymer matrix and hence, fungicide release and accumulation in soil.

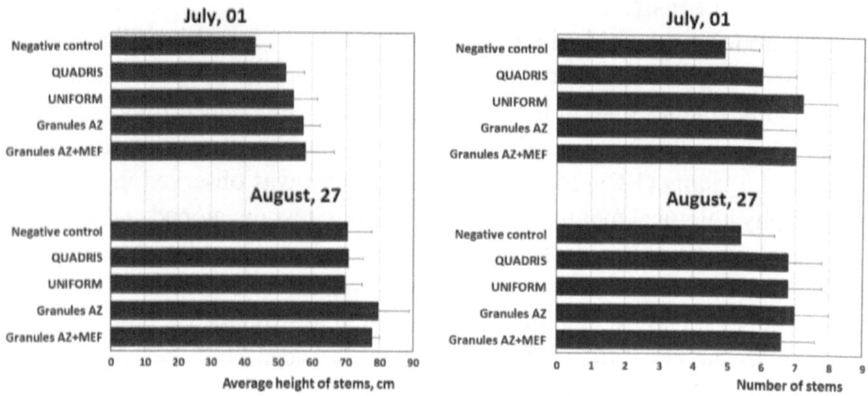

FIGURE 9.5 Biometrics of potato cv. Krasnoyarskiy ranniy grown in the field with different fungicide forms: full-sized plants (July 1, 2021), early physiological maturity (August 27, 2021).

In June, when the weather was cool, the total counts of microorganisms were low, $(4.6\pm0.3)\times10^6$ and $(14.7 \pm 1.2)\times10^3$ CFUs/g soil for bacteria and fungi, respectively. The abundance of fungi in the negative control was similar to the counts at the time of potato planting. In that period, the use of embedded fungicides did not cause any significant change in the counts of fungi in potato rhizosphere relative to the negative control. The dry and hot weather in July also had an unfavorable effect on the development of soil microflora. In the rhizosphere of intact plants, the abundance of fungi remained the same as in the previous month: 10,000–15,000 CFUs/g soil. *Alternaria* constituted about 2–3% and *Fusarium* 8–10% of the total counts of fungi. In the second half of August, however, the combination of high average daily temperature (19–20°C) and abundant rainfall (42 mm) induced a dramatic increase in the counts of microorganisms. In the rhizosphere of intact plants and in the groups of the plants grown on soil treated with Quadris and Uniform, the counts of fungi increased to 30,000–35,000 CFUs/g soil, and that number was almost twice higher than in July and early August. In the soil with embedded fungicides, the counts of fungi were significantly lower than in the negative and positive controls (18,000–20,000 CFUs/g soil), indicating that fungicides were released to soil from the granules in that period of higher activity of microorganisms.

In all soil samples, the fungal communities were dominated by *Penicillium* species; in the samples treated with commercial fungicides, their percentage increased to 61–68% while the percentage of *Trichoderma* fungi decreased considerably (by a factor of 2–4 compared to the rhizosphere soil of intact plants). That was an unfavorable development, as *Trichoderma* fungi are natural antagonists of phytopathogenic fungi. Moreover, by the end of the growing season, the fungicidal activity of commercial formulations against *Alternaria* and *Fusarium* became weaker, and they constituted 6–8% of the fungal community, which was an insignificantly smaller percent than in the rhizosphere of intact plants (10–12%). In contrast to the positive control groups, in the treatments, the abundance of *Alternaria* and *Fusarium* species in soil was below the detection limit, and they were not identified when soil suspensions were plated onto wort agar, suggesting that embedded fungicides remained active towards those potato pathogens. This is an important fact because at the end of the growing season, potato tubers are highly susceptible to diseases caused by these pathogens.

Not only were the weather conditions of the field season of 2021 unfavorable for the development of soil microflora and phytopathogenic fungi but they also affected biological effectiveness of embedded fungicides.

The first examination of potato plants was carried out when all plants were in the flowering stage (the last 10 days of July). Early blight spots were revealed on the leaves of the plants in the negative control and in the positive control group with Quadris. No signs of early blight were found on the plants in the groups with embedded fungicides and in the positive control group with Uniform. In the flowering stage, about 10% of the intact plants of potato cv. Krasnoyarskiy ranniy were affected by late blight. Both free and embedded fungicides were effective against the pathogen of this disease in this potato variety. The embedded AZ and AZ + mefenoxam demonstrated the best protective effect–the damaged surface area of the plants was no greater than 5–6%. In the groups with the corresponding commercial fungicides, the damaged area was 9–10%, and these differences were statistically significant. The inconsiderable damage done by pathogens to potato plants in that period was evidently associated with the dry and hot weather, which is unfavorable for the development of fungi.

The second examination revealed signs of late blight and early blight in plants of all groups, but those signs were the most pronounced in the negative controls (intact plants). The severity of early blight damage to plant leaves in the negative control rated on a 1–9 scale was 7, and the damaged leaf area was estimated as 20–25%. The severity of damage in the positive control and treatments was similar, about 10–15%. The second examination revealed plants affected by late blight. Late blight is the most dangerous potato disease. It does the greatest harm to early potato varieties, which have low resistance to this pathogen. Late blight, causing premature death of leaves and stems, leads to crop shortages, and zoospores that infect tubers cause significant losses during storage.

Between two examinations, late blight spread precipitously, but the severity of damage varied between low and medium. The signs of the disease were detected on all plants (100% disease prevalence), but the severity of damage varied. Late blight was controlled most effectively in the treatment with AZ + mefenoxam (the damaged leaf area was 12%); in the treatment with AZ, the damaged leaf area was somewhat larger–18%. In the negative control, the area affected by late blight was about 30% while in the positive control groups, it was about 21–22%. In the positive controls, with a single application of Quadris and Uniform solutions to soil, fungicides could be removed from soil by rainwater and as a result of exposure to both abiogenic and biogenic processes, primarily as a result of metabolism by microflora. This is the difference between how fungicides that are active ingredients of commercial formulations are applied to soil and how they behave in soil and the prolonged action of fungicides embedded in the matrix based of poly-3-hydroxybutyrate, which is slowly degraded in soil.

Neither examination revealed black scurf signs on the plants although seed tubers had been infected by that pathogen. The basidial stage of this fungus is usually visible on the plants growing in warm and humid conditions in the second half of summer. In this period, light-colored mycelium can be seen at the base of the stems of the affected clones. Other symptoms of this disease are rolling of the upper leaves along the vein, anthocyanin coloration on the edges of the leaves, and formation of aerial tubers in the axils of the leaves in the lower part of the stems. Examination of the plants on August 27, 2021, did not reveal the above signs on potato plants.

One week before harvesting, the plant tops were mechanically removed. Harvesting and crop accounting were carried out on September 7, 2021. To assess the damage to tubers by diseases after harvesting, a single sample was taken from each plant group. After washing and drying, each tuber was examined for the presence of damage by late blight, black scurf, and scabs (Figure 9.6). For black scurf and scab, the severity of diseases was determined as a percentage of the affected surface area. For late blight, the percentage of damage was not taken into account, since such tubers are not suitable for storage and further use. The number of such tubers was used to determine the prevalence of the disease. All tubers infected with late blight die, and thus, the damaged area was not determined for them, but the prevalence of disease was.

FIGURE 9.6 Photographs of potatoes (cv. Krasnoyarskiy ranniy) grown in the field with different fungicide forms.

The number of affected tubers (disease prevalence) is shown in Table 9.1. In the groups with fungicides, the prevalence of tuber damage by black scurf, scabs, and late blight was significantly reduced. In assessing the damage to tubers by scab, the total damage caused by common scab, corky scab, and powdery scab was determined. In the negative control, 16% of tubers were affected by black scurf, 47% by scab, and 2% by late blight. In the groups with Quadris and Uniform, 38 and 35% of tubers, respectively, were affected by scab, and that was more than in the groups with both types of embedded fungicides (25%). The prevalence of black scurf on the tubers of the experimental groups was no more than 1%; no signs of late blight were found; a similar effect was recorded in both positive control groups.

TABLE 9.1 Prevalence of Diseases on Tubers of Potato cv. Krasnoyarskiy Ranniy Grown in the Field with Different Fungicide Forms (Percentage of Damaged Tubers)

Disease	Treatments/Controls				
	Control (–)	Control (+) Commercial Fungicides		Treatment (Embedded Fungicides)	
		Quadris	Uniform	Azoxystrobin Granules	Azoxystrobin + Mefenoxam Granules
Black scurf	16	5.4	1.3	1	1
Scab	47	38	35	25	25
Late blight	2	0	0	0	0

Thus, the maximum number of tubers that were not affected by scabs, black scurf, and late blight was obtained in the treatments with pre-emergence application of embedded fungicides to the soil.

Evaluation of the structure of the yield obtained in the present study made it possible to estimate the response of potato cv. Krasnoyarskiy ranniy to the use of free and embedded fungicides (Figure 9.7).

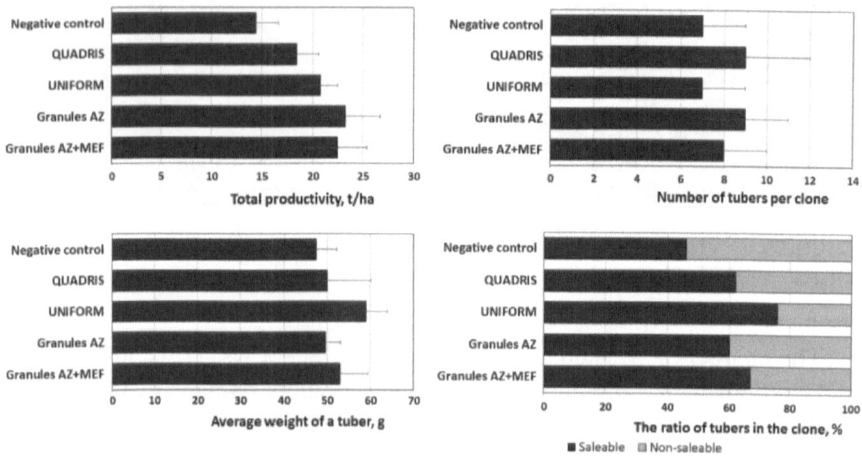

FIGURE 9.7 The structure of the yields of potato cv. Krasnoyarskiy ranniy in groups with different fungicide forms (in the field).

While weights of saleable tubers were comparable (above 40 g) in the groups with two forms of fungicides (free fungicides as components of commercial formulations and embedded fungicides), in most plant groups, another parameter (average weight of tubers) was higher due to the different saleable to non-saleable tuber ratio. In the negative control, the number of non-saleable tubers (weighing less than 40 g) was the highest (54%); in the groups with Uniform and the embedded combination of two fungicides, it was the lowest: 24 and 33%, respectively. The number of tubers in the clone varied from 7 to 9, but these differences were not significant. Significant differences were found in the total potato yield, which was significantly higher in the treatment groups compared with the groups with commercial fungicides in the positive control. The lowest yield (14.4 t/ha) was in the group of intact plants. Compared with this group, the yield increase due to the use of Quadris and Uniform was 27.7 and 44.4%, respectively; and in the treatments with embedded azoxystrobin (AZ) and AZ + mefenoxam, the yield increase was even higher: 61.8 and 56.3%, respectively.

The effects of prolonged-action embedded fungicides and free fungicides on the quality of the potato yield were assessed by studying the chemical composition of tubers (Table 9.2).

TABLE 9.2 The Chemical Composition of Tubers of Potato cv. Krasnoyarskiy Ranniy in the Groups of Plants with Different Fungicide Forms (n = 4)

Treatment/Control	Dry Matter (%)	Reducing Sugars (%)	Starch (%)	Nitrates (mg/kg)
Control "–"	25.9	0.81	18.2	157.7
	Control (+) Commercial Fungicides			
Quadris	22.0	0.89	16.3	148.7
Uniform	23.6	0.70	17.8	140.0
	Treatment-Embedded Fungicides			
Azoxystrobin	24.1	0.89	18.3	139.3
Azoxystrobin + mefenoxam	23.5	0.87	17.8	125.3
p	0.040*	0.002*	0.050*	0.030*

The quality of tubers depends on a number of parameters, including varietal characteristics, climatic conditions, cultivation technologies, etc. The dry matter in potatoes is on average 70% starch, 7% fiber, 5% ash, and 10% protein. For the industrial processing of potatoes, an important parameter is the content of dry matter in tubers in combination with a decreased content of reducing sugars. For food processing, potato varieties must contain at least 22% dry matter. The content of solids during processing into fried products affects the consumption of oil and stock material, consistency (taste), and the yield of finished products per unit area and in the production of puree, the consumption of raw materials and the yield of finished products. The starch content also determines the nutritional value of tubers and the efficiency of processing into starch.

A study of the chemical composition of tubers of potato cv. Krasnoyarskiy ranniy showed that the dry matter content in all groups was about 22–23%, that is, it met the requirements established for food varieties (Table 9.2). The content of reducing sugars in the tubers of both varieties was close in all groups and amounted to 0.89–1.0%. The content of starch in tubers changed slightly, within 16–18%, across the groups. An important indicator of the quality and safety of potatoes is the content of nitrates in tubers; their content in potato tubers is limited by the maximum allowable concentration (MAC) equal to 250 mg/kg fresh weight. The accumulation of nitrates in potato tubers is influenced by many controllable and uncontrollable factors, including the duration of the growing period and the genetic characteristics of varieties, the level of nitrogen nutrition, the timing and density of planting, the quality of seed material, and other agricultural practices. Nitrates in the tubers of the potato variety used in the present study did not exceed the MAC and varied within 125–158 mg/kg (Table 9.2). In the groups of plants grown without fungicides, nitrate content was the highest (158 mg/kg). The use of embedded fungicides reduced the content of nitrates in tubers by 9–12% compared to the negative control ($p = 0.030$). The most pronounced effect was achieved in the group with embedded AZ (139 mg/kg). Anyway, the nitrate contents of tubers in all experimental groups did not exceed the MAC, and the chemical composition of the tubers grown in the presence of experimental fungicide formulations met the requirements for food potato varieties.

The current study demonstrated a significant positive effect of the use of embedded fungicides to control pathogens that spread on the vegetative organs and tubers of potato cv. Krasnoyarskiy ranniy. There are no analogs of slow-release fungicide formulations produced using poly-3-hydroxybutyrate, a degradable biopolymer of the PHA family; there are very few works on using embedded fungicides to control potato diseases. Therefore, it is difficult to discuss the results obtained. At the same time, the choice of the fungicides studied in this work is justified, since AZ, mefenoxam (metalaxyl), and DIF are widely used in crop production for treating tubers before planting and for spraying potato plants [20, 21, 38, 44, 73, 75]. The antifungal activity of the Uniform commercial formulation, combining two fungicides, which was used in the positive control, was comparable to the antifungal activity of the embedded fungicides. This is consistent with the literature data showing the effectiveness of this formulation for pre-sowing soil treatment and suppression of *Pythium* rot, black scurf, black dot, and silver scurf, which improved the growth and yield of potatoes [27–29]. The data on embedded controlled-release fungicides include descriptions of 130- to 238-nm microparticles loaded with AZ and embedded in a matrix of silica, carboxymethyl chitosan or polylactide/glycolide, which are more§ environmentally friendly and effective for suppressing *Fusarium* and *Phytophthora* potato pathogens compared to spraying plants with aqueous solutions of the free fungicide [87, 89, 92]. A study by Liang et al. [32] described blended mulch films made from various proportions of chitosan and hydroxypropyl methylcellulose loaded with metalaxyl; it was shown that the release rate of metalaxyl within 72 hours was 70.6%; such films suppressed the development of late blight in laboratory experiments, reducing the prevalence of plant disease by 13.3%. In another study [7], a PHA copolymer–[P(3HB/4HB)] was used as mulch film; films loaded with the prothioconazole systemic fungicide showed high bioactivity against soil-borne *Sclerotium rolfsii* Sacc. That is, biodegradable mulching films containing prothioconazole effectively suppressed soil-borne pathogenic fungi causing plant diseases. The positive result of pre-emergence soil application of the embedded Quadris fungicide (active ingredient AZ) was described in the work of Smagin et al. [63]. The fungicide was loaded into prolonged-action gel structures based on acrylates with silver ions. Field trials of the embedded fungicide to protect potato varieties Red Scarlett, Lady Claire, and Gala were carried out in humid regions of European Russia (Serebryany Bor, Moscow, and Krasnodar Territory) and arid regions (Uzbekistan) under atmospheric precipitation and irrigation on various soils. The formulations were highly effective against many pathogens (common scab, black dot, fusarium disease, late blight, rubber, and bacterial rots). The use of this fungicide increased the yield of potatoes in some experiments from 6 to 15 t/ha while saving water for irrigation by 1.3–2.0 times.

The results obtained in the present work and the literature data indicate the considerable significance of research aimed at the development and application of ecofriendly slow-release fungicide formulations to protect potatoes from diseases.

9.4 CONCLUSION

The present study, for the first time, investigated the use of slow-release fungicide formulations prepared by embedding fungicides in degradable matrix of microbial polymer poly-3-hydroxybutyrate blended with birch wood flour. The biological activity of embedded fungicides with different mechanisms of action (AZ, AZ + mefenoxam, and DIF) was compared with the activity of their commercial analogs (Quadris, Uniform, and Score) towards plant pathogens causing potato diseases. *In vitro* and *in vivo* studies were performed; the *in vivo* study was carried out using potato cv. Krasnoyarskiy ranniy. In the *in vitro* study, the embedded fungicides showed a pronounced antifungal effect, suppressing the development of colonies of the most harmful potato pathogens causing black scurf, early blight, and late blight. In the laboratory pot experiment and in the field experiment with potato cv. Krasnoyarskiy ranniy, the biological activity of the embedded fungicides was superior to that of their commercial analogs. The embedded fungicides had a positive effect on the germination, growth, and development of potatoes, more effectively controlling pathogens damaging plants and tubers and providing a higher yield. The results obtained showed that the experimental slow-release fungicide formulations prepared by embedding fungicides in the P(3HB)/wood flour degradable matrix had an advantage over their commercial analogs and were effective in protecting potatoes from pathogens and in increasing potato yields.

FUNDING

This work was financially supported by Project "Agropreparations of the new generation: a strategy of construction and realization" (Agreement No.: 075-15-2021-626) in accordance with Resolution No. 220 of the Government of the Russian Federation of April 9, 2010, "On measures designed to attract leading scientists to the Russian institutions of higher learning."

KEYWORDS

- **commercial analogs**
- **degradable matrix**
- **disease susceptibility**
- **fungicides**
- ***in vitro* and *in vivo* pathogen suppression**
- **microbiological study**
- **P(3HB)/wood flour**
- **slow-release formulations**

REFERENCES

1. Agustien, A., Sari, A., Fitria, A., & Djamaan, A., (2016). Manufacture of a slow-release herbicide of methyl-metsulfuron using biopolymer of poly (3-hydroxybutyrate) as matrix. *Der. Pharma. Chem., 8*(7), 105–111.
2. Ali, M., Parmar, A., Niedbała, G., Wojciechowski, T., El-Yazied, A. A., El-Gawad, H., Nahhas, N., et al., (2021). Improved shelf-life and consumer acceptance of fresh-cut and fried potato strips by an edible coating of garden cress seed mucilage. *Foods, 10*(7), 1536.
3. Barrera-Mendez, F., Miranda-Sanchez, D., Sanchez-Rangel, D., Bonilla-Landa, I., Rodriguez-Haas, B., Monribot-Villanueva, J. L., & Olivares-Romero, J. L., (2019). Propiconazole nanoencapsulation in biodegradable polymers to obtain pesticide-controlled delivery systems. *J. Mex. Chem. Soc., 63*(1), 50–60.
4. Belov, G. L., Belosokhov, A. F., Kutuzova, I. A., Statsyuk, N. V., Chudinova, E. M., Alexandrova, A. V., Kokaeva, L. Y., & Elansky, S. N., (2018). *Colletotrichum coccodes* in potato and tomato leaves in Russia. *J. Plant Diseases Prot., 125*(3), 311–317.
5. Biswas, S. K., Shankar, U., Kumar, S., Kumar, A., Kumar, W., & Lal, K., (2015). Impact of bio-fertilizers for the management of spot blotch disease and growth and yield contributing parameters of wheat. *J. Pure Appl. Microbiol., 9*(4), 3025–3030.
6. Brasovean, I., & Oroian, V., (2009). Florian integrated control of potato diseases. *Pro-Environment Promediu, 2*(4), 230–234.
7. Chen, G., Cao, L., Cao, C., Zhao, P., Li, F., Xu, B., & Huang, Q., (2021). Effective and sustained control of soil-borne plant diseases by biodegradable polyhydroxybutyrate mulch films embedded with fungicide of prothioconazole. *Molecules, 26*(3), 762.
8. Chen, G. Q., Chen, X. Y., Wu, F. Q., & Chen, J. C., (2020). Polyhydroxyalkanoates (PHA) toward cost competitiveness and functionality. *Advanced Industrial and Engineering Polymer Research, 3*(1), 1–7.
9. Rakhimol, K. R., Thomas, S., Volova, T., & Jayachandran, K., (2020). *Controlled Release of Pesticides for Sustainable Agriculture* (p. 266). Springer, Cham.
10. Cooke, L. R., Schepers, H. T. A. M., Hermansen, A., Bain, R. A., Bradshaw, N. J., Ritchie, F., Shaw, D. S., et al., (2011). Epidemiology and integrated control of potato late blight in Europe. *Potato Research, 54*(2), 183–222.
11. Damalas, C. A., & Eleftherohorinos, I. G., (2011). Pesticide exposure, safety issues, and risk assessment indicators. *Int. J. Environ. Res. Public Health, 8*(5), 1402–1419.
12. Davoren, M. J., & Schiestl, R. H., (2018). Glyphosate-based herbicides and cancer risk: A post-IARC decision review of potential mechanisms, policy and avenues of research. *Carcinogenesis, 39*(10), 1207–1215.
13. Dmitriev, E. A., (1995). *Mathematical Statistics in Soil Science* (p. 319). M.: Publishing house of Moscow State University.
14. Duellman, K. M., Price, W. J., Lent, M. A., Christian, C. L., Bertram, M. C., & Nolte, P., (2021). Fungicide seed treatment improves performance of single-drop whole and cut seed potatoes. *Am. J. Potato Res., 98*(4), 315–327.
15. GOST 28561-90. By-products of Fruits and vegetables. *Methods for the Determination of Dry Matter or Moisture*.
16. GOST 33996-2016 Interstate Standard" Seed Potatoes, (2017).
17. GOST 7194-81. Fresh potatoes. *Acceptance Rules and Quality Determination Methods*.

18. Grillo, R., De Melo, N. F. S., De Lima, R., Lourenço, R. W., Rosa, A. H., & Fraceto, L. F., (2010). Characterization of atrazine-loaded biodegradable poly (hydroxybutyrate-co-hydroxyvalerate) microspheres. *J. Polym. Environ., 18*(1), 26–32.
19. Grillo, R., Pereira, A. D. E. S., De Melo, N. F. S., Porto, R. M., Feitosa, L. O., Tonello, P. S., Filho, N. L. D., et al., (2011). Controlled release system for ametryn using polymer microspheres: Preparation, characterization and release kinetics in water. *J. Hazard. Mater, 186*(2, 3), 1645–1651.
20. Kassaw, A., Abera, M., & Belete, E., (2021). The response of potato late blight to potato varieties and fungicide spraying frequencies at meket, Ethiopia. *Cogent Food & Agriculture, 7*(1). 1870309.
21. Khadka, R. B., Chaulagain, B., Subedi, S., Marasini, M., Rawal, R., Pathak, N., & Sharma-Poudyal, D., (2020). Evaluation of fungicides to control potato late blight (*Phytophthora infestans*) in the plains of Nepal. *J. Phytopathol., 168*(5), 245–253.
22. Khan, H., Kaur, S., Baldwin, T. C., Radecka, I., Jiang, G., Bretz, I., Duale, K., Adamus, G., & Kowalczuk, M., (2020). Effective control against broadleaf weed species provided by biodegradable PBAT/PLA mulch film embedded with the herbicide 2-methyl-4-chlorophenoxyacetic acid (MCPA). *ACS Sustain. Chem. Eng., 8*(13), 5360–5370.
23. Kiselev, E. G., Boyandin, A. N., Zhila, N. O., Prudnikova, S. V., Shumilova, A. A., Baranovskiy, S. V., Shishatskaya, E. I., et al., (2020). Constructing sustained-release herbicide formulations based on poly-3-hydroxybutyrate and natural materials as a degradable matrix. *Pest Manag. Sci., 76*(5), 1772–1785.
24. Koller, M., & Mukherjee, A., (2020). Polyhydroxyalkanoates–linking properties, applications, and end-of-life options. *Chem. Biochem. Eng., 34*(3), 115–129.
25. Kubsad, D., Nilsson, E. E., King, S. E., Sadler-Riggleman, I., Beck, D., & Skinner, M. K., (2019). Assessment of glyphosate induced epigenetic transgenerational inheritance of pathologies and sperm epimutations: Generational toxicology. *Sci. Rep., 9*(1), 1–17.
26. Kumar, S., Bhanjana, G., Sharma, A., Sidhu, M. C., & Dilbaghi, N., (2014). Synthesis, characterization and on field evaluation of pesticide loaded sodium alginate nanoparticles. *Carbohydr. Polym., 101*, 1061–1067.
27. Kuznetsova, M. A., Kozlovsky, B. E., Rogozhin, A. N., Smetanina, T. I., Spiglazova, S. Yu., Derenko, T. A., & Filippov, A. V., (2010). Late blight and potato Alternaria: A program of protective actions. *Potatoes and Vegetables, 3*, 27–29. In Russian.
28. Kuznetsova, M. A., Rogozhin, A. N., Smetanina, T. I., & Denisenkov, I. A., (2017). Protection of potatoes from rhizoctonia, anthracnose and silvery scab. *Potatoes and Vegetables, 4*, 27–29. In Russian.
29. Kuznetsova, M. A., Rogozhin, A. N., Smetanina, T. I., Demidova, V. N., Denisenkov, I. A., & Stashuk, N. V., (2018). UNIFORM®: Reducing the harmfulness of potato pite rot. *Potatoes and Vegetables, 3*, 30–33. In Russian.
30. Kwiecien, I., Adamus, G., Jiang, G., Radecka, I., Baldwin, T. C., Khan, H. R., Johnston, B., et al., (2018). Biodegradable PBAT/PLA blend with bioactive MCPA-PHBV conjugate suppresses weed growth. *Biomacromol., 19*(2), 511–520.
31. Li, S., Wu, Q., Sun, Q., Coffin, S., Gui, W., & Zhu, G., (2019). Parental exposure to tebuconazole causes thyroid endocrine disruption in zebrafish and developmental toxicity in offspring. *Aquatic Toxicology, 211*, 116–123.
32. Liang, W., Zhao, Y., Xiao, D., Cheng, J., & Zhao, J., (2020). A biodegradable water-triggered chitosan/hydroxypropyl methylcellulose pesticide mulch film for sustained control of *Phytophthora sojae* in soybean (*Glycine max* L. Merr.). *J. Clean. Prod., 245*, 118943.

33. Liefrink, S. R., (2005). *Inspection of Dutch Seed Potatoes, the Importance of Certified Seed Potatoes* (p. 36). NIVAP Netherlands publisher.
34. Lobo, F. A., De Aguirre, C. L., Silva, M. S., Grillo, R., De Melo, N. F. S., De Oliveira, L. K., De Morais, L. C., et al., (2011). Poly (hydroxybutyrate-co-hydroxyvalerate) microspheres loaded with atrazine herbicide: Screening of conditions for preparation, physico-chemical characterization, and *in vitro* release studies. *Polym. Bull., 67*(3), 479–495.
35. Majumder, S., Shakil, N. A., Kumar, J., Banerjee, T., Sinha, P., Singh, B. B., & Garg, P., (2016). Eco-friendly PEG-based controlled release nano-formulations of mancozeb: Synthesis and bioefficacy evaluation against phytopathogenic fungi *Alternaria solani* and *Sclerotium rolfsii*. *J. Environ. Sci. Health, Part B, 51*(12), 873–880.
36. Manfo, F. P. T., Harthé, C., Nantia, E. A., et al., (2019). Bisphenol A differentially affects male reproductive function biomarkers in a reference population and agro pesticides users from Djutitsa, Cameroon. *Toxicol. Industr. Health, 35*(4), 324–335.
37. Mineau, P., & Whiteside, M., (2013). Pesticide acute toxicity is a better correlate of US grassland bird declines than agricultural intensification. *PloS One, 8*(2). e57457.
38. Muzhinji, N., Woodhall, J. W., Truter, M., & Van, D. W. J. E., (2018). Variation in fungicide sensitivity among *Rhizoctonia* isolates recovered from potatoes in South Africa. *Plant Disease, 102*(8), 1520–1526.
39. Fraceto, L., Grillo, R., Avila, D., Oloveria, H., & Lima, R., (2020). *Nanopesticides–From Research and Development to Mechanisms of Action and Sustainable Use in Agriculture* (p. 353). Springer Nature.
40. Netrusov, A. I., Egorova, M. A., & Zakharchuk, L. M., (2005). *Praktikum Po Mikrobiologii (Practical Course in Microbiology)* (p. 608). Moscow, Akademiya. In Russian.
41. Nguyen, D. B., Rose, M. T., Rose, T. J., Morris, S. G., & Van, Z. L., (2016). Impact of glyphosate on soil microbial biomass and respiration: A meta-analysis. *Soil Biol. Biochem., 92*, 50–57.
42. Nishimoto, R., (2019). Global trends in the crop protection industry. *J. Pestic. Sci.*, D19–101.
43. Núñez-Zofío, M., Larregla, S., & Garbisu, C., (2011). Application of organic amendments followed by soil plastic mulching reduces the incidence of *Phytophthora capsici* in pepper crops under temperate climate. *J. Crop Prot., 30*(12), 1563–1572.
44. Odilbekov, F., Edin, E., Mostafanezhad, H., Coolman, H., Grenville-Briggs, L. J., & Liljeroth, E., (2019). Within-season changes in *Alternaria solani* populations in potato in response to fungicide application strategies. *Eur. J. Plant Pathol., 155*(3), 953–965.
45. Pascual, J. A., Rozas, M. M., Rojo, M. A. D., & Ros, M., (2019). Soil fungicides to control soil-borne diseases of Mediterranean crops grown under greenhouse. *ISHS Acta Horticulturae 1268: XI International Symposium on Protected Cultivation in Mild Winter Climates and I International Symposium on Nettings and Screens in Horticulture* (pp. 179–186).
46. Paśmionka, I., & Kotarba, K., (2015). Possible application of effective microorganisms in environmental protection. *Kosmos, Problemy Nauk Biologicznych, 64*(1), 173–184. In Polish.
47. Patel, S., & Sangeeta, S., (2019). Pesticides as the drivers of neuropsychotic diseases, cancers, and teratogenicity among agro-workers as well as general public. *Environ. Sci. Pollut. Res., 26*(1), 91–100.
48. Patent RF for the invention No. 2733295, (2020). *"Long-acting Pesticidal Agent for Soil Application."* Priority from 23.03.2020.

49. *Polyhydroxyalkanoate (PHA) Market by Type, Manufacturing Technology & Application,* (2017). Global Forecast to 2021. Research and Markets. https://www.marketsandmarkets.com/Market-Reports/pha-market-395.html (accessed on 10 July 2023).
50. Popov, S. Y., Dorozhkina, L. A., & Kalinin, V. A., (2003). *Fundamentals of Chemical Plant Protection* (p. 208). Moscow, Art-Lyon. In Russian.
51. Prudnikova, S., Streltsova, N., & Volova, T., (2021a). The effect of the pesticide delivery method on the microbial community of field soil. *Environ. Sci. Pollut. Res., 28*(7), 8681–8697.
52. Prudnikova, S. V., Churakov, A. A., Ovsyankina, S. V., & Khizhnyak, S. V., (2021b). Isolation and identification of autochthonous pathogens of potato diseases common in the regions of Siberia. In: *Materials of the IV International Conference "Biotechnology of New Materials-Environment-Quality of Life"* (pp. 174–177). Krasnoyarsk, Russia. In Russian.
53. Pszczółkowski, P., & Sawicka, B., (2018). The effect of application of biopreparations and fungicides on the yield and selected parameters of seed value of seed potatos. *Acta Agroph., 25*(2), 239–255.
54. Rakitskiy, V. N., (2011). *Handbook on Pesticides (Toxicological and Hygienic Characteristics)* (Vol. 1, p. 960). Moscow, Publishing house Agrorus. In Russian.
55. Roberts, R. G., (2005). *Alternaria yaliinficiens* sp. Nov. on Ya Li pear fruit: From interception to identification. *Plant Dis., 89*(2), 134–145.
56. Roshani, B., Tavanai, H., Morshed, M., & Khajehali, J., (2017). Controlled release of thiram pesticide from poly (L-lactic acid) nanofibers. *J. Textile Institute, 108*(9), 1504–1509.
57. Sander, M., (2019). Biodegradation of polymeric mulch films in agricultural soils: Concepts, knowledge gaps, and future research directions. *Environ. Sci. Technol., 53*(5), 2304–2315.
58. Sandhya, Kumar, S., Kumar, D., & Dilbaghi, N., (2017). Preparation, characterization, and bio-efficacy evaluation of controlled release carbendazim-loaded polymeric nanoparticles. *Environ. Sci. Pollut. Res, 24*(1), 926–937.
59. Savenkova, L., Gercberga, Z., Muter, O., Nikolaeva, V., Dzene, A., & Tupureina, V., (2002). PHB-based films as matrices for pesticides. *Proc. Biochem., 37*(7), 719–722.
60. Schepers, H., Andrivon, D., Gaucher, D., Kapsa, J., Lebecka, R., Nielsen, B., Evenhuis, B., & Ruocco, M., (2009). Results of the potato case study in the EU-network of excellence, ENDURE. *Proceedings of Eleventh Euro Blight Workshop* (Vol. 13, pp. 31–36). Pro-Special Report.
61. Singh, B., Sharma, D. K., & Dhiman, A., (2013). Environment friendly agar and alginate-based thiram delivery system. *Toxicol. Environ. Chem., 95*(4), 567–578.
62. Singh, M., Biswas, S. K., Nagar, D., Lal, K., & Singh, J., (2017). Impact of bio-fertilizer on growth parameters and yield of potato. *Int. J. Curr. Microbiol. Appl. Sci., 6*(5), 1717–1724.
63. Smagin, A., Sadovnikova, N., & Smagina, M., (2019). Synthetic gel structures in soils for sustainable potato farming. *Sci. Rep., 9*(1), 1–15.
64. Smagin, A. V., Sadovnikova, N. B., Vasenev, V. I., & Smagina, M. V., (2018). Biodegradation of some organic materials in soils and soil constructions: Experiments, modeling and prevention. *Materials, 11*(10), 1889.
65. Smith, C. M., Vera, M. K. M., & Bhandari, R. K., (2019). Developmental and epigenetic effects of roundup and glyphosate exposure on Japanese medaka (*Oryzias latipes*). *Aquatic Toxicology, 210*, 215–226.
66. *State Catalog of Pesticides and Agrochemicals Permitted for Use on the Territory of the Russian Federation* (2021). Moscow, Ministry of Agriculture of Russia. In Russian.

67. Suave, J., Dall'Agnol, E. C., Pezzin, A. P. T., Meier, M. M., & Silva, D. A. K., (2010). Biodegradable microspheres of poly (3-hydroxybutyrate)/poly (ε-caprolactone) loaded with malathion pesticide: Preparation, characterization, and *in vitro* controlled release testing. *J. Appl. Polym. Sci., 117*(6), 3419–3427.
68. Sudesh, K., (2010). *Practical Guide to Microbial Polyhydroxyalkanoates*. Smitthes, London, UK.
69. Sutton, D., Fothergill, A., & Rinaldi, M., (2001). *Guide to Pathogenic and Opportunistic Fungi (Trans. from English)*. Mir, Moscow, In Russian.
70. Sychev, V. G., Shapoval, O. A., Mozharova, I. P., Verevkina, T. M., Mukhina, M. T., Korshunov, A. A., Ponomareva, A. S., et al., (2018). *Guidelines for the Registration Tests of Agrochemicals in Agriculture: Production and Practice* (p. 220). Moscow, FGBNU "Rosinformagrotech," In Russian.
71. Tarrahi, R., Fathi, Z., Seydibeyoğlu, M. Ö., Doustkhah, E., & Khataee, A., (2020). Polyhydroxyalkanoates (PHA): From production to nanoarchitecture. *Int. J. Biol. Macromol, 146*, 596–619.
72. Thomas, S., Shumilova, A. A., Kiselev, E. G., Baranovsky, S. V., Vasiliev, A. D., Kuzmin, E. G., Nemtsev, I., et al., (2020). Thermal, mechanical and biodegradation studies of biofiller based poly-3-hydroxybutyrate biocomposites. *Int. J. Biol. Macromol., 155*, 1373–1384.
73. Tiwari, R. K., Kumar, R., Sharma, S., Sagar, V., Aggarwal, R., Naga, K. C., Lal, M. K., et al., (2020). Potato dry rot disease: Current status, pathogenomics and management. *3 Biotech, 10*(11), 1–18.
74. Tleuova, A. B., Wielogorska, E., Talluria, P., Štěpánek, F., Elliott, C. T., & Grigoriev, D. O., (2020). Recent advances and remaining barriers to producing novel formulations of fungicides for safe and sustainable agriculture. *J. Controlled Rel., 326*, 468–481.
75. Tsror, L., Erlich, O., Hazanovsky, M., Marshak, G., Segev, G., & Zig, U., (2021). Fungicide field treatments to control potato leak caused by *Pythium ultimum*. *Am. J. Potato Re., 98*(2), 115–121.
76. Venugopal, N. V. S., & Sainadh, N. V. S., (2016). Novel polymeric nanoformulation of mancozeb–an eco-friendly nanomaterial. *Int. J. Nanosci., 15*(4), 1650016.
77. Volova, T. G., Shishatskaya, E. I., & Sinskey, A. J., (2013). *Degradable Polymers: Production, Properties, Applications* (p. 380). Nova Science Pub Inc, New York.
78. Volova, T. G., Prudnikova, S. V., & Zhila, N. O., (2018). Fungicidal activity of slow-release P(3HB)/TEB formulations in wheat plant communities infected by *Fusarium moniliforme*. *Environ. Sci. Pollut. Res., 25*(1), 552–561.
79. Volova, T., Prudnikova, S., Boyandin, A., Zhila, N., Kiselev, E., Shumilova, A., Baranovskiy, S., et al., (2019). Constructing slow-release fungicide formulations based on poly(3-hydroxybutyrate) and natural materials as a degradable matrix. *J. Agric. Food Chem., 67*(33), 9220–9231.
80. Volova, T., Baranovsky, S., Petrovskaya, O., Shumilova, A., & Sukovatyi, A., (2020a). Biological effects of the free and embedded metribuzin and tribenuron-methyl herbicides on various cultivated weed species. *J. Environ. Sci. Health, Part B, 55*(11), 1009–1019.
81. Volova, T. G., Shumilova, A. A., Zhila, N. O., Sukovatyi, A. G., Shishatskaya, E. I., & Sabu, T., (2020b). Efficacy of slow-release formulations of metribuzin and tribenuron methyl herbicides for controlling weeds of various species in wheat and barley stands. *ACS Omega, 5*(39), 25135–25147.
82. Volova, T. G., Shishatskaya, E. I., Prudnikova, S. V., Zhila, N. O., & Boyandin, A. N., (2020c). *New Generation Formulations of Agrochemicals: Current Trends and Future Priorities* (p. 250). Apple Academic Press, Toronto.

83. Volova, T. G., Demidenko, A. V., Kurachenko, N. L., Baranovsky, S. V., Petrovskaya, O. D., & Shumiliva, A. A., (2021). Efficacy of embedded metribuzin and tribenuron-methyl herbicides in field-grown vegetable crops infested by weeds. *Environ. Sci. Pollut. Res., 28*(1), 982–994.
84. Wang, C., Wang, F., Zhang, Q., & Liang, W., (2016). Individual and combined effects of tebuconazole and carbendazim on soil microbial activity. *Eur. J. Soil Biol., 72*, 6–13.
85. World Potato Statistics, (2018). The potato sector (FAOSTAT). http://www.potatopro.com/world/potatostatistics (accessed on 10 July 2023).
86. Xu, L., Cao, L. D., Li, F. M., Wang, X. J., & Huang, Q. L., (2014). Utilization of chitosan-lactide copolymer nanoparticles as controlled release pesticide carrier for pyraclostrobin against *Colletotrichum gossypii* Southw. *J. Disp. Sci. Technol., 35*(4), 544–550.
87. Xu, C., Cao, L., Zhao, P., Zhou, Z., Cao, C., Li, F., & Huang, Q., (2018). Emulsion-based synchronous pesticide encapsulation and surface modification of mesoporous silica nanoparticles with carboxymethyl chitosan for controlled azoxystrobin release. *Chem. Eng. J, 348*, 244–254.
88. Yang, X., Li, S., Wang, Z., Lee, S. M., Wang, L. H., & Wang, R., (2018). Constraining the teratogenicity of pesticide pollution by a synthetic nanoreceptor. *Chem. Asian J, 13*(1), 41–45.
89. Yao, J., Cui, B., Zhao, X., Zhi, H., Zeng, Z., Wang, Y., Sun, C., et al., (2018). Antagonistic effect of azoxystrobin poly (lactic acid) microspheres with controllable particle size on *Colletotrichum higginsianum* Sacc. *Nanomaterials, 8*(10), 857.
90. Zhang, M., Teng, Y., Xu, Z., Wang, J., Christie, P., & Luo, Y., (2016). Cumulative effects of repeated chlorothalonil application on soil microbial activity and community in contrasting soils. *J. Soils Sediments, 16,* 1754–1763.
91. Zhang, M., Wang, W., Zhang, Y., Teng, Y., & Xu, Z., (2017). Effects of fungicide iprodione and nitrification inhibitor 3,4-dimethylpyrazole phosphate on soil enzyme and bacterial properties. *Sci. Total Environ., 599,* 254–263.
92. Zhang, M., Zhu, Z. Q., Yuan, S., Wang, S. Y., Yang, C. Y., Dwivedi, P., Si, T., & Xu, R. X., (2019). One-step microencapsulation and spraying of pesticide formulations for improved adhesion and sustained release. *J. Microencapsul., 36*(7), 649–658.
93. Zhevora, S. V., Zeyruk, V. N., Belov, G. L., Vasilyeva, S. V., Derevyagina, M. K., Anisimov, B. V., Starovoitov, V. I., et al., (2019). *Advanced Methods for Diagnosing Potato Pathogens: A Scientific and Analytical Review* (p. 92). FGBNU "Rosinformagrotech," Moscow, In Russian.
94. Zhu, F., Duan, T. T., Liao, G. H., Cao, L. D., Cao, C., Li, F. M., Chen, C., & Huang, Q. L., (2018). Preparation of fenoxanil micro-capsule. *Proceedings of the Academic Annual Conference Sponsored by the Chinese Society of Plant Protection in 2018* (pp. 213–219). Shaanxi, China.
95. Zoteeva, N. M., & Kosareva, O. S., (2021). Evaluation of resistance of potato varieties from the VIR collection to *Phytophthora infestans* in laboratory studies. *Bull. Plant Prot., 104*(2), 113–119. In Russian.

CHAPTER 10

Parameters That Influence the Fermentation Submerged Process: Bioprocess Development in Biotechnology

MARICELA ESMERALDA-GUZMÁN,[1] J. BUENROSTRO-FIGUEROA JUAN,[2] CRISTÓBAL NOÉ AGUILAR,[1] A. ASCACIO-VALDÉS JUAN,[1] and LEONARDO SEPÚLVEDA[1]

[1]*Bioprocess and Bioproducts Research Group, Food Research Department, School of Chemistry, Autonomous University of Coahuila, Saltillo, Coahuila, México*

[2]*Research Center in Food and Development A.C., Cd. Delicias, Chihuahua, Mexico*

ABSTRACT

Submerged fermentation processes are presented as an alternative bioprocess that can be applied in different industries to obtain bioactive molecules of high-added value. In submerged fermentation processes, various elements are considered for the economic production of the desired products, such as substrates and their conversion, product selectivity, productivity, and energy consumption. These elements are influenced by physical and chemical factors like temperature, agitation, and aeration, pH, media composition, among others. Therefore, it is essential to review the factors that affect submerged fermentation for its optimization and control at an industrial level. This chapter reviews previous work related to the physical and chemical parameters that intervene and affect fermentation in submerged fermentation.

Bioresources and Bioprocess in Biotechnology for a Sustainable Future. Leonardo Sepúlveda Torre, Juan Carlos Contreras-Esquivel, Ann Rose Abraham, & A. K. Haghi (Eds.)

10.1 INTRODUCTION

Fermentation describes the processes that transform one material into another simply through live organisms like yeasts, fungi, and bacteria, from which specific products are obtained [1]. These fermented products vary depending on the substrate and microorganism selected. Fermentation has been used since ancient times to transform raw materials into products, especially in the food industry, such as cheese, yogurt, some sauces, bread, wine, and beer, mainly [2]. In addition, fermentation has been described as an anaerobic process of incomplete oxidation [1], in the industry, these processes can be subdivided by the addition of oxygen in the medium or the lack of this, finding aerobic and anaerobic fermentations.

1. **Aerobic Processes:** Oxygen is important for the growth and development of microorganisms.
2. **Anaerobic Processes:** These processes must be performed in the absence of oxygen.

Most fermentations can be classified into two types: Solid state fermentation, which can be defined as the process that takes place in a solid matrix which lacks free water; however, it requires moisture for microorganisms' growth [3]. Some examples of products derived from this type of fermentation are baked products, cocoa processing, and some traditional foods such as soy sauce. On the other hand, submerged fermentation, in which microorganisms are in contact with the substrate in an aqueous environment with abundant water availability, facilitates, and promotes the growth and development of microorganisms [4]. This type of fermentation is used to ferment vegetables, yogurt, brewing, and wine. In this review, we will focus on the physicochemical parameters that influence submerged fermentation processes, besides addressing the transport phenomena involved in mass and energy transfer in this type of process.

10.2 FACTORS THAT INFLUENCE THE SUBMERGED FERMENTATION

Several factors can influence submerged fermentation, like temperature, pH, dissolved oxygen, agitation, and mixing, and composition of the culture medium. Variations of these parameters can directly influence the fermentation rate, affecting the organoleptic and physicochemical properties of the desired products (altering the production of sensory compounds, generating unwanted compounds) [5]. The medium should offer sources of carbon, oxygen, nitrogen, and oligonutrients for the proper growth and development of the microorganism and understanding fermentation biochemistry is critical to the proper development of an appropriate medium for fermentation. Micronutrients are present in smaller concentrations; however, they can influence the cell growth rate and productivity [6]. Trace elements are important in cellular growth because they act as coenzymes to catalyze reactions, help primary metabolism, and regulate secondary metabolism. Some of these elements are iron, zinc, and magnesium [7].

10.2.1 TEMPERATURE

Temperature is one of the most influential factors in submerged fermentation, so its measurement and control is essential since the temperature of the culture is going to influence cell growth, the formation of desired products and oxygen solubility, also temperature directly affects production costs [8]. If the culture is at temperatures below the optimal, the microorganisms' growth will be affected by slowing their metabolism, consequently the rate of cellular production and the synthesis of products will be affected. On the other hand, if temperatures exceed the optimal, not only will the death or inactivity of enzymes occur, the expression of proteins and the synthesis of the compounds will also be affected, decreasing the performance and quality of the desired product [9].

The Arrhenius equation is commonly used to model the effect of temperature on chemical reactions and cell growth and death [10]. Cell growth rate increases when temperature rises to its optimal, on the contrary, when temperature exceeds its optimal, growth rate decreases, and thermal death occurs. The net rate of cell growth is proposed using the following equation.

$$\frac{dx}{dt} = (\mu - kd)X \quad (1)$$

where; dx/dt is the change between cellular concentration regarding time, μ represents cell growth rate and kd is the cell death rate. Both μ, kd are temperature-depending and according to the Arrhenius equation can be expressed as:

$$\mu = A * e^{(-\frac{Ea}{RT})} \quad (2)$$

$$kd = A' * e^{(-\frac{Ed}{RT})} \quad (3)$$

where; Ea and Ed are activation energies for cell growth and thermal death, respectively.

Table 10.1 shows reported studies on the influence of temperature in submerged cultivation.

TABLE 10.1 Studies on the Influence of Temperature in Submerged Fermentation

Fermented Material	Microorganism	References
Baijiu	*Lactobacillus* and *Saccharomyces*	[11]
Kombucha tea	Microbial electrolysis cell	[12]
Wine	*Saccharomyces Cerevisiae* and not *Cerevisiae.*	[13]
Milk	*Lactococcus*	[14]
Kombucha tea	Tea fungus, acetic acid bacteria and yeasts	[15]
Maize straw and cabbage waste	*Lactobacillus* and *Leuconostoc*	[16]
Merlot wines	*Saccharomyces cerevisiae*. Five different strains.	[17]
Loose tea	*Eurotium cristatum*	[18]
KB medium	*Pseudomonas fluorescens*	[19]
Sugarcane	*Saccharomyces cerevisiae*	[20]

10.2.2 pH

pH control and regulation in cell cultures is essential for microorganism development [21]. Monitoring and controlling this parameter are important because each microorganism needs a specific pH for its growth and development. An inadequate pH in the culture media can influence the production of metabolites, affecting metabolism and cell growth [22]. Sepulveda et al. conducted a study on the factors that influence the accumulation of ellagic acid in a submerged fermentation process using *Aspergillus niger* GH1. Among the parameters evaluated, it was found that pH influences the production of ellagic acid, at a low pH, the accumulation of ellagic acid increased. The authors attributed this result to the presence of enzymes that degrade ellagitannins at low pH [23]. Walaszczyk et al. carried out a study to evaluate the influence of pH in a range between 3 and 7 to obtain oxalic acid by submerged fermentation using *Aspergillus niger* and sucrose as a carbon source. The results showed that pH 6 is optimal for obtaining oxalic acid [24]. On the other hand, Ahmad et al. conducted a study to investigate the effect of pH on crops development and the production of steviol glycides from *Stevia rebaudiana*. The results showed that medium to high pH favor biomass production, however, a low pH favors the production of glycosides [25].

10.2.3 DISSOLVED OXYGEN

Oxygen transfer often limits the optimal performance in biological systems because it is poorly soluble in aqueous media and is of paramount importance for aerobic cultures as microorganisms use oxygen to carry out cell growth, directly affecting the production and performance of fermentation products. If dissolved oxygen concentration falls below the critical level, cell metabolism could be altered. However, this is often the goal in fermentation processes to obtain products from the secondary metabolism of microorganisms. In addition, if the concentration of dissolved oxygen in the medium is higher than the critical level, it may not affect biomass production, however, it may contribute to forming products [26]. Waldir et al. studied the effect of aeration on the fermenting activity of *Brettanomyces intermedius* RIVE 2-2-2 in a mixed culture with *Saccharomyces cerevisiae* RIVE 15-1-416, these strains were cultured in apple juice to evaluate the production of chemical compounds of sensory importance. The best results of cider organoleptic quality regarding taste and smell were obtained using static culture fermentation [27]. Molnár et al. conducted a study to evaluate the influence of oxygen concentration on dicarboxylic acid yield by submerged fermentation using *Aspergillus terreus*. The results obtained show that 2% of dissolved oxygen was sufficient to obtain the highest biomass concentration, however, by raising the oxygen concentration to 30% the products obtain increase [28]. Ha et al. investigated the fermentation parameters to produce surfactin by submerged fermentation using *Bacillus subtilis* Y9. The results show that the rate of oxygen absorption is positively related to surfactin production [29].

10.2.4 AGITATION

In fermentation processes, operations depend to a high degree on agitation and mixing of fluids. Agitation refers to forcing a fluid by mechanical means to acquire a circulatory movement inside the fermenter, and mixing involves two or more individual phases being randomly distributed to each other. Marques et al. studied the effects of agitation rates on extractive fermentation of clavulanic acid based on a two-phase aqueous system using *Streptomyces variabilis*. In this study, it was reported that at medium levels of agitation (950 rpm), the oxygen transfer in the system was more efficient, concluding that agitation contributes to the transfer of oxygen in the system [30].

In bioreactors, the mixture is essential to ensure the adequate supply of nutrients and prevent the accumulation of toxic metabolites. An inadequate mixing can cause modifications to the cell culture, leading to the production of by-products, decreasing the performance of the system [31]. In addition, it must be considered the fluid hydrodynamics, fluid rheology, impeller type, energy input and output, and bioreactor size as these factors can influence the mixing process.

The control of temperature, pH, and substrate concentration depends to a large extent on mixing. Although it is easy to control agitation and mixing in small-scale bioreactors; it becomes a major challenge in industrial processes and often this is a limitation for the development of these systems. In industrial bioreactors, inadequate mixing leads to undesirable concentration gradients, causing a decrease in the rate of mass and heat transfer.

In biological systems that are sensitive to shear, such as fermentations with filamentous fungi, mixing cannot be improved simply by increasing agitation because it can damage the cells by cutting. The presence of biomass has been reported to significantly reduce mixing efficiency, affecting heat and mass transfer.

10.3 MASS TRANSFER

Mass transfer is crucial in many biochemical processes. This transfer occurs when the component has a mixture migrates from one phase to another due to the difference in concentration between phases [32]. An important example of mass transfer in a submerged fermentation system is the addition of oxygen. The difference in concentration between the air bubbles generated and the culture medium favors the transfer of oxygen to the liquid medium; if mass transfer is slow, the rate of cellular metabolism will depend on the oxygen supply in the gas phase because oxygen is important in aerobic systems and is insoluble in aqueous solutions [33].

10.3.1 PHASE RULE

To predict the concentration of a single reaction in two phases in balance, experimental balance data are required. In addition, if the phases are not in balance, the mass transfer rate is proportional to the driving force, which is the deviation from the balance. Variables that affect the phases balance are temperature, pressure, and concentration [32].

The two-phase balance is given by the following equation:

$$F = C - P + 2 \tag{4}$$

where; P is the number of phases in the balance, C is the total number of components in the phases and F is the number of system variants.

The phase rule does not indicate partial pressure pA in balance with the selected X_A this value must be obtained by experimental methods.

Often the balance between partial pressure in the gas phase and X_A can be expressed by Henry's law equation.

$$pA = HX_A \tag{5}$$

where; H represents Henry's constant, pA and X_A represent the partial pressure and concentration of the gas, respectively.

10.3.2 MOLECULAR DIFFUSION

Molecular diffusion can be defined as the transport of individual molecules in a fluid through individual displacements and molecule disorder [32]. Diffusion is defined as the transport of molecules from one region to another due to their difference in concentration. This transport always occurs from the region of high concentration to low concentration [34].

Fick's law equation can be represented as follows for a binary mixture of a compound A and B

$$J_{Az} = -CD_{AB}\frac{dx}{dz} \tag{6}$$

where; J_{AZ} is the mass flow of component A in z direction; CD_{AB} is the diffusion coefficient of component A in B; and dx/dz is the differential concentration of A comprising z distance.

The diffusion coefficient is a property of the materials that can be found in the literature, this represents how easy compound A diffuses in compound B, so its value depends on both components in the mixture. In addition, the value of CD_{AB} may depend on the temperature and pressure when working with gases.

Molecular diffusion is important for mass transfer in submerged fermentation, its importance lies in the mixture of the culture broth, the reaction in the solid phase and the transfer of mass through the phase limits.

10.3.3 MASS TRANSFER MECHANISM

Various mechanisms have been suggested to describe the theory of mass transfer in interface. The three best known are the theory of the two films, the theory of penetration and the theory of superficial renewal [35].

1. **Theory of the Two Films:** This assumes that there are fictitious films on the sides of the interface, representing the resistance to mass transfer, which is carried out by molecular diffusion. This model concludes that the mass transfer coefficient K_L is proportional to the diffusivity D_{AB} and inversely proportional to the thickness of the film Z_f as:

$$K_L = \frac{D_{AB}}{Z_f} \tag{7}$$

2. **Penetration Theory:** It assumes that turbulent swirls travel from most of the phase to the interface where they remain for a time (exposure time, T_e). The solute is supposed to penetrate a given swirl during its stay on the interface through an unstable molecular diffusion process. This model predicts that the mass transfer coefficient is directly proportional to the square root of molecular diffusivity [36].

$$K_L = 2(\frac{D_{AB}}{\pi te})^{1/2} \tag{8}$$

3. **Theory of Surface Renewal:** It proposes that there is an infinite range of ages for surface elements and the function of distribution of the age of the surface (t) can be expressed as:

$$\varnothing(\mathrm{t}) = se^{-st} \tag{9}$$

where; "s" is the fractional rate of surface renewal. This theory also predicts the mass transfer coefficient is proportional to the square root of molecular diffusivity [37].

$$\mathrm{KL} = (sD_{AB})^{1/2} \tag{10}$$

For the three theories, it is necessary to know an unknown parameter, effective film thickness (Z_f), exposure time (te), or fractional rate of surface renewal (s). Little is known about their properties, so all theories are incomplete. However, these theories help us visualize the mass transfer mechanism at the interface and to know the exponential dependence on molecular diffusivity of the mass transfer coefficient.

10.3.3.1 TRANSFER OF LIQUID-LIQUID MASS

In solvent extraction, in the post-fermentation stage, two immiscible or partially miscible liquid phases are put in contact. One or more of the compounds present in one of the phases is transferred to the other. In fermentation broths, organic solvents are used for the isolation of alkaloids, steroids, and antibiotics [38].

10.3.3.2 GAS-LIQUID MASS TRANSFER

Gas-liquid mass transfer is of great importance in aerobic fermentation systems due to the oxygen requirement in cell culture. The transfer of gas-liquid mass that takes place in a submerged fermentation process usually depends on the formation of bubbles due to agitation and aeration of the system [39].

10.3.3.3 SOLID-LIQUID MASS TRANSFER

Mass transport in liquids and solids usually originates due to a combination of convection (fluid mobilization) and diffusion [40]. Mass transfer within solid particles is generally not convection assisted; therefore, the only mechanism for mass transfer between particles in solid is molecular diffusion [32].

10.4 DEMAND FOR AERATION AND OXYGEN

In the bioreactor, oxygen transfer is affected by different factors that intervene on the bioreactor hydrodynamics like the physical properties of both the culture medium and the gas used to provide oxygen, the geometric parameters of the bioreactor, the operating conditions at which the process is developed, and the type of microorganism cultured [41].

Dissolved oxygen is a limiting nutrient in culture with high oxygen demand, which may be due to the growth rate of the microorganism, a high biomass production or the rheological properties of the culture media resisting mass transfer [42].

10.4.1 FACTORS AFFECTING CELL OXYGEN DEMAND

The rate of oxygen transfer from the gas phase to the liquid phase is determined by the rate at which the cells are present in the culture, consuming the oxygen. The factors that affect oxygen transfer are the geometric and operational characteristics of the fermentation vessels, the composition of the culture medium, viscosity, surface tension, the oxygen concentration in the gas stream, the morphology of the microorganism and the surface area provided by the bubbles [43]. In batch cultivation the rate of oxygen absorption varies over time, because the cell concentration increases during batch culture and the uptake of the total oxygen rate is proportional to the number of cells. In Eqn. (11), *Qo* represent the rate of oxygen consumption per volume of broth and *qo* the specific rate of oxygen absorption:

$$Qo = qoX \tag{11}$$

where; *X* is cellular concentration. Typical units for *qo* are gmol $g^{-1}s^{-1}$ and for *Qo* gmol l^{-1} s^{-1}.

In a well-mixed suspension system, the mass balance for oxygen is represents by the following equation:

$$\frac{dCo}{dt} = kLa\left(Co^* - Co\right) - QoX \tag{12}$$

where; *kLa, Co**, *Co*, and *Qo* are the volumetric mass transfer coefficient, saturated oxygen concentration, oxygen concentration in the liquid and the specific rate of liquid absorption, respectively.

In stationary state the above equation can be solved to obtain the oxygen concentration in the liquid.

$$Co = Co^* - \frac{QoX}{kLa} \tag{13}$$

where; *Co** is constant at a fixed air pressure; *Co* is determined by three factors: the specific rate of oxygen absorption (*Qo*) that is determined by the biological system, the cell concentration (*X*); and the volumetric mass transfer coefficient (*kLa*). In biological systems, a high oxygen shortage is expected at a high cell density, so it is important to determine the oxygen demand of the microorganism to supply the appropriate amount that does not influence the formation of fermentation products.

To be used by the cells, the oxygen molecules must cross various transport obstacles. Eight steps are described for mass transfer in oxygen transport from inside the bubble to the intracellular reaction site. These steps are:

- Transfer from inside the bubble to the gas-liquid interface.
- Movement through the gas-liquid interface.
- Broadcast through the relatively stagnant liquid film surrounding the bubble.

- Transport through the liquid in bulk.
- Diffusion through the relatively stagnant liquid film surrounding the cells.
- Movement through the liquid-cellular interface.
- If the cells are in a solid floculum, group or particle, their diffusion through solid to the individual cell.
- Transport through cytoplasm to the reaction site.

When the cells are dispersed in the fermentation broth, the transport of oxygen through the film surrounding the bubble becomes the principal resistance to oxygen transfer, limiting the speed of the process and controlling the overall rate of mass transfer [26].

10.4.2 BUBBLES

The bubbles dispersed in the liquid medium are of great importance in the efficiency of gas-liquid mass transfer. The principal characteristic of air bubbles in fermenters is their size. In a gas volume, there will be more interfacial area for mass and heat transfer in small bubbles instead of some large ones, in addition small bubbles contribute to the transfer due to their increase speeds which are slow making them stay longer in the liquid so that the oxygen dissolves [35].

10.5 HEAT GENERATION

All fermentation processes are exothermic, that is, they generate heat, this is mainly because the cells generate significant amounts of heat to avoid temperature increases, so the heat generated by the cellular metabolism must be removed [44] and for this reason the fermenters must have a cooling system to prevent the increase and accumulation of heat that denatures the enzymes involved in this process. Fermentations are commonly limited by heat transfer, as it is difficult to remove this heat through an exchange.

10.5.1 HEAT TRANSFER MECHANISM

In a liquid medium, there are three ways to explain heat migration.

1. **Conduction:** The heat conduction occurs between the particles of two bodies in contact, without matter flow.

2. **Convection:** Heat transfer through the movement of a fluid mass, so the convection is limited to gases and liquids.
3. **Phase Change:** In heat transfer mechanisms, a phase change can occur in saturated systems, these changes happen through vaporization and condensation of the system.

10.5.2 HEAT TRANSFER

Heat transfer depends directly on the surface area, the temperature difference, the properties, and the turbulence in fluids. Temperature changes have significant variations in heat transfer and water loss. Heat transfer is regulated by Fourier's law which is defined as:

$$\frac{q}{A} = -k\frac{dt}{dz} \tag{14}$$

where; q is the amount of heat generated by volume unit; A represents the cross-sectional area normal to the direction of the heat flow in m^2; k is the thermal conductivity; T the temperature in k in $W/m\ m^2$; dt/dx is the temperature gradient in x direction. The negative sign of the equation is included because the temperature decreases when is in the same direction that the heat flow.

The transport of energy in liquids is quite simple. Molecules have a continuous and disordered movement and collide with each other exchanging energy in a linear moment. If a molecule transports from a high temperature region to a lower temperature region, it carries kinetic energy to that region and gives it in by colliding with low-energy molecules. However, in liquid systems, molecular force fields have a considerable effect on energy exchange. Since there is no molecular theory for liquid systems, most correlations to predict their conductivity are empirical. The thermal conductivity of liquids can vary with temperature. This variation can be expressed with the following linear equation.

$$k = a + bT \tag{15}$$

where; a and b are empirical constants. The thermal conductivity of liquids is pressure independent. The physical mechanism of energy conduction in gases is very similar to liquids, molecules exchange energy from vibratory motion, however in gases the movement of molecules is very messy.

The conduction of heat or energy through the solids is verified by two mechanisms, first the heat, like electricity, is driven by free electrons that move in a structural metal network, in the second the heat is driven by the transmission of vibrational energy between adjacent atoms. Heat transfer improves with increased agitation and aeration.

10.6 PROCESSES IN INDUSTRIAL FERMENTATION

Industrial fermentations can be carried out in different processes, including batch, feed, and continuous. In batch processing, an initial amount of substrate and microorganism is added, the fermentation process is conserved during a time and then the products are collected. In the food industry there are fermentations that extend even to years, such as some wines.

A batch-fed process has a continuous feed of substrate and microorganism. The volume of the broth increases with each addition of the medium and the product is collected after the batch time. In a continuous process, the fermentation broth is continuously fed in the fermenter and the fermented product is continuously extracted, so the volume of fermentation remains constant.

10.7 BIOREACTORS IN SUBMERGED FERMENTATION

A bioreactor is a vessel or system in which biochemical reactions are carried out to convert a substrate into a product [45]. Conversion occurs by the action of a biocatalyzer which can be enzymes, microorganisms, animal cells and plant cells. Bioreactors that produce microbial cells and animal and plant cells are known as fermenters, which can also produce chemicals or biotransform substrates in other compounds. A fermenter can operate using a specific cell type (single-septic operation) or a mixed population of different cell types. There are several reactors to carry out submerged fermentation. Below are the most common ones.

Table 10.2 presents some studies in which the use of bioreactors in submerged fermentation are reported.

10.7.1 STIRRED TANK BIOREACTOR

The stirred tank bioreactor is a cylindrical vessel which has an engine with one or more agitators, this bioreactor is widely used by various industries and is generally used at steady state [56].

TABLE 10.2 Reported Studies Using Bioreactors in Submerged Fermentation

Reactor Type	Products	References
Air-lift bioreactor	Chitosanase	[46]
Stirred tank	Scopoletin, tomentin, and sphaeralcic acid	[47]
Bubble column	Eurycomanone and polysaccharides	[48]
Fixed-Bed	lacasa, manganese peroxidase, endoxylanase, β-glucosidase and total cellulases	[49]
Air-lift bioreactor	Biomass	[50]
Stirred tank reactor	Xylanase	[51]
Bubble column reactor	Biomass	[52]
Fluidized bed reactor	Ethanol	[53]
Stirred tank reactor	Pullulan	[54]
Fluidized bed reactor	Ethanol	[55]

10.7.2 BUBBLE COLUMN

A column of bubbles consists of a puddle of gas-bubbled liquid. The structure of the bubble column comprises a cylindrical vessel with a gas distributor at the bottom for the gas bubbles to feed into the reactor [57].

10.7.3 AIR LIFT BIOREACTOR

These reactors are designed so that the air maintains the agitation in the breeding ground. The bioreactor is equipped with either an inner draft tube or an external loop, by which the bioreactor volume is divided into a gassed and unguessed region. This generates a vertically circulating flow that improves oxygen transfer and equalizes shear forces [58].

10.7.4 FLUIDIZED BED REACTOR

These are like bubble columns, because the bed of solids or catalytic particles are supported by gas with an expanded cross section near the top. Fresh or recirculated liquid is pumped into the bottom of the container at a speed sufficient to fluidize the solids or keep them in suspension [59] (Figure 10.1).

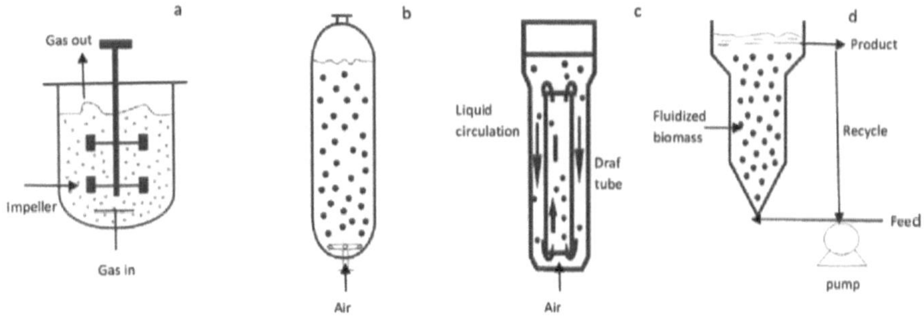

FIGURE 10.1 (a) Stirred tank reactor; (b) bubble column reactor; (c) air lift reactor; and (d) fluidized bed reactor.

10.8 GROWTH OF FUNGI IN FERMENTATION IN SUBMERGED FERMENTATION

Fungi have played an important role in biotechnological processes for their ability to produce secondary metabolites, mainly in the food and pharmaceutical industry [60]. Today, filamentous fungi are used as cell factories to produce many compounds of industrial relevance, such as organic acids, proteins, enzymes, exopolysaccharides, and secondary metabolites [61].

Filamentous fungi are characterized by their ability to secrete substances in their growth medium, which are of interest to some industries. Some fungi are fundamental to produce enzymes at an industrial level. Some of them are *Aspergillus niger*, *Aspergillus oryzae*, and *Trichoderma reesei* [62].

Fungi morphology can occur in various forms, from hyphae, scattered mycelia, pellets, or mycelial masses called granules. The way they grow depends largely on the physical and chemical factors in the breeding ground [63]. The growth kinetics of microorganisms is associated with differences in morphology that occur in crops. The morphological macro-characteristics of fungi have an important effect on the rheological properties of liquid culture. These influence the production and excretion of various microbial metabolites [64]. It is important to keep track of the growth and development of microorganisms within bioprocesses, this involves monitoring and controlling crops by optimizing process conditions. Therefore, a control is required on the factors that can affect growth, like carbon source/concentration, pH, ion content, density of spore inoculum, addition of surfactants, oxygen enrichment, agitation [65], osmolarity [66], addition of insoluble microparticles [67], or change from stirred tank to wave mixed bioreactor equipment [68], amongst others. On the other hand, the form of granules is presented as an attractive method since its to decrease the viscosity in the culture medium without modifying its rheological properties. In this way, the medium is considered a Newtonian fluid which is characterized by having a good transfer of heat and mass. In addition, the formation of granules facilitates the separation of the microorganism and fermentation extract in industrial systems.

Growth morphology characteristics and phase interactions play an important role in mixing. In the study reported by Lucatero et al. the influence of growth morphology of *Trichoderma harzianum* in submerged fermentation, the bubble sizes observed with the dispersed mycelium were smaller than with the granules, in addition it was shown that the location of the bubble was influenced by the morphology of the biomass, which is why it is important to control the morphology of the microorganism in a submerged fermentation system [69].

In general, the growth of fungal strains in submerged fermentation is subjected to three main phases:

1. **Micro-Morphological Growth:** Swelling of spores, germination of spores, extension, and branching of hyphae.
2. **Macro-Morphological Growth:** The formation of the hyphae network or the formation of fungal pellets.
3. **Fungal Cell Autolysis:** Phase of cell death and product formation.

10.8.1 MICRO-MORPHOLOGICAL GROWTH

In biotechnological processes like fermentation, a short germination phase is desired to decrease the growing time. The germination process directly influences the efficiency of fermentation [70]. There are several environmental factors that can influence swelling and germination of spores, such as temperature, humidity, and carbon dioxide (CO_2), besides the concentration of spores affects the germination time.

10.8.2 MACRO-MORPHOLOGICAL GROWTH

In submerged fermentation, microorganisms can growth form diverse macro-morphologies, as a result, hyphae, compact pellets, or intermediates of these growth types named loose clumps can be observed, resulting from the interaction between spore and mycelial in moist substrates [71]. Pellet formation is conventionally non-coagulative types [72].

10.8.3 FUNGAL CELL AUTOLYSIS

Fungal autolysis is considered a dynamic phase of cell death. This process is complicated, because it involves interactions between growing parameters, such as lack of nutrients in the medium and shear cutting. In submerged fermentation, autolysis can interfere in the formation of products. Fungi are presented as ideal microorganisms in bioprocessing; however, it is important to know the physiology, morphology, and biochemistry of them. This will allow us to exploit even more this type of culture [73].

10.9 CONCLUSION

Submerged fermentation is a bioprocess widely used in various industries to obtain enzymes, food, secondary metabolites, proteins, among many other products. That is why it becomes important to study the components and factors that can intervene and affect these bioprocesses. Among the most important factors that affect submerged fermentation are temperature, pH, agitation, and mixing, dissolved oxygen, carbon sources and oligonutrients, these parameters can affect the development and growth of the microorganism as well as the transfer of mass and energy in the system, affecting the expected yields in the process. The research and optimization of systems in submerged cultivation will allow the development of new and better processes with yields comparable and even higher than those obtained through chemical techniques.

KEYWORDS

- **bioprocess**
- **bioreactors**
- **bubble column**
- **fermentation**
- **heat transfer**
- **industrial fermentation**
- **macro/micro-morphological growth**
- **mass transfer**
- **stirred tank bioreactor**

REFERENCES

1. Mazzeo, L., & Piemonte, V., (2020). Fermentation and biochemical engineering: Principles and applications. In: Basile, A., Centi, G., De Falco, M. M., & Iaquaniello, G., (eds.), *Studies in Surface Science and Catalysis* (pp. 261–285). Elsevier.
2. Wacher, R. C., (2014). Former food biotechnology: Fermented foods. *Revista Digital Universitaria – UNAM, 15*, 1–14.
3. Lizardi-Jiménez, M. A., & Hernández-Martínez, R., (2017). Solid state fermentation (SSF): Diversity of applications to valorize waste and biomass. *3 Biotech, 7*, 44.
4. Chisti, Y., (2014). Fermentation (industrial)|basic considerations. In: Batt, C. A., & Tortorello, M. L., (eds.), *Encyclopedia of Food Microbiology* (2nd edn., pp. 751–761). Academic Press.
5. Arellano, M., Pelayo, C., Ramírez, J., & Rodríguez, I., (2008). Characterization of kinetic parameters and the formation of volatile compounds during the tequila fermentation by wild yeast isolated from agave juice. *J. Ind. Microbiol. Biotechnol., 35*, 835–841.
6. Dou, X., Lu, X. H., Lu, M. Z., Yu, L. S., Xue, R., & Ji, J. B., (2013). The effects of trace elements on the lipid productivity and fatty acid composition of *Nannochloropis oculate*. *J. Renew Energy, 671545*, 1–6.
7. Humbird, D., & Qiang, F., (2016). Scale-up considerations for biofuels. In: Carrie, A. E., & Trinh, C. T., (eds.), *Biotechnology for Biofuel Production and Optimization* (513–537). Elsevier.
8. Converti, A., Casazza, A. A., Ortiz, E. Y., Perego, P., & Del Borghi, M., (2009). Effect of temperature and nitrogen concentration on the growth and lipid content of *Nannochloropsis oculata* and *Chlorella vulgaris* for biodiesel production. *Chem. Eng. Process.: Process Intensif., 48*, 1146–1151.
9. Wang, S. J., & Zhong, J. J., (2007). Bioreactor engineering. In: Yang, S. T., (ed.), *Bioprocessing for Value-Added Products from Renewable Resources* (pp. 131–161). Elsevier.
10. Ross, T., & Nichols, D. S., (2014). Ecology of bacteria and fungi in foods | INFLUENCE of temperature. In: Batt, C. A., & Tortorello, M. L., (eds.), *Encyclopedia of Food Microbiology* (2nd edn., pp. 602–609). Academic Press.
11. Zhang, H., Wang, L., Wang, H., Yang, F., Chen, L., Hao, F., Lv, X., Du, H., & Xu, Y., (2021). Effects of initial on microbial community succession rate and volatile flavors during *Baijiu* fermentation process. *Food Res Int., 141*, 109887.
12. Wang, Y. Z., Zhang, L., Xu, T., & Ding, K., (2017). Influence of initial anolyte pH and temperature on hydrogen production through simultaneous saccharification and fermentation of lignocellulose in microbial electrolysis cell. *Int. J. Hydrog. Energy, 42*, 22663–22670.
13. Alonso-del-Real, J., Lairón-Peris, M., Barrio, E., & Querol, A., (2017). Effect of temperature on the prevalence of *Saccharomyces* non *cerevisiae* species against a *S. cerevisiae* wine strain in wine fermentation: Competition, physiological fitness, and influence in final wine composition. *Front. Microbiol., 8*, 1–15.
14. Moonga, H. B., Schoustra, S. E., Linnemann, A. R., Van, D. H. J., Shindano, J., & Smid, E. J., (2021). Influence of fermentation temperature on microbial community composition and physicochemical properties of mabisi, a traditionally fermented milk. *LWT – Food Sci. Technol., 136*, 110350.
15. Neffe-Skocińska, K., Sionek, B., Ścibisz, I., & Kołożyn-Krajewska, D., (2017). Acid contents and the effect of fermentation condition of Kombucha tea beverages on physicochemical, microbiological and sensory properties. *CYTA J. Food, 15*, 601–607.

16. Ren, H., Feng, Y., Liu, T., Li, J., Wang, Z., Fu, S., Zheng, Y., & Peng, Z., (2020). Effects of different simulated seasonal temperatures on the fermentation characteristics and microbial community diversities of the maize straw and cabbage waste co-ensiling system. *Sci. Total Environ., 708*, 135113.
17. Massera, A., Assof, M., Sari, S., Ciklic, I., Mercado, L., Jofré, V., & Combina, M., (2021). Effect of low temperature fermentation on the yeast-derived volatile aroma composition and sensory profile in merlot wines. *LWT – Food Sci. Technol., 142*, 111069.
18. Yao, Y., Wu, M., Huang, Y., Li, C., Pan, X., Zhu, W., & Huang, Y., (2017). Appropriately raising fermentation temperature beneficial to the increase of antioxidant activity and gallic acid content in *Eurotium cristatum*-fermented loose tea. *LWT – Food Sci. Technol., 82*, 248–254.
19. Bisutti, I., & Stephan, D., (2020). Influence of fermentation temperature and duration on survival and biocontrol efficacy of *Pseudomonas fluorescens* Pf153 freeze-dried cells. *J. Appl. Microbiol., 128*, 232–241.
20. Ccopa, R. E., Yamakawa, C. K., Saad, M. B. W., Atala, D. I. P., Ambrosio, W. B., Bonomi, A., Nolasco, J. J., & Rossell, C. E. V., (2017). Effect of temperature on sugarcane ethanol fermentation: Kinetic modeling and validation under very-high-gravity fermentation conditions. *Biochem. Eng. J., 119*, 42–51.
21. Zhu, L., Zong, M., & Wu, H., (2008). Efficient lipid production with *Trichosporon ferment* and its use for biodiesel preparation. *Bioresour Technol., 99*, 7881–7885.
22. Fei, Q., Chang, H. N., Shang, L., Kim, N., & Kang, J., (2011). The effect of volatile fatty acids as a sole carbon source on lipid accumulation by *Cryptococcus albidus* for biodiesel production. *Bioresour Technol., 102*, 2695–2701.
23. Sepúlveda, L., Buenrostro-Figueroa, J. J., Ascacio-Valdés, J. A., Aguilera-Carbó, A. F., Rodríguez-Herrera, R., Contreras-Esquivel, J. C., & Aguilar, C. N., (2014). Submerged fermentation for production of ellagic acid from pomegranate husk by *Aspergillus Niger* GH1. *Int. Appl. Mycol., 26*, 27–35.
24. Walaszczyk, E., Podgórski, W., Janczar-Smuga, M., & Dymarska, E., (2018). Effect of medium pH on chemical selectivity of oxalic acid biosynthesis by *Aspergillus Niger* W78C in submerged batch cultures with sucrose as a carbon source. *Chem. Zvesti., 72*, 1089–1093.
25. Ahmad, N., Rab, A., Ahmad, N., & Fazal, H., (2018). Differential pH-induced biosynthesis of steviol glycosides and biochemical parameters in submerge root cultures of *Stevia rebaudiana* (Bert.). *Sugar Tech., 20*, 734–744.
26. Stanbury, P. F., Whitaker, A., & Hall, S. J., (2003). Aeration and agitation. In: *Principles of Fermentation Technology* (p. 357). Butterworth-Heinemann, Oxford.
27. Dxh. Estela-Escalante, W., Rychtera, M., Melzoch, K., Torres-Ibáñez, F., Calixto-Cotos, R., Bravo-Araníbar, N., Memenza-Zegarra, M. E., & Chávez-Guzmán, Y. M., (2014). Effect of aeration on the production of volatile compounds by mixed culture of *Brettanomyces intermedius* and *Saccharomyces cerevisiae* during cider fermentation. *Tip. Rev. Espec. Cienc. Quím. Biol., 17*, 5–14.
28. Molnár, Á. P., Németh, Z., Kolláth, I. S., Fekete, E., Flipphi, M., Ág, N., Soós, Á., et al., (2018). High oxygen tension increases itaconic acid accumulation, glucose consumption, and the expression and activity of alternative oxidase in *Aspergillus terreus*. *Appl. Microbiol. Biotechnol., 102*, 8799–8808.
29. Ha, S., Kim, H. M., Chun, H. H., Hwang, I. M., Lee, J. H., Kim, J. C., Kim, I. S., & Park, H. W., (2018). Effect of oxygen supply on surfactin production and sporulation in submerged fermentation of *Bacillus subtilis*Y9. *Appl. Sci., 8*, 1–10.

30. Viana, M. D. A., Santos-Ebinuma, V. C., Pessoa-Júnior, A., Porto, A. L., Torres, B. R., & Converti, A., (2016). Effect of aeration and agitation on extractive fermentation of clavulanic acid by using aqueous two-phase system. *Biotechnol Prog., 32*, 1444–1452.
31. Al Taweel, A. M., Shah, Q., & Aufderheide, B., (2012). Effect of mixing on microorganism growth in loop bioreactors. *Int. J. Chem. Eng., 984827*, 1–12.
32. Geankoplis, C. J., (2009). Mass transfer principles. In: *Transport Processes and Principles of Separation Processes Include Unit Operations*. Mexico: Editorial group patria.
33. Doran, P. M., (2013). Mass transfer. In: Doran, P. M., (ed.), *Bioprocess Engineering Principles* (2nd edn., pp. 379–444). Academic Press.
34. Zhao, W., Cheng, Y., Pan, Z., Wang, K., & Liu, S., (2019). Gas diffusion in coal particles: A review of mathematical models and their applications. *Fuel, 252*, 77–100.
35. Lee, J. M., (2009). Agitation and aereation. In: Lee, J. M., (ed.), *Biochemical Engineering.* Department of chemical engineering. Washington State University.
36. Higbie, R., (1931). The rate of absorption of a pure gas into a still liquid during short periods of exposure. *Trans. AIChE, 31*, 365–389.
37. Danckwerts, P. V., (1951). Significance of liquid-film coefficients in gas absorption. *Ind. Eng. Chem., 43*,1460–1467.
38. Lobo-Oehmichen, R. A., (2007). *Principios De Transferencia De Masa Segunda Edición* (pp. 1–499). Universidad Autónoma Metropolitana.
39. Amaral, A., Gillot, S., Garrido-Baserba, M., Filali, A., Karpinska, A. M., Benedek, G. P., De Groot, C., et al., (2019). Modelling gas–liquid mass transfer in wastewater treatment: When current knowledge needs to encounter engineering practice and vice versa. *Water Sci. Technol., 80*, 607–619.
40. Kashid, M. N., Renken, A., & Kiwi-Minsker, L., (2011). Gas–liquid and liquid–liquid mass transfer in microstructured reactors, *Chem. Eng. Sci., 66*, 3876–3897.
41. Buitrago, H. G., Otálvaro, A. M., & Duarte, P. G., (2013). Evaluation of oxygen transfer in a conventional bioreactor with external aerator. *Rev. Colomb. Biotecnol., 15*, 106–114.
42. Casas, J. A., Santos, V., & García-Ochoa, F., (2000). Xanthan gum production under several operational conditions: Molecular structure and rheological properties. *Enzyme Microb. Technol. 26*, 282–291.
43. Soler, A., & Buitrago, H. G., (2010). Evaluating oxygen transfer in a *Lactococcus lactis cultures* using an external aeration fermentation system (EAFS). *Rev. Colomb. Biotecnol., 12*, 124–138.
44. Doran, P. M., (2013). Heat transfer. In: Doran, P. M., (ed.), *Bioprocess Engineering Principles* (2nd edn., pp. 333–377). Academic Press.
45. Yusuf, C., & Moo-Young, M., (2003). Bioreactors. In: Meyers, R. A., (eds.), *Encyclopedia of Physical Science and Technology* (3rd edn., pp. 247–271). Academic Press.
46. Suyotha, W., Cheirsilp, B., Yano, S., Matsuba, S., & Konno, H., (2021). Production of chitosanase by *Lentzea* sp. OUR-I1 using acid-pretreated shrimp shell in an air-lift bioreactor and the feasibility of utilizing the residual biomass. *Waste Biomass Valorization, 12*, 2445–2458.
47. Pérez-Hernández, J., Nicasio-Torres, M. P., Sarmiento-López, L. G., & Rodríguez-Monroy, M., (2019). Production of anti-inflammatory compounds in suspension of *Sphaeralcea angustifolia* cells cultured in stirred tank bioreactor, *Eng Life Sci., 19*, 196–205.
48. Fan, M. Z., An, X. L., Cui, X. H., Jiang, X. L., Piao, X. C., Jin, M. Y., & Lian, M. L., (2021). Production of eurycomanone and polysaccharides through adventitious root culture of *Eurycoma longifolia* in a bioreactor, *Biochem. Eng. J., 171*, 108013.

49. Montoya, S., Patiño, A., & Sánchez, O., (2021). Production of lignocellulolytic enzymes and biomass of *Trametes versicolor* from agro-industrial residues in a novel fixed-bed bioreactor with natural convection and forced aeration at pilot scale. *Processes*, *9*, 1–19.
50. Ho, T. T., Le, K. C., Kim, S. W., & Park, S. Y., (2021). Culture condition optimization and FT-IR analysis of *Polygonum multiflorum* Thunb. adventitious root cultures grown in an air-lift bioreactor system. *Plant Cell Tiss. Organ Cult.*, *144*, 371–381.
51. Abdella, A., Segato, F., & Wilkins, M. R., (2020). Optimization of process parameters and fermentation strategy for xylanase production in a stirred tank reactor using a mutant *Aspergillus nidulans* strain. *Biotechnol. Rep.*, *26*, e00457.
52. Saberi, A., Jalili, H., Nikfarjam, A., Koohsorkhi, J., Jarmoshti, J., & Bizukojc, M., (2020). Monitoring of *Aspergillus terreus* morphology for the lovastatin production in submerge culture by impedimetry. *Biochem. Eng. J.*, *159*, 107615.
53. Antunes, F. A. F., Chandel, A. K., Brumano, L. P., Terán, H. R., Peres, G. F. D., Ayabe, L. E. S., Sorato, V. S., et al., (2018). A novel process intensification strategy for second-generation ethanol production from sugarcane bagasse in fluidized bed reactor. *Renew. Energ.*, *124*, 189–196.
54. Singh, R. S., Saini, G. K., & Kennedy, J. F., (2021). Pullulan production in stirred tank reactor by a color-variant strain of *Aureobasidium pullulans* FB-1. *Carbohydrate Polymer Technologies and Applications*, *2*, 100086.
55. Antunes, F. A. F., Chandel, A. K., Dos Santos, J. C., Milessi, T. S. S., Peres, G. F. D., & Da Silva, S. S., (2017). Hemicellulosic ethanol production in fluidized bed reactor from sugar cane bagasse hydrolysate: Interplay among carrier concentration and aeration rate. *ACS Sustain Chem. Eng.*, *5*, 8250–8259.
56. Shokrkar, H., Ebrahimi, S., & Zamani, M., (2018). A review of bioreactor technology used for enzymatic hydrolysis of cellulosic materials. *Cellulose*, *25*, 6279–6304.
57. Dhotre, M. T., Ekambara, K., & Joshi, J. B., (2004). CFD simulation of sparger design and height to diameter ratio on gas hold-up profiles in bubble column reactors. *Exp. Therm. Fluid Sci.*, *28*, 407–421.
58. Bai, F. W., Zhao, X. Q., & Xu, J., (2011). Bioengineering at the interface between science and society. In: Moo-Young, M., (ed.), *Comprehensive Biotechnology* (2nd edn., pp. 477–489). Academic Press.
59. Bello, M. M., Raman, A. A. A., & Purushothaman, M., (2016). Applications of fluidized bed reactor in wastewater treatment–A review of the major design and operational parameters. *J. Clean. Prod.*, *141*, 1492–1514.
60. Shu, C. H., (2007). Fungal fermentation for medicinal products. In: Yang, S. T., (ed.), *Bioprocessing for Value-Added Products from Renewable Resources* (pp. 447–463). Elsevier.
61. Nielsen, J. C., & Nielsen, J., (2017). Development of fungal cell factories for the production of secondary metabolites: Linking genomics and metabolism. *Synth. Syst. Biotechnol.*, *2*, 5–12.
62. Frisvad, J. C., Møller, L. L. H., Larsen, T. O., Kumar, R., & Arnau, J., (2018). Safety of the fungal workhorses of industrial biotechnology: Update on the mycotoxin and secondary metabolite potential of *Aspergillus Niger*, *Aspergillus oryzae*, and *Trichoderma reesei*. *Appl. Microbiol. Biotechnol.*, *102*, 9481–9515.
63. Papagianni, M., (2004). Fungal morphology and metabolite production in submerged mycelial processes. *Biotechnol. Adv.*, *22*, 189–259.

64. El-Enshasy, H. A., (2007). Filamentous fungal cultures–process characteristics, products, and applications. In: Yang, S. T., (ed.), *Bioprocessing for Value-Added Products from Renewable Resources* (pp. 225–261). Elsevier.
65. Papagianni, M., (2007). Advances in citric acid fermentation by *Aspergillus Niger*: Biochemical aspects, membrane transport and modeling. *Biotechnol. Adv., 25*, 244–263.
66. Wucherpfennig, T., Hestler, T., & Krull, R., (2011). Morphology engineering—Osmolality and its effect on *Aspergillus Niger* morphology and productivity. *Microb. Cell Fact., 10*, 1–15.
67. Wucherpfennig, T., Lakowitz, A., Driouch, H., Krull, R., & Wittmann, C., (2012). Customization of *Aspergillus Niger* morphology through addition of talc micro particles. *J. Vis. Exp*., e4023.
68. Kurt, T., Marbà-Ardébol, A. M., Turan, Z., Neubauer, P., Junne, S., & Meyer, V., (2018). Rocking *Aspergillus*: Morphology-controlled cultivation of *Aspergillus Niger* in a wave-mixed bioreactor for the production of secondary metabolites. *Microb Cell Fact., 17*, 1–19.
69. Lucatero, S., Larralde-Corona, C. P., Corkidi, G., & Galindo, E., (2003). Dispersion of oil and air in a simulated fermentation broth depending on the morphology of the mycelium, *Biotechnol Prog., 19*, 285–292.
70. James, S., Nwabueze, T. U., Ndife, J., Onwuka, G. I., & Ata'Anda, U. M., (2020). Influence of fermentation and germination on some bioactive components of selected lesser legumes indigenous to Nigeria. *J. Food Agric. Environ., 2*, 100086.
71. Cairns, T. C., Zheng, X., Zheng, P., Sun, J., & Meyer, V., (2019). Moulding the mould: Understanding and reprogramming filamentous fungal growth and morphogenesis for next generation cell factories. *Biotechnol. Biofuels, 12*, 1–18.
72. Veiter, L., Rajamanickam, V., & Herwig, C., (2018). The flamentous fungal pelle–relationship between morphology and productivity. *Appl. Microbiol. Biotechnol., 102*, 2997–3006.
73. White, S., McIntyre, M., Berry, D. R., & McNeil, B., (2002). The analysis of industrial filamentous fungi. *Crit. Rev. Biotechnol., 22*, 1–14

CHAPTER 11

Novel Extraction Technologies for the Recovery of Bioactive Compounds from Citrus By-Products: Recent Findings

SOUJANYA GADI,[1] SAMUEL PÉREZ-VEGA,[2]
RAFAEL MINJARES-FUENTES,[3] LOURDES MORALES-OYERVIDES,[1]
JUAN CARLOS CONTRERAS-ESQUIVEL,[1] and JULIO MONTAÑEZ[1]

[1]*Faculty of Chemical Sciences, Universidad Autonoma de Coahuila, Saltillo, Coahuila, Mexico*

[2]*Faculty of Chemical Sciences, Universidad Autonoma de Chihuahua, Chihuahua, Mexico*

[3]*Faculty of Chemical Sciences, Universidad Juárez del Estado de Durango, Gómez Palacio, Durango, Mexico*

ABSTRACT

Citrus is one of the most cultivated and consumed fruit crops worldwide. These fruits are either eaten fresh or processed to produce juices and marmalades and as a result, large quantities of waste are generated. Citrus residues are a potential source of various added-value compounds like polyphenols, pectin, essential oils (EOs), etc., and the traditional extraction methods used to obtain them are usually time-consuming and use large quantities of toxic solvents. This chapter aims to discuss the effectiveness and operating conditions of novel extraction technologies like ultrasound-assisted extraction (UAE), microwave-assisted extraction (MAE), pulsed electric field technology (PEF), supercritical carbon dioxide (SC-CO_2), subcritical water extraction (SWE), enzyme-assisted extraction (EAE), and ohmic heating extraction (OHE),

Bioresources and Bioprocess in Biotechnology for a Sustainable Future. Leonardo Sepúlveda Torre, Juan Carlos Contreras-Esquivel, Ann Rose Abraham, & A. K. Haghi (Eds.)

etc., that affect the recovery of bioactive compounds from citrus residues. Valorization of citrus waste helps not only to reduce the environmental impact but also improves the global economy. Also, emerging technologies have proven to be a greener and more sustainable alternative to conventional extraction methods in terms of better extraction yields and reduced extraction times.

11.1 INTRODUCTION

Citrus juices represent an important market in the worldwide beverage sector. Among the citrus juices, orange juice is the most commercialized item, followed by grapefruit and lemon. However, during juice extraction process, huge amounts of residues such peels, juices sacs, and seeds are generated. These residues contain a variety of bioactive compounds such as polyphenols EOs, pectin, carotenoids, vitamin C, etc., that can be extracted and re-inserted as natural food additives [1–3]. Polyphenolic compounds, such as flavonoids, and phenolic acids, are known to be potential sources of antioxidants, and many authors have reported their therapeutic activity for treating diseases such as diabetes, cancer, cardiovascular, and liver diseases, and more. Because of these properties, the bioactive compounds present in citrus by-products have a great potential as natural and organic supplements for different sector such food, cosmetic, nutraceutical, and pharmaceutical [4–9]. Extraction is considered a crucial step for the recovery of these bioactive compounds from citrus byproducts. Traditional extraction methods like maceration, solvent extraction, hydro-distillation, etc., have the disadvantages of using large quantities of solvents, prolonged extraction times and low yields [10]. Because of these drawbacks, conventional extraction methods are gradually being replaced by "green technologies" using friendly-solvents and short-processing times, thus they have a positive impact on the environment. The application of novel extraction methods such as microwave-assisted (MAE), ultrasound-assisted (UAE), pulsed electric field technology (PEF), supercritical carbon dioxide (SC-CO_2), subcritical water extraction (SWE), and EAE, Ohmic heating extraction (OHE), etc., to obtain bioactive compounds from citrus residues has been reported in the literature. The main target of emerging technologies arises sustainability, eco-friendliness, higher yields, better quality, reduce solvent usage, energy consumption and the duration of extraction [11–15]. However, still some work like techno-economic viability and life cycle analysis before its industrial implementation must be addressed. In the present chapter, a general overview, and recent findings regarding the use of emerging technologies for bioactive compounds from citrus by-products is presented (Figure 11.1).

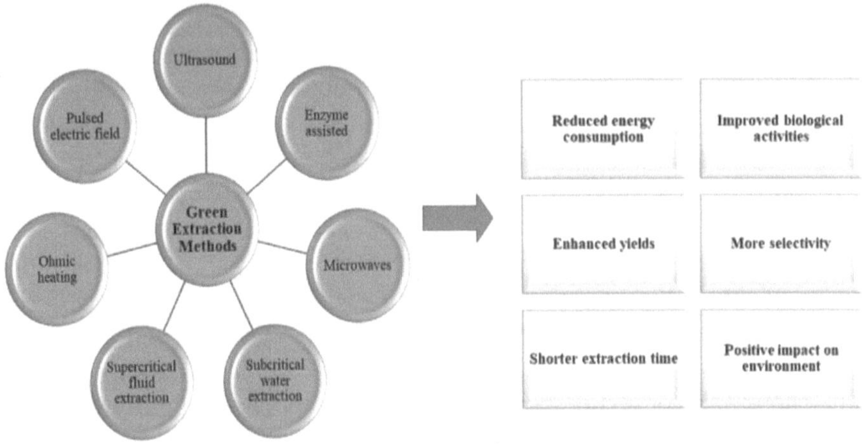

FIGURE 11.1 Novel extraction technologies and their advantages.

11.2 ULTRASOUND ASSISTED-EXTRACTION

Ultrasound assisted extraction (UAE) is a novel technology with potential industrial application in food sector. Ultrasound is based on the cavitation phenomenon generating micro-streaming followed by collapsing; thus, it triggers a series of physical effects such as shear force, turbulence, shock waves and micro-jets [16]. These effects generate fragmentation and sono-poration, leading to disruption of the cell wall structure, improving solvent penetration, and allowing the release of intracellular matter. UAE is usually carried out using an ultrasonic bath or a probe-type ultrasound system, both attached to transducers that generate sound waves [16–19]. UAE is an attractive technology reducing processing time, easy operation, diminishes use of solvents and reduces energy, thus its use in the recovery of bioactive compounds from agro-industrial by-products have risen in recent years.

Recovery of bioactive compounds from agro-industrial residues using UAE is widely reported in the literature, demonstrating advantages (yield and extraction time) over conventional extraction methods [20–22]. Saini et al. [23] compared the efficiency of polyphenol extraction from kinnow mandarin (*Citrus reticulata)* and mousambi (*Citrus limetta*) peels utilizing maceration and ultrasound-assisted extraction, observing that total poly-phenol content and antioxidant activity were higher in ultrasound extracts of both the peel varieties [23]. Also, Wang et al. [24] did a comparative study of UAE and conventional extraction to recover polymethoxyflavones from Sichuan red orange (*Citrus reticulata* 'Blanco') peels, reporting that yields of tangeretin and nobiletin obtained using UAE were 1.5 times higher than the conventional solvent extraction [24]. Nishad et al. used the Box-Behnken design to optimize the ultrasound-assisted extraction of phenolic compounds from grapefruit peels (*Citrus paradisi* L.) and compared the extraction yield to a traditional method. The total polyphenol and flavonoid content were more in ultrasound extracts when compared to conventional solvent extracts (TPC: 21.17 and 15.28 mg GAE/g; TFC: 2.76 and 1.39 mg QE/g, respectively). An increase in ethanol concentration up to 70% increased the TPC yield, but at higher concentrations, the yield was lower [20]. In another study, the pectin obtained from tangerine (*Citrus reticulata*) and grapefruit (*Citrus paradisi*) peels using UAE was compared with conventional extraction. The extracted pectin showed a typical pseudoplastic fluid with shear thinning behavior. Tangerine peel pectin had higher phenolic and protein content, while grapefruit peel had higher viscosimetric molecular weight. The results revealed that the pectin yield was higher, and the extraction time was shorter in the UAE (tangerine: 13.46%, 15 min; grapefruit: 26.05%, 30 min) than in the conventional extraction (tangerine: 10.06%, grapefruit: 24.08% in 90 minutes) [25].

Jagannath and Biradar reported that when compared to Soxhlet extraction, UAE was better in terms of recovery of phenolic compounds from lemon (Citrus limon L.) peels (TPC: 76.22 and 88.06 mg GAE/100 g; TFC: 5.47 and 13.92 mg CE/100 g, respectively). Antioxidant activity and retention of vitamin C were also higher in UAE extracts. The temperature, followed by the time were the parameters that had a significant effect on the extraction of bioactive compounds from using ultrasound [26].

Although citrus seeds account for a minor portion of the fruits, they are a powerhouse of various chemical compounds with medicinal properties. Flavanone glycosides, alkaloids, fatty acids, tocopherols, and triterpenoids (such as limonin and obacunone) are some of the phytochemicals commonly found in citrus seeds [4, 27]. The flavonoids obtained from the ultrasonic extraction of lemon seeds (200 W power, 95% aqueous ethanol, 50°C for 60 min) had hepatoprotective effects [27].

The ultrasound extraction efficiency can be affected by various factors like extraction temperature, time, power, amplitude, particle size, the polarity of the solvent, frequency, duty cycle, etc. [28].

Table 11.1 illustrates a summary of the use of UAE and critical parameters for enhancing extraction of bioactive compounds.

TABLE 11.1 Ultrasound-Assisted Extraction of Citrus Bioactive Compounds

Citrus Source	Critical Factors	Optimal Conditions	Yield	References
Citrus reticulata Blanco cv. Sainampueng)	Extraction time, temperature, and ultrasonic power.	48°C and 56.71 W for 40 min.	Extraction yield: 26. 5%, TPC: 152.5 mg GAE/g, hesperidin: 64.35 mg/g.	[29]
Citrus depressa Hayata	Solvent types, extraction time, and fruit maturity.	50% ethanol, 50 minutes. unripe peels	Nobiletin: 12.5 mg/g tangeretin: 7 mg/g rutin: 6.27 mg/g.	[30]
Citrus reticulata	Temperature, time, and amplitude	33% amplitude, 6.16 mL/g liquid-solid ratio, 43°C and extraction time of 34 min..	Lutein yield: 2.97 mg/100 g.	[31]
Lime peels	Ethanol concentration, ultrasonic amplitude, and extraction time.	38% amplitude, 55% ethanol concentration, and extraction time of 4 min.	TPC: 54 ± 1 mg GAE/g DPPH assay: 19 ± 0.4 µM Trolox/g ABTS assay: 456 ± 2 µM Trolox/g	[32]
Citrus limetta	Solvent type, extraction time, particle size, stirring speed, solid to solvent ratio, extraction temperature, ultrasound duty cycle, and power.	S/L ratio: 1:10 Particle size: 2 mm Agitation speed: 300 rpm Ultrasonic power: 80 W 50% duty cycle Temperature: 60°C	d-limonene: 32.9 mg/g	[28]
Lemons Mandarins	Temperature, sonication time, S/L ratio	S/L: 1:30 g/mL Temperature: 75°C Sonication time: 45 min	Lemon pectin yield: 10.11% Mandarin pectin yield: 11.29%	[33]
Citrus sinensis cv. Malta	Amplitude; Liquid-solid ratio Extraction time	70.89% amplitude, 40 ml/g solvent–solid ratio and extraction time: 35 min.	TPC: 15.9 mg GAE/g and TFC: 1.04 mg QE/g.	[34]

TABLE 11.1 *(Continued)*

Citrus Source	Critical Factors	Optimal Conditions	Yield	References
Citrus sinensis	Power, extraction time and ethanol concentration.	Ultrasonic power: 400 W, Time: 30 min, 50% aqueous ethanol	TPC: 105.96 mg GAE/100 g Carotenoids: 0.63 mg β-carotene/100 g	[35]
Citrus aurantium L.	Ultrasound power, extraction time, and pH	Ultrasound power: 150 W for 10 min using citric acid at pH 1.5	Pectin yield: 28.07 ± 0.67%	[21]
Citrus reticulata Blanco	S/L ratio, power, frequency, and extraction time.	Solid-liquid ratio: 1/100 Ultrasonic power: 320 W Frequency: 40 kHz Time: 30 min	Triterpene limonin: 11.16 mg/g Hesperidin: 0.04 mg/g	[36]
Citrus sinensis	Extraction time, extraction temperature, and ethanol concentration.	Extraction time: 44 min; temperature: 50°C ethanol concentration: 57.7%.	TPC: 292.16 μg GAE/g and TFC: 191.15 mg CE/g DPPH: 5.199 μg/ml and ABTS: 2.96 μg/ml.	[37]

11.3 ENZYME ASSISTED EXTRACTION (EAE)

EAE involves the use of enzymes to cause hydrolytic disruption of cell walls and membranes and increase their permeability for the efficient extraction of biologically active compounds [38]. The commonly used enzymes for EAE are cellulases, hemicellulases, pectinases, tannases, proteases, and amylases. Parameters such as enzyme type and concentration, particle size, extraction temperature, incubation time, and pH play a critical role in the yield of bioactive compounds recovery. This green extraction process commonly uses water or buffer as the solvent medium, is carried out at low temperatures, and requires low energy. Another advantage of EAE is that it uses mild extraction conditions, thus preventing the degradation of some other phytochemicals [38–40].

Some studies have reported that EAE has a better extraction yield of phenolic compounds than conventional extraction. Enzymatic treatment of citrus peels with pectinase and cellulase (6 U/g DW, pH 5.0 and 40°C) increased the extraction efficiency of phenolic compounds. When compared to the control extracts, the extraction yield was approximately 30% higher in the case of flavedo; while in albedo, the recovery of polyphenols improved by 60% [41]. Barbosa et al. [42] reported that treatment of citrus peels with tannase and β-glucosidase in an acetate buffer (20 mM at pH 5.0) at 40°C for 24 hours caused deglycosylation of flavanones, resulting in extracts rich in naringenin and hesperetin. The aglycone-rich extracts also prevented the growth and adhesion of the pathogen, *Salmonella typhimurium*, in the abdomen. The enzymatic extracts were better than the conventional ethanolic extracts, with improved polyphenol extraction and biological activity [42]. Ruviaro et al. [43] reported that citrus by-products subjected to EAE with pectinase, cellulase, tannase, and β-glucosidase resulted in hydrolysis of flavanone glycosides (hesperidin and naringin) releasing their aglycone forms (hesperetin and naringenin). A positive correlation was observed between the enzyme concentration and TPC yield. The highest content of hesperetin (462.2 ± 31.7 mg/100 gm DW) was obtained upon treating the citrus waste with β-glucosidase (20 U/g DW) [43]. Recently, Nishad et al. [44] optimized the recovery of antioxidant polyphenols from grapefruit (*C. paradisi* cv. Redblush) peels using the complex Viscozyme (pH 4.8 and 60°C), demonstrating that extraction time had a significant effect on total phenolic and flavonoids content yields. At optimum extraction conditions (extraction time of 4.8 hours, 0.9% enzyme concentration, and S/L ratio of 1: 40 g/ml), total phenolic and flavonoids content were 31.8 mg GAE/g and 3.3

mg QE/g, respectively [20]. In similar study, the same research group [30] reported that EAE enhanced by 2.6 times the recovery of total phenolic and flavonoid content in the orange (*C. sinensis* cv. Malta) peel extracts, being extraction time and enzyme concentration the factors with significant effects.

11.4 MICROWAVE ASSISTED EXTRACTION (MAE)

The mechanism of MAE is based on the electrophoretic transfer of ions and electrons. The irradiation of the food matrices by non-ionizing electromagnetic waves with frequencies ranging from 300 MHz to 300 GHz, also known as microwaves, generates an electric field by initiating ionic conduction. This event further creates a dipole rotation causing molecular motions in the water and electrolytes in the biomass and heats up the materials [44]. The moisture content of the sample matrix is one of the important factors in microwave heating. As the water evaporates, the cell membranes disrupt due to an increase in the intracellular pressure and thereby release the target compounds [44–46]. The efficiency of MAE depends on several factors, including the microwave power, polarity, and volume of extracting solvent, sample material and particle size, extraction temperature, time, and number of extraction cycles [47].

Microwave-assisted extraction technique has been effective for the extraction of polyphenols from citrus by-products. Rodsamran & Sothornvit [48] optimized the MAE of phenolic compounds from lime peel residues (*Citrus aurantifolia*), resulting in a total polyphenol content (TPC) of 53 mg GAE/g, under the optimized conditions (55 ± 5% ethanol concentration, microwave power of 140 W for 45 s and 8 repeats of the extraction). Ateş et al. [47] reported the recovery of bioactive compounds (TPC and TFC) by-MAE from mandarin leaf waste; under optimized conditions (microwave power of 275 W, 2 g of mandarin leaf, and extraction time of 45 seconds), the maximum TPC and TFC were 17.2 mg GAE/g and 1.72 mg CE/g, respectively. Also, these extracts rich in TPC and TFC were highly correlated to antioxidant capacity.

Rajput et al. [64] investigated the effect of various during MAE of phenolic antioxidants from kinnow (*Citrus reticulata*) peels. Under optimal conditions (microwave power of 300 W, 80% ethanol, liquid: solid ratio of 50 ml/g after 6 min of extraction), the maximum phenolic content (2.48 mg GAE/g peel) had antioxidant activities of 63.20% and 6.38 mmol Fe^{2+}/100 g peels for DPPH and FRAP assays, respectively.

MAE have also been applied for the recovery of added-value compounds from citrus like pectin. Rodsamran and R. Sothornvit [32] extracted pectin from lime peels (*Citrus aurantifolia*) through MAE, reporting higher yields (15.91 ± 0.52%) in comparison with control experiments (no-MAE) [48]. Su et al. [49] reported that the use of MAE and the surfactant Tween 80 (8 g/L) has increased the yield of pectin extracted from satsuma mandarin (*Citrus unshiu*) peel by 17% when compared to MAE alone [49]. Karbuz et al. [33] found MAE to be more effective and efficient method than UAE for extraction of pectin from lemon and mandarin peels. However, prolonged microwave exposure caused deterioration of pectin chains. By-using nitric acid as a solvent, application of 360 W of microwave power for 1 min obtained maximum pectin yield (9.71% from lemon peels and 7.6% from mandarin peels) [33].

Also, MAE have been applied in the recovery of EO from citrus residues. Zhang et al. [50] studied the combined effect of simultaneous extraction of EO by-using ultrasonic/microwave-assisted hydro distillation from *Citrus medica L. var. sarcodactylis*, concluding that microwave power had a significant effect on EO yield. With an extraction time of 15 minutes, increasing the MW power from 500 W to 800 W resulted in an increase in EO yield from 0.36 ± 0.02% to 0.55 ± 0.03% [50]. A similar effect of the microwave power was reported by Auta et al. [51] during microwave-assisted hydro-diffusion extraction of d-limonene from orange and lemon peels. For an optimal extraction time of 10 min, an increase in microwave power from 200 to 1000 W led to a corresponding increase in extraction efficiency to 2.4 and 1.33% for orange and lemon peels, respectively [51].

Razzaghi et al. [52] evaluated six different methods for the extraction of EOs from sweet orange peels (*Citrus sinensis*). These include traditional processes such as steam distillation (SD) and microwave-assisted techniques such as microwave steam diffusion (MSDf), microwave steam distillation (MSD), microwave hydraulic diffusion, and gravity (MHG), solvent-free microwave extraction (SFME), and steam hydraulic diffusion (SHDF). Among the extraction strategies evaluated, MHG demonstrated the fastest extraction cycle, yielding 6.1 mL of EO from 400 grams of orange peels in less than an hour. On the other hand, it took more than 100 minutes to extract 4 mL of EO using SD. Also, the use of microwaves alone, or in combination with steam increased the power efficiency compared to SD [52]. Song et al. [53] compared the efficiency of traditional and microwave heating to obtain EOs from citrus peels. The maximum limonene yield (37.82%) was obtained by microwave heating at a power of 450 W for 10 min. According to their findings, in the case of MAE, the yield of limonene per unit of energy consumption was 18.6 times higher than that of conventional heating [53].

Therefore, MAE is not only an energy-efficient extraction method, but its short extraction time helps avoiding the degradation of heat-sensitive compounds [44, 54].

11.5 SUPERCRITICAL FLUID EXTRACTION (SCFE)

Super critical fluid extraction has been used for the selective recovery of phytochemicals from food waste and by-products [55, 56] This procedure uses fluids in the state above their critical temperature (T_c) and critical pressure (P_c) where they exhibit both gas-like and liquid-like properties (high diffusivity, lower densities, viscosities, and surface tensions) and can easily penetrate the solid matrices when compared to conventional extraction solvents. Near the critical point, the density of the solvent can be changed by varying the pressure and temperature. This process enhances the solubility of the target compounds, enabling selective and faster extraction [14, 57].

Carbon dioxide (CO_2) is the commonly used supercritical solvent because of it is non-toxic and its inert nature and low critical properties (T_c: 31.1C, P_c: 73.8 bar). Other supercritical solvents reported in the literature are ethane, propane, and dimethyl ether; these solvents have critical points comparable with CO_2 and have a higher polarity index compared to CO_2, facilitating the efficient recovery of bioactive compounds from plant matrices [58]. The extraction pressure, temperature, solvent flow rate, extraction time, and physical structure of the matrix are the main parameters influencing the efficiency of supercritical carbon dioxide extraction (SC-CO_2) [59, 60]. Ateş et al. [61] reported that extraction temperature had a significant effect on the total polyphenol and flavonoid content of the mandarin leaf extracts. The optimal conditions for SFE were 50°C, 200 bar, and a flow rate of 0.39 mL/min [61].

Giovanna et al. [62] reported that the ripening stage of the fruit and particle size of the sample had also a significant effect during the recovery of d-limonene from lemon (*Citrus limon* L.) peels. The samples collected in December demonstrated the highest content of d-limonene (88% w/w). They also observed that at optimal conditions (pressure:15 MPa; temperature: 40°C; CO_2 flow rate: 8 l/min, extraction time:40 min), a decrease in particle size improved SFE rate and yield [62]. Barrales et al. [63] used supercritical CO_2 extraction to recover the EO from orange peels (at 40°C, 35 MPa, and solvent to feed ratio of 32 kg CO_2/kg dry peel). The extraction yield was 0.48%, and the main component identified in the EO was α-terpineol, followed by d-limonene [63]. Rajput et al. [64] optimized the supercritical fluid extraction of EOs from kinnow mandarin peels and reported that SCFE at a temperature of 43°C, pressure of 297 bar for 120 min resulted in a maximum extraction yield of 1.57%, TPC of 41.22 mg GAE/g and DPPH activity of 79.94% [64].

One of the limitations during the use of SC-CO_2 is the inadequate solubility of polar functional groups in supercritical CO_2. The addition of polar solvents like ethanol, methanol, or water is recommended to overcome this problem. These added solvents are called co-modifiers, co-solvents, or entrainers [65]. Several works have demonstrated the improvements of SCFE added with co-solvents [66].

Ndayishimiye et al. [67] reported that adding co-solvent during the oil extraction by SCFE from citrus seeds, peels, and mixtures increased the yields (23.04%, 1.91%, and 12.73%, respectively). The oils extracted using SC-CO_2 + ethanol showed significantly higher TPC values (27.31 to 43.64 mg GAE/g) than those obtained using SC-CO_2 alone (12.17 and 27.94 mg GAE/g). The increase in phenolic content in extracts was attributed to the polarity of ethanol. Moreover, the addition of ethanol as a modifier enhanced the antioxidant and antimicrobial activities of the extracted oils [67].

As it was previously mentioned, one of the advantages of supercritical fluids is their selectivity toward bioactive compounds extraction. Franco-Arnedo et al. [68] compared the efficiency of Soxhlet extraction with SCFE + ethanol as co-solvent on the extraction of antioxidant bioactive compounds from tangerine (*Citrus reticulata* var. Arrayana) peels. The extraction yield was higher in Soxhlet extracts (34.85%) than the supercritical extracts (1.42–34.76%). However, the total polyphenol and flavanone content was higher in the case of SCFE. This pattern was also observed in the antioxidant activity of the peel extracts (24.02% for Soxhlet extraction and 89.22% for SCFE) [68]. In a recent study, sequential SC-CO_2 extractions were carried out to obtain extract different bioactive compounds from mandarin peels using acetone and methanol as co-solvents. Limonene was the main compound in the peel extracts. With the use of SC-CO_2-acetone (7%), 0.45–3.97 μg of β-carotene/g of dried peel were extracted, meanwhile with SC-CO_2-methanol (7%) the hesperidin yield was in the range of 1.6–1.8% [69]. Yaqoob et al. [70] reported that the tangerine peel extracts obtained after SCFE demonstrated highest total polyphenol content (27.91 ± 0.05 mg GAE/g) and total flavonoids content (10.31 ± 0.39 mg/g) when compared to ultrasonication method [70]. Rosa et al. [71] evaluated the yields and compositions of citrus seed oils obtained by *n*-hexane Soxhlet extraction and SCFE. Compared to the traditional extraction method, the recovery yield was lower in SCFE, but the chemical profile of the oils was very similar. Linoleic acid (35–42%), oleic acid (22–28%), palmitic acid (21–25%), alpha-linolenic acid (4–10%), and stearic acid (4–7%) were the prominent fatty acids, while monolinolein and dilinolein derivatives are the main triacylglycerols in oil extracted by SCFE [71].

Also, raw material conditioning plays an important role during SCFE. The effects of drying methods and SCFE parameters on the phenolic content and antioxidant activity of mandarin (*Clementina orogrande*) peel oil were evaluated by Quispe-Fuentes et al. [72]. Convection drying gave better results than vacuum drying. Maximum TPC (19.5 ± 1.8 mg GAE/g oil) was obtained using SC-CO_2 + 10% ethanol at 40°C and 300 bar pressure. Hesperidin, naringin, and rutin were the prominent phenolic compounds in mandarin peel oil. Their results demonstrated that the extraction conditions for maximum antioxidant activity (DPPH: 33.1 ± 3.2 μmol Trolox/g; ORAC: 2860 ± 276 μmol Trolox/g) were under SC-CO_2 + 10% ethanol at 40°C and 400 bar pressure [72].

Jokić et al. [73] used SC-CO_2 to obtain nonpolar EO compounds from orange (*Citrus sinensis* L.) peels. Limonene was the predominant compound found in the supercritical extracts (up to 89%) and extraction yield was significantly affected by the pressure. Under the optimal conditions (pressure 30 MPa and temperature of 40°C), the predicted extraction yield was 3.13 g/100 g and limonene 51.69% [73]. Regarding pigments recovery through SCFE from citrus residues, Priyadarsani et al. [74] reported the extraction of lycopene from ripe grapefruit peels. At optimum conditions of 70°C temperature, 305 bar pressure, and 35 g/min CO_2 flow rate for 135 min, the extraction efficiency was 91.03 ± 1.86% [74]. In a similar study, SCFE was carried out using SC-CO_2 and rice bran oil as co-solvent (flow rate of 35 g/min and 3% w/w, respectively). The authors used response surface methodology to optimize the extraction parameters and maximize the recovery of lycopene and γ-oryzanol retention in the extract. At optimal conditions (325 bar, 64°C, for 143 min), nearly 70% of lycopene was recovered, and 97% of γ-oryzanol was retained in the extracts. Moreover, rice bran oil as a co-solvent helped delay oxidative degradation during extraction and storage.

As such, SFE is a potential alternative to traditional solvent extraction due to its high extraction efficiency, prevention of oxidation reactions, and elimination of residual solvents in the final extract.

11.6 SUBCRITICAL WATER EXTRACTION (SWE)

SWE, also known as pressurized liquid extraction, uses superheated water as an extraction solvent. At temperatures around its boiling point (100°C) and critical point (374°C) and high pressures ranging from 4–20 MPa, water exists in a liquid form, also known as a subcritical state [75]. In this state, water behaves like regular organic solvents due to dielectric constant decrease, allowing enhanced mass transfer, increase the solubility of the non-polar analytes, and an increase in the extraction yields [76, 77]. The efficiency of SWE depends on several factors like temperature, pressure, solvent flow rate, extraction time, sample particle size, the presence of co-solvents, etc. [77–80]. Liew et al. [81] evaluated the effect of different extraction parameters on the isolation of low methoxyl pectin from pomelo (*Citrus grandis* (L.) Osbeck) peels. It was observed that extraction yield was significantly affected by the temperature. As the water temperature increases, the strength of the hydrogen bond weakens, thus the dielectric constant decreases, subsequently extraction efficiency was enhanced. Pectin yield of 19.6% was obtained under optimum extraction conditions (120°C and at a pressure of 30 bar) [81].

A similar effect of temperature on the SWE extraction yield of flavanones from defatted orange peel was reported by Lachos-Perez et al. [82]. The maximum extraction yield (10.63 ± 0.55%), and hesperidin (20 mg/g) and naringin (2.33 mg/g defatted peel) were obtained at an extraction temperature of 150°C and a flow rate of 10 mL/min. The total phenolic compounds, total flavonoids and the antioxidant activity of SWE extracts were higher than the conventional extracts [82]. Kim & Lim [77] reported that the extraction rate of citrus peel bioactive compounds was directly proportional to the extraction temperature and water flow rate. The extraction yield of the citrus phytochemicals increased when the temperature increased from 120°C to 160°C, however temperatures beyond 160°C, thermal hydrolysis occurred [77]. SWE is a promising green extraction technique for isolation of bioactive compounds because of the nontoxic nature, safety, and easy availability of water [83].

11.7 PULSED ELECTRIC FIELD TECHNOLOGY (PEF)

The pulsed electric field technology is a non-thermal food processing technique that has been widely used in the past few years to extract bioactive compounds from plant substrates. PEF works on the principle of "electroporation" or "electropermeabilization," where short pulses of moderate-high electric field strength disrupt the cell membranes and increase their permeability [84]. PEF treatment improves the mass transfer and increases the extraction yield in a short duration (micro- to milliseconds). Factors like electric field strength, number of the pulses, pulse duration, pulse shape, pulse polarity, pulse frequency, sample size, and conductivity of the medium determine the efficiency of the PEF treatment [85–88]. Peiro et al. [89] reported that PEF treatment of lemon peels with high-intensity pulses (7 kV/cm) improved the extraction yield of polyphenols by 300%. Hesperidin and eriocitrin were identified as the main flavanones in the peel extracts (84 mg /100 g FW and 176 mg /100 g FW, respectively) [89]. The impact of PEF treatment on the extraction yield of naringin from pomelo peels (*Shantian* Variety) was determined by Niu et al. [90] at an electric field strength of 2 to 10 kV/cm and with 10 to 50 pulses. At an electric field strength of 4 kV/cm and 30 pulses, the extraction yield, and the antioxidant activity of naringin increased by 20% and 70%, respectively, compared to the samples without PEF treatment. The authors also demonstrated that the electrical pulses changed the structure of naringin without causing its degradation [90]. Kantar et al. used PEF (3–10 kV/cm) as a pretreatment to the conventional extraction procedure (50% ethanol for 60 min) and compared the total polyphenolic content of juice and peel extracts (flavedo/albedo) of different citrus fruits. The results showed that PEF treatment of orange peels at 10 kV/cm improved the phenolic concentration from 12 to 22 mg GAE/g dry weight [91].

PEF is a sustainable technique with significantly shorter extraction times, and high energy efficiency, and typically operates at room temperature, thus preventing the degradation of thermosensitive phytochemicals.

11.8 OHMIC-HEATING EXTRACTION (OHE)

Ohmic heating, otherwise known as direct Joule heating, is a process where alternating electric current is passed through the semiconductive materials to achieve rapid and homogenous heating [92, 93]. In the presence of an electric field, the ions present in the food materials move towards the electrodes with opposite charges and collide with each other, creating electrical resistance and increasing their kinetic energy, thereby resulting in volumetric heating [94]. Processing parameters like electrical conductivity, electrical field strength, the surface area of the electrodes, etc., influence the heating rate and efficiency in an ohmic process [95]. The application of ohmic heating for extraction purposes has gained a lot of attention in the last few years. Studies have reported that OHE is an energy efficient, time-saving process improving the extraction quality and yield of bioactive compounds compared to conventional extraction methods [96, 97]. Saberian et al. [98] evaluated the effect of the voltage gradient, temperature, and extraction time on the pectin yield from orange juice waste. They reported that at optimal extraction conditions (15 V/cm, temperature of 90°C for 30 min), the pectin yield of ohmic extracts was higher than conventional extracts (14.32 and 13.53 g/100 g dry matter, respectively) [98]. Ohmic heating assisted extraction of pectin from lemon peel powder using citric acid at different temperatures (70–90°C), solid: liquid ratios (1:30, 1:40, 1:50 g/ml), and extraction holding times (0–30 min), were analyzed by Çilingir et al. [99]. The highest pectin yield (16.91%) was obtained at 90°C, 1:40 solid: liquid ratio after 30 minutes. They also reported that OHE was more energy efficient than the traditional pectin extraction method (total energy consumption: 109.9 ± 2.3 kJ and 155.25 ± 11.6 kJ, respectively) [99]. Sabanci et al. [100] reported that the pectin yield of various citrus peels and the total energy consumed during the extraction procedure increased upon increasing the processing time of ohmic heating. In the case of orange peels, upon increasing the holding time from 0 to 180 minutes, the pectin yield increased from 9.68 ± 0.70% to 18.32 ± 0.75, and the energy consumption increased from 39 kJ to approximately 260 kJ [100]. Thus, ohmic heating extraction is a sustainable alternative to conventional extraction methods to obtain added-value compounds from citrus residues.

11.9 COMBINATION OF EXTRACTION TECHNOLOGIES

The novel technologies have demonstrated lot of advantages over traditional extraction procedures like reduced extraction times, limited use of toxic solvents to name a few. Several authors have reported the use of these new technologies as a pre-treatment to conventional methods or simultaneously used two green technologies to make the most of the agro-industrial residues. The synergistic effect of the extraction methods helps to further improve the extraction yields without compromising the biological activities of the extracted compounds. For example EAE may not be sufficient to completely hydrolyze the cellular matrix, but sonication can have complementary effect to maximize the extraction process [101]. A variety of bioactive compounds have been extracted from plant matrices using enzyme-assisted supercritical fluid extraction and the extraction yield was better when compared to individual extraction methods. Combination of these extraction methods helps in targeting both polar and non-polar bioactive compounds [102]. Some of the recent studies which combined the extraction technologies are generalized in Table 11.2.

TABLE 11.2 Combination of Extraction Methods to Obtain Citrus Bioactive Compounds

Extraction Method	Source	Target Compound	Extraction Conditions	References
UAE+MAE	Pomelo peels	Pectin	Ultrasound power: 50 W Microwave power: 360 W Time: 15 min Liquid-solid ratio: 27 mL/g	[103]
UAE+EAE	Pomelo peels (*Citrus grandis limonia* Osbeck species)	Polyphenols	2% Pectinex Ultra SP-L Water-solid ratio: 40 ml/g Incubation temperature: 50°C Ultrasonic energy: 40 kHz Time: 60 min	[101]
UAE + Pressure (Manosonication)	Eureka lemon (*Citrus limon* Burm.), Guanxi pomelo (*Citrus maxima* Osbeck), grapefruit (*Citrus paradisi* Macf.), and navel orange (*Citrus sinensis* var.)	Pectin	Pressure: 200 kPa Temperature: 25°C Amplitude: 25% Time: 20 min	[104]
UAE + Hydro-distillation	Bergamot *(Citrus medica* l. var. sarcodactylis)	Essential oil	Ultrasonic power: 800 W Sonication time: 20 to 60 minutes	[105]
UAE + MAE + Hydro-distillation	Bergamot (*Citrus medica* L. var. sarcodactylis	Essential oil	Microwave power: 800 W Ultrasonic power: 50 W Time: 15 min	[50]
Manosonication	*Citrus unshiu* Marc. peels	Pectin	Pressure: 250 kPa Temperature: 42°C Amplitude: 40% Time: 20 min	[106]

TABLE 11.2 *(Continued)*

Extraction Method	Source	Target Compound	Extraction Conditions	References
UAE+SC-CO_2	Orange peels (*Citrus sinensis* L.)	Limonene Hesperidin	Optimum conditions: Pressure: 30 MPa Temperature: 40°C UAE temperature: 68°C Time: 45 min Solvent–solid ratio: 41 mL/g Aqueous ethanol: 61%	[73]
UAE + Acid hydrolysis	Orange peels	Essential oil Pectin Bacterial cellulose	1.21% H_2SO_4 concentration 5.75% solid loading Extraction time: 34.2 minutes	[107]
MAE+HD	Orange leaves	Essential oil	Water-solid ratio: 3.46 mL/g Time: 100.47 min Microwave power: 471.58 W	[108]
EAE+HD	Sweet orange peels (*Citrus sinesis* L.)	Essential oil	3.9% Viscozyme L® Liquid/solid ratio of 4.0 mL/g Temperature: 55°C Time: 3.8 hours	[109]
SWE+PEF	*Citrus unshiu* peels	Hesperidin and narirutin	Electric field strength: 3 kV/cm Pulse frequency: 50 Hz Pulse width: 25 μs Time: 60–120 s Pressure:10 MPa 110–190°C; t: 3–15 min	[110]

TABLE 11.2 *(Continued)*

Extraction Method	Source	Target Compound	Extraction Conditions	References
Ohmic heating + hydro distillation	*Citrus limon*	Pectin	Voltage gradient: 14.2 V/cm	[111]
		Essential oil	Liquid/solid ratio: 8.7:1	
			Extraction time: 58.4 min	
PEF+UAE	*Pericarpium Citri Reticulatae*	Flavonoids, hesperidin, nobiletin, and tangeretin	Ultrasonic power: 340 W	[112]
			Ultrasonic time: 30 min	
			Electric field strength: 3 kV/cm and 81 pulses	

11.10 CONCLUSIONS

The novel extraction techniques discussed above have been utilized for the recovery of bioactive compounds from citrus by-products. These environment-friendly extraction methods are a potential alternative to traditional extraction methods and can also be used as a pre-treatment. Moreover, the combination of more than one technique increases the extraction efficiency without compromising the biological activity of the extracted compounds. The efficiency of each of these extraction methods is dependent on different processing parameters, which need to be evaluated and optimized to obtain maximum results. The selection of the extraction method depends on the type of residue and the target compounds. Since sustainability is one of the main objectives of these green technologies, emphasis should be given to energy consumption, environmental impact and use of green solvents instead volatile organic compounds (VOC´s). Therefore, scale-up studies are needed in terms of the use of novel emerging technologies for improving efficiency, processing time and costs in the recovery of bioactive compounds from citrus wastes for further implementation at industrial scale.

KEYWORDS

- **bioactive compounds**
- **citrus by-products**
- **enzyme assisted extraction**
- **microwave assisted extraction**
- **novel extraction technologies**
- **pulsed electric field technology**
- **subcritical water extraction**
- **supercritical fluid extraction**
- **ultrasound assisted-extraction**

REFERENCES

1. Paw, M., Begum, T., Gogoi, R., Pandey, S. K., & Lal, M., (2020). Chemical composition of *Citrus limon* L. Burmf peel essential oil from North East India. *J. Essent. Oil Bear. Plants, 23*(2), 337–344.

2. Elkhatim, K. A., Elagib, R. A. A., & Hassan, A. B., (2018). Content of phenolic compounds and vitamin C and antioxidant activity in wasted parts of Sudanese citrus fruits. *Food Sci. Nutr., 6*(5), 1214–1219.
3. Yabalak, E., Erdoğan, E. E. A., & Nazlı, M. D., (2021). Evaluation of Citrus reticulata essential oil: Chemical composition and antibacterial effectiveness incorporated gelatin on E. coli and S. aureus. *Int. J. Environ. Health Res.*, 1–10.
4. Zayed, A., Badawy, M. T., & Farag, M. A., (2021). Valorization and extraction optimization of Citrus seeds for food and functional food applications. *Food Chem., 355*, 129609.
5. Giovagnoli-Vicuña, C., Briones-Labarca, V., Bernal, G., De La Fuente-Ortega, E., & Macuer, J., (2021). Ultrasound and high hydrostatic pressure extractions on antioxidant capacity, antiproliferative and apoptosis effects in gastric cancer cells by lemon extract treatment. *Nat. Prod. Res.*, 1–5.
6. Zareiyan, F., & Khajehsharifi, H., (2021). Bioactive compounds analysis in ethanolic extracts of *Citrus maxima* and *Citrus sinensis* exocarp and mesocarp. *Nat. Prod. Res.*, 1–4.
7. M'hiri, N., Ioannou, I., Ghoul, M., & Mihoubi, B. N., (2017). Phytochemical characteristics of citrus peel and effect of conventional and nonconventional processing on phenolic compounds: A review. *Food Rev. Int., 33*(6), 587–619.
8. Mahato, N., Sharma, K., Sinha, M., & Cho, M. H., (2018). Citrus waste derived nutra-/pharmaceuticals for health benefits: Current trends and future perspectives. *J. Funct. Foods, 40*(2017), 307–316.
9. Mohsin, A., et al., (2021). Advances in sustainable approaches utilizing orange peel waste to produce highly value-added bioproducts. *Crit. Rev. Biotechnol.*, 1–20.
10. Panwar, D., Saini, A., Panesar, P. S., & Chopra, H. K., (2021). Unraveling the scientific perspectives of citrus by-products utilization: Progress towards circular economy. *Trends Food Sci. Technol., 111*, 549–562.
11. Imeneo, V., Romeo, R., De Bruno, A., & Piscopo, A., (2022). Green-sustainable extraction techniques for the recovery of antioxidant compounds from 'citrus limon' by-products. *J. Environ. Sci. Heal. Part B*, 1–13.
12. Panwar, D., Panesar, P. S., & Chopra, H. K., (2021). Recent trends on the valorization strategies for the management of citrus by-products. *Food Rev. Int., 37*(1), 91–120.
13. Liang, Z., Zhang, P., & Fang, Z., (2022). Modern technologies for extraction of aroma compounds from fruit peels: A review. *Crit. Rev. Food Sci. Nutr., 62*(5), 1284–1307.
14. Anticona, M., Blesa, J., Frigola, A., & Esteve, M. J., (2020). High biological value compounds extraction from citrus waste with non-conventional methods. *Foods, 9*(6).
15. Putnik, P., et al., (2018). Novel food processing and extraction technologies of high-added value compounds from plant materials. *Foods, 7*(7), 1–16.
16. Chemat, F., Rombaut, N., Sicaire, A. G., Meullemiestre, A., Fabiano-Tixier, A. S., & Abert-Vian, M., (2017). Ultrasound assisted extraction of food and natural products. Mechanisms, techniques, combinations, protocols and applications: A review. *Ultrason. Sonochem., 34*, 540–560.
17. Belwal, T., et al., (2018). A critical analysis of extraction techniques used for botanicals: Trends, priorities, industrial uses and optimization strategies. *TrAC Trends Anal. Chem., 100*, 82–102.
18. Vernès, L., Abert-Vian, M., El Maâtaoui, M., Tao, Y., Bornard, I., & Chemat, F., (2019). Application of ultrasound for green extraction of proteins from spirulina. Mechanism, optimization, modeling, and industrial prospects. *Ultrason. Sonochem., 54*, 48–60.

19. Vernès, L., Vian, M., & Chemat, F., (2020). Ultrasound and microwave as green tools for solid-liquid extraction. *Liq. Extr.*, 355–374.
20. Nishad, J., Saha, S., Dubey, A. K., Varghese, E., & Kaur, C., (2019). Optimization and comparison of non-conventional extraction technologies for *Citrus paradisi* L. peels: A valorization approach. *J. Food Sci. Technol.*, *56*(3), 1221–1233.
21. Hosseini, S. S., Khodaiyan, F., Kazemi, M., & Najari, Z., (2019). Optimization and characterization of pectin extracted from sour orange peel by ultrasound assisted method. *Int. J. Biol. Macromol.*, *125*, 621–629.
22. Soquetta, M. B., et al., (2019). Ultrasound extraction of bioactive compounds from *Citrus reticulata* peel using electrolyzed water. *J. Food Process. Preserv.*, *43*(12), 1–9.
23. Saini, A., Panesar, P. S., & Bera, M., (2019). Comparative study on the extraction and quantification of polyphenols from citrus peels using maceration and ultrasonic technique. *Curr. Res. Nutr. Food Sci.*, *7*(3), 678–685.
24. Wang, Q., et al., (2018). HPLC–DAD–ESI–MS2 analysis of phytochemicals from Sichuan red orange peel using ultrasound-assisted extraction. *Food Biosci.*, *25*, 15–20.
25. Polanco-Lugo, E., et al., (2019). Citrus pectin obtained by ultrasound-assisted extraction: Physicochemical, structural, rheological and functional properties. *CYTA–J. Food*, *17*(1), 463–471.
26. Jagannath, A., & Biradar, R., (2019). Comparative evaluation of Soxhlet and ultrasonics on the structural morphology and extraction of bioactive compounds of lemon (*Citrus limon* L.) peel. *J. Food Chem. Nanotechnol.*, *5*(3), 56–64.
27. Yang, M., et al., (2020). Preventive effect of lemon seed flavonoids on carbon tetrachloride-induced liver injury in mice. *RSC Adv.*, *10*(22), 12800–12809.
28. Khandare, R. D., Tomke, P. D., & Rathod, V. K., (2021). Kinetic modeling and process intensification of ultrasound-assisted extraction of d-limonene using citrus industry waste. *Chem. Eng. Process.–Process Intensif.*, *159*, 108181.
29. Nipornram, S., Tochampa, W., Rattanatraiwong, P., & Singanusong, R., (2018). Optimization of low power ultrasound-assisted extraction of phenolic compounds from mandarin (*Citrus reticulata* Blanco cv. Sainampueng) peel. *Food Chem.*, *241*(2017), 338–345.
30. Chien, W. J., Saputri, D. S., & Lin, H. Y., (2022). Valorization of Taiwan's *Citrus depressa* Hayata peels as a source of nobiletin and tangeretin using simple ultrasonic-assisted extraction. *Curr. Res. Food Sci.*, *5*, 278–287.
31. Saini, A., Panesar, P. S., & Bera, M. B., (2021). Valuation of *Citrus reticulata* (kinnow) peel for the extraction of lutein using ultrasonication technique. *Biomass Convers. Biorefinery*, *11*(5), 2157–2165.
32. Rodsamran, P., & Sothornvit, R., (2019). Extraction of phenolic compounds from lime peel waste using ultrasonic-assisted and microwave-assisted extractions. *Food Biosci.*, *28*, 66–73.
33. Karbuz, P., & Tugrul, N., (2021). Microwave and ultrasound assisted extraction of pectin from various fruits peel. *J. Food Sci. Technol.*, *58*(2), 641–650.
34. Nishad, J., Saha, S., & Kaur, C., (2019). Enzyme- and ultrasound-assisted extractions of polyphenols from *Citrus sinensis* (cv. Malta) peel: A comparative study. *J. Food Process. Preserv.*, *43*(8), 1–13.
35. Montero-Calderon, A., Cortes, C., Zulueta, A., Frigola, A., & Esteve, M. J., (2019). Green solvents and ultrasound-assisted extraction of bioactive orange (*Citrus sinensis*) peel compounds. *Sci. Rep.*, *9*(1), 1–8.
36. Zheng, G. D., et al., (2020). Identification and determination of chemical constituents of *Citrus reticulata* semen through ultra-high performance liquid chromatography combined with Q exactive orbitrap tandem mass spectrometry. *J. Sep. Sci.*, *43*(2), 438–451.

37. Shehata, M. G., Abd El Aziz, N. M., Youssef, M. M., & El-Sohaimy, S. A., (2021). Optimization conditions of ultrasound-assisted extraction of phenolic compounds from orange peels using response surface methodology. *J. Food Process. Preserv.*, *45*(10), 1–10.
38. Gligor, O., Mocan, A., Moldovan, C., Locatelli, M., Crişan, G., & Ferreira, I. C. F. R., (2019). Enzyme-assisted extractions of polyphenols–a comprehensive review. *Trends Food Sci. Technol.*, *88*, 302–315.
39. Cai, H., et al., (2021). Novel extraction methods and potential applications of polyphenols in fruit waste: A review. *J. Food Meas. Charact.*, *15*(4), 3250–3261.
40. Nadar, S. S., Rao, P., & Rathod, V. K., (2018). Enzyme assisted extraction of biomolecules as an approach to novel extraction technology: A review. *Food Res. Int.*, *108* (2017), 309–330.
41. Multari, S., Licciardello, C., Caruso, M., Anesi, A., & Martens, S., (2021). Flavedo and albedo of five citrus fruits from Southern Italy: Physicochemical characteristics and enzyme-assisted extraction of phenolic compounds. *J. Food Meas. Charact.*, *15*(2), 1754–1762.
42. De Barbosa, P. P. M., Ruviaro, A. R., Martins, I. M., Macedo, J. A., LaPointe, G., & Macedo, G. A., (2021). Enzyme-assisted extraction of flavanones from citrus pomace: Obtention of natural compounds with anti-virulence and anti-adhesive effect against *Salmonella enterica* subsp. enterica serovar typhimurium. *Food Control, 120*, 107525.
43. Ruviaro, A. R., De Barbosa, P. P. M., & Macedo, G. A., (2019). Enzyme-assisted biotransformation increases hesperetin content in citrus juice by-products. *Food Res. Int.*, *124*, 213–221.
44. Alvi, T., Asif, Z., & Iqbal, K. M. K., (2022). Clean label extraction of bioactive compounds from food waste through microwave-assisted extraction technique: A review. *Food Biosci.*, *46*, 101580.
45. Hu, W., et al., (2019). Microwave-assisted extraction, physicochemical characterization and bioactivity of polysaccharides from *Camptotheca acuminata* fruits. *Int. J. Biol. Macromol.*, *133*, 127–136.
46. Marić, M., Grassino, A. N., Zhu, Z., Barba, F. J., Brnčić, M., & Rimac, B. S., (2018). An overview of the traditional and innovative approaches for pectin extraction from plant food wastes and by-products: Ultrasound-, microwaves-, and enzyme-assisted extraction. *Trends Food Sci. Technol.*, *76*, 28–37.
47. Gomez, L., Tiwari, B., & Garcia-Vaquero, M., (2020). Emerging extraction techniques: Microwave-assisted extraction. *Sustain. Seaweed Technol.*, 207–224.
48. Rodsamran, P., & Sothornvit, R., (2019). Microwave heating extraction of pectin from lime peel: Characterization and properties compared with the conventional heating method. *Food Chem.*, *278*, 364–372.
49. Su, D. L., et al., (2019). Efficient extraction and characterization of pectin from orange peel by a combined surfactant and microwave assisted process. *Food Chem.*, *286*, 1–7.
50. Zhang, H., et al., (2019). Effect of simultaneous ultrasonic and microwave assisted hydrodistillation on the yield, composition, antibacterial and antibiofilm activity of essential oils from *Citrus medica* L. var. sarcodactylis. *J. Food Eng.*, *244*, 126–135.
51. Auta, M., et al., (2018). Optimization of citrus peels D-limonene extraction using solvent-free microwave green technology. *Chem. Eng. Commun.*, *205*(6), 789–796.
52. Razzaghi, S. E., et al., (2019). Operational efficiencies of six microwave-based extraction methods for orange peel oil. *J. Food Eng.*, *241*, 26–32.

53. Song, Z., et al., (2022). Study on the microwave extraction process and product distribution of essential oils from citrus peel. *Chem. Eng. Process.–Process Intensif.*, *171*, 108726.
54. Teigiserova, D. A., Tiruta-Barna, L., Ahmadi, A., Hamelin, L., & Thomsen, M., (2021). A step closer to circular bioeconomy for citrus peel waste: A review of yields and technologies for sustainable management of essential oils. *J. Environ. Manage.*, *280*, 111832.
55. Yaqoob, M., Aggarwal, P., & Purandare, N., (2020). Extraction of phenolic compounds by supercritical fluid extraction. In: Inamuddin, & Asiri, A. M., (eds.), *Advanced Nanotechnology and Application of Supercritical Fluids* (pp. 125–139). Cham: Springer International Publishing.
56. Mitani, R., et al., (2021). Extraction of nobiletin and tangeretin with antioxidant activity from peels of *Citrus poonensis* using liquid carbon dioxide and ethanol entrainer. *Sep. Sci. Technol.*, *56*(2), 290–300.
57. Manjare, S. D., & Dhingra, K., (2019). Supercritical fluids in separation and purification: A review. *Mater. Sci. Energy Technol.*, *2*(3), 463–484.
58. Cvjetko, B. M., Vidović, S., Radojčić, R. I., & Jokić, S., (2018). New perspective in extraction of plant biologically active compounds by green solvents. *Food Bioprod. Process.*, *109*, 52–73.
59. Yang, L., et al., (2021). Supercritical extraction and antioxidant activity of major ingredients in *Puerariae lobatae* root, *Pinus massoniana* needle, Citrus reticulata peel and their mixture. *J. CO_2 Util.*, *48*(2020), 101518.
60. Del Valle, J. M., Calderón, D., & Núñez, G. A., (2019). Pressure drop may negatively impact supercritical CO_2 extraction of citrus peel essential oils in an industrial-size extraction vessel. *J. Supercrit. Fluids, 144*, 108–121.
61. Ateş, F., Şahin, S., İlbay, Z., & Kırbaşlar, I., (2019). A green valorization approach using microwaves and supercritical CO_2 for high-added value ingredients from mandarin (*Citrus deliciosa* Tenore) leaf waste. *Waste and Biomass Valorization, 10*(3), 533–546.
62. Giovanna, C., Alessandra, L., Giuseppe, M., Sanzo, D., Chakraborty, S., & Calabrò, V., (2019). Process–intensified waste valorization and environmentally friendly d–limonene extraction. *Euro-Mediterranean J. Environ. Integr.*, *4*(1).
63. Barrales, F. M., et al., (2018). Recovery of phenolic compounds from citrus by-products using pressurized liquids—An application to orange peel. *Food Bioprod. Process.*, *112*, 9–21.
64. Rajput, S., Kaur, S., Panesar, P. S., & Thakur, A., (2022). Supercritical fluid extraction of essential oils from *Citrus reticulata* peels: Optimization and characterization studies. *Biomass Convers. Biorefinery*, 0123456789.
65. Vidovic, S., Vladić, J., Nastic, N., & Jokić, S., (2020). Subcritical and supercritical extraction in food by-product and food waste valorization. *Innov. Food Process. Technol. A Compr. Rev.*, 705–721.
66. Romano, R., De Luca, L., Aiello, A., Rossi, D., Pizzolongo, F., & Masi, P., (2022). Bioactive compounds extracted by liquid and supercritical carbon dioxide from citrus peels. *Int. J. Food Sci. Technol.*, 3826–3837.
67. Ndayishimiye, J., Lim, D. J., & Chun, B. S., (2018). Antioxidant and antimicrobial activity of oils obtained from a mixture of citrus by-products using a modified supercritical carbon dioxide. *J. Ind. Eng. Chem.*, *57*, 339–348.
68. Franco-Arnedo, G., Buelvas-Puello, L. M., Miranda-Lasprilla, D., Martínez-Correa, H. A., & Parada-Alfonso, F., (2020). Obtaining antioxidant extracts from tangerine

(*C. reticulata* var. Arrayana) peels by modified supercritical CO_2 and their use as protective agent against the lipid oxidation of a mayonnaise. *J. Supercrit. Fluids,* 165.

69. Tsitsagi, M., Ebralidze, K., Chkhaidze, M., Rubashvili, I., & Tsitsishvili, V., (2018). Sequential extraction of bioactive compounds from tangerine (*Citrus unshiu*) peel. *Ann. Agrar. Sci.*, *16*(2), 236–241.
70. Yaqoob, M., Aggarwal, P., & Babbar, N., (2022). Extraction and screening of kinnow (*Citrus reticulata* L.) peel phytochemicals, grown in Punjab, India. *Biomass Convers. Biorefinery.*
71. Rosa, A., et al., (2019). Supercritical CO_2 extraction of waste citrus seeds: Chemical composition, nutritional and biological properties of edible fixed oils. *Eur. J. Lipid Sci. Technol.*, *121*(7), 1–33.
72. Quispe-Fuentes, I., Uribe, E., López, J., Contreras, D., & Poblete, J., (2022). A study of dried mandarin (*Clementina orogrande*) peel applying supercritical carbon dioxide using co-solvent: Influence on oil extraction, phenolic compounds, and antioxidant activity. *J. Food Process. Preserv.*, *46*(1), e16116.
73. Jokić, S., Molnar, M., Cikoš, A. M., Jakovljević, M., Šafranko, S., & Jerković, I., (2020). Separation of selected bioactive compounds from orange peel using the sequence of supercritical CO_2 extraction and ultrasound solvent extraction: Optimization of limonene and hesperidin content. *Sep. Sci. Technol.*, *55*(15), 2799–2811.
74. Priyadarsani, S., Patel, A. S., Kar, A., & Dash, S., (2021). Process optimization for the supercritical carbondioxide extraction of lycopene from ripe grapefruit (Citrus paradisi) endocarp. *Sci. Rep.*, *11*(1), 1–8.
75. Plaza, M., & Marina, M. L., (2019). Pressurized hot water extraction of bioactives. *TrAC Trends Anal. Chem.*, *116*, 236–247.
76. Plaza, M., & Turner, C., (2015). Pressurized hot water extraction of bioactives. *TrAC Trends Anal. Chem.*, *71*, 39–54.
77. Kim, D. S., & Lim, S. B., (2020). Kinetic study of subcritical water extraction of flavonoids from *Citrus unshiu* peel. *Sep. Purif. Technol.*, *250*, 117259.
78. Halim, N. A. A., Abidin, Z. Z., Siajam, S. I., Hean, C. G., & Harun, M. R., (2020). Screening of factors influencing the yield of *Citrus hystrix* leaves essential oil extracted via pressurized hot water extraction based on resolution V fractional factorial design. *J. Food Process Eng.*, *43*(11).
79. Ko, M. J., Kwon, H. L., & Chung, M. S., (2016). Pilot-scale subcritical water extraction of flavonoids from satsuma mandarin (*Citrus unshiu* Markovich) peel. *Innov. Food Sci. Emerg. Technol.*, *38*, 175–181.
80. Gbashi, S., Adebo, O. A., Piater, L., Madala, N. E., & Njobeh, P. B., (2017). Subcritical water extraction of biological materials. *Sep. Purif. Rev.*, *46*(1), 21–34.
81. Liew, S. Q., Teoh, W. H., Tan, C. K., Yusoff, R., & Ngoh, G. C., (2018). Subcritical water extraction of low methoxyl pectin from pomelo (*Citrus grandis* (L.) Osbeck) peels. *Int. J. Biol. Macromol.*, *116*, 128–135.
82. Lachos-Perez, D., et al., (2018). Subcritical water extraction of flavanones from defatted orange peel. *J. Supercrit. Fluids, 138*, 7–16.
83. Cheng, Y., Xue, F., Yu, S., Du, S., & Yang, Y., (2021). Subcritical water extraction of natural products. *Molecules, 26*(13), 1–38.
84. Martínez, J. M., Delso, C., Maza, M., Álvarez, I., & Raso, J., (2018). Utilizing pulsed electric field processing to enhance extraction processes. *Ref. Modul. Food Sci.*, 1–7.
85. Poojary, M. M., Lund, M. N., & Barba, F. J., (2020). *Pulsed Electric Field (PEF) as an Efficient Technology for Food Additives and Nutraceuticals Development.* INC.

86. Lakka, A., Bozinou, E., Makris, D. P., & Lalas, S. I., (2021). Evaluation of pulsed electric field polyphenol extraction from *Vitis vinifera, Sideritis scardica* and *Crocus sativus. ChemEngineering, 5*(2).
87. Siddeeg, A., et al., (2019). Pulsed electric field-assisted ethanolic extraction of date palm fruits: Bioactive compounds, antioxidant activity and physicochemical properties. *Processes, 7*(9).
88. Redondo, D., Venturini, M. E., Luengo, E., Raso, J., & Arias, E., (2018). Pulsed electric fields as a green technology for the extraction of bioactive compounds from thinned peach by-products. *Innov. Food Sci. Emerg. Technol., 45*, 335–343.
89. Peiró, S., Luengo, E., Segovia, F., Raso, J., & Almajano, M. P., (2019). Improving polyphenol extraction from lemon residues by pulsed electric fields. *Waste and Biomass Valorization, 10*(4), 889–897.
90. Niu, D., Ren, E. F., Li, J., Zeng, X. A., & Li, S. L., (2021). Effects of pulsed electric field-assisted treatment on the extraction, antioxidant activity and structure of naringin. *Sep. Purif. Technol., 265*, 118480.
91. El Kantar, S., et al., (2018). Pulsed electric field treatment of citrus fruits: Improvement of juice and polyphenols extraction. *Innov. Food Sci. Emerg. Technol., 46*, 153–161.
92. Castro, I., Teixeira, J. A., Salengke, S., Sastry, S. K., & Vicente, A. A., (2004). Ohmic heating of strawberry products: Electrical conductivity measurements and ascorbic acid degradation kinetics. *Innov. Food Sci. Emerg. Technol., 5*(1), 27–36.
93. Kumar, T., (2018). A review on ohmic heating technology: Principle, applications and scope. *Int. J. Agric. Environ. Biotechnol., 11*(4).
94. Loypimai, P., Moongngarm, A., Chottanom, P., & Moontree, T., (2015). Ohmic heating-assisted extraction of anthocyanins from black rice bran to prepare a natural food colorant. *Innov. Food Sci. Emerg. Technol., 27*, 102–110.
95. Gavahian, M., & Chu, R., (2021). Design, development, and performance evaluation of an ohmic extractor to valorize fruit by-products based on Taguchi method: Reduced energy consumption and enhanced total phenolics. *J. Food Process Eng.*, 1–11.
96. Silva, V. L. M., Santos, L. M. N. B. F., & Silva, A. M. S., (2017). Ohmic heating: An emerging concept in organic synthesis. *Chem.–A Eur. J., 23*(33), 7853–7865.
97. Khan, M. K., Ahmad, K., Hassan, S., Imran, M., Ahmad, N., & Xu, C., (2018). Effect of novel technologies on polyphenols during food processing. *Innov. Food Sci. Emerg. Technol., 45*, 361–381.
98. Saberian, H., Hamidi-Esfahani, Z., Ahmadi, G. H., & Barzegar, M., (2017). Optimization of pectin extraction from orange juice waste assisted by ohmic heating. *Chem. Eng. Process. Process Intensif., 117*, 154–161.
99. Çilingir, S., Goksu, A., & Sabanci, S., (2021). Production of pectin from lemon peel powder using ohmic heating-assisted extraction process. *Food Bioprocess Technol., 14*(7), 1349–1360.
100. Sabanci, S., Çevik, M., & Göksu, A., (2021). Investigation of time effect on pectin production from citrus wastes with ohmic heating assisted extraction process. *J. Food Process Eng., 44*(6), e13689.
101. Van, H. P., Yen, N. N. H., Ting, L. Y., Lan, P. N. T., & Al Loman, A., (2020). Chemical composition and biological activities of extracts from pomelo peel by-products under enzyme and ultrasound-assisted extractions. *J. Chem.*
102. Patil, P. D., Patil, S. P., Kelkar, R. K., Patil, N. P., Pise, P. V., & Nadar, S. S., (2021). Enzyme-assisted supercritical fluid extraction: An integral approach to extract bioactive compounds. *Trends Food Sci. Technol., 116*, 357–369.

103. Liu, Z., Qiao, L., Yang, F., Gu, H., & Yang, L., (2017). Brönsted acidic ionic liquid-based ultrasound-microwave synergistic extraction of pectin from pomelo peels. *Int. J. Biol. Macromol., 94*, 309–318.
104. Hu, W., Chen, S., Wu, D., Zhu, K., & Ye, X., (2021). Manosonication assisted extraction and characterization of pectin from different citrus peel wastes. *Food Hydrocoll., 121*, 106952.
105. Xing, C., et al., (2019). Chemical composition and biological activities of essential oil isolated by HS-SPME and UAHD from fruits of bergamot. *LWT, 104*, 38–44.
106. Hu, W., Ye, X., Chantapakul, T., Chen, S., & Zheng, J., (2020). Manosonication extraction of RG-I pectic polysaccharides from citrus waste: Optimization and kinetics analysis. *Carbohydr. Polym., 235*, 115982.
107. Karanicola, P., et al., (2021). Ultrasound-assisted dilute acid hydrolysis for production of essential oils, pectin and bacterial cellulose via a citrus processing waste biorefinery. *Bioresour. Technol., 342*, 126010.
108. Dao, T. P., et al., (2019). Modeling and optimization of the orange leaves oil extraction process by microwave-assisted hydro-distillation: The response surface method based on the central composite approach (RSM-CCD model). *Rasayan J. Chem., 12*(2), 666–676.
109. Waheed, A., Akram, S., Ashraf, R., Mushtaq, M., & Adnan, A., (2020). Kinetic model and optimization for enzyme-assisted hydrodistillation of d-limonene-rich essential oil from orange peel. *Flavour Fragr. J., 35*(5), 561–569.
110. Hwang, H. J., Kim, H. J., Ko, M. J., & Chung, M. S., (2021). Recovery of hesperidin and narirutin from waste *Citrus unshiu* peel using subcritical water extraction aided by pulsed electric field treatment. *Food Sci. Biotechnol., 30*(2), 217–226.
111. Tunç, M. T., & Odabaş, H. İ., (2021). Single-step recovery of pectin and essential oil from lemon waste by ohmic heating assisted extraction/hydrodistillation: A multi-response optimization study. *Innov. Food Sci. Emerg. Technol., 74*, 102850.
112. Xin-An, Z., & Id, O. (2019). *Pulsed Electric Field-Ultrasonic Assisted Extraction Combined with Macroporous Resin for the Preparation of Flavonoids from* (pp. 1, 2).

CHAPTER 12

Isothermal Nucleic Acid Amplification Techniques: New Horizons in Microbial Food Safety Applications

A. ARUN PRINCE MILTON,[1] G. BHUVANA PRIYA,[2] SANDEEP GHATAK,[1] SAMIR DAS,[1] and M. CHENDU BHARAT PRASAD[1]

[1]*ICAR Research Complex for NEH Region, Umiam, Meghalaya, India*

[2]*College of Agriculture, Central Agricultural University (Imphal), Kyrdemkulai, Meghalaya, India*

ABSTRACT

Foodborne pathogens present major challenges to food safety. Foodborne diseases are one of the most prevailing health issues in the world with interference in the health of the population and the development of society [68]. In the food industry, to ensure microbial food safety, hazard analysis and critical control points (HACCP) and good hygiene practices (GHP) are implemented as a preventive strategy [12]. However, the function of pathogen detection is vital, as it is the key to the identification and prevention of problems. It is essential to control them and for that, early or rapid detection is very crucial [64]. Nucleic acid-based detection assays have become popular and one of the most regularly used platforms to routinely detect foodborne pathogens. Although PCR dominates molecular diagnostics, isothermal amplification methods are gaining momentum recently. There are a number of isothermal amplification methods deployed for the detection of foodborne pathogens. In this chapter, special attention has been given to loop-mediated isothermal amplification (LAMP), rolling circle amplification (RCA), saltatory rolling circle amplification (SRCA), nucleic acid sequence-based

Bioresources and Bioprocess in Biotechnology for a Sustainable Future. Leonardo Sepúlveda Torre, Juan Carlos Contreras-Esquivel, Ann Rose Abraham, & A. K. Haghi (Eds.)

amplification (NASBA), single primer isothermal amplification (SPIA), recombinase polymerase amplification (RPA), polymerase spiral reaction (PSR), helicase dependent amplification (HDA), cross-priming amplification (CPA), competitive annealing mediated isothermal amplification (CAMP), strand displacement amplification (SDA), and strand exchange amplification (SEA). A detailed appraisal of all the above techniques has been made taking into account the complexity of the assay, instrumentation/reagent necessities, detection limit/time, culture independence, repeatability, portability, etc.

12.1 INTRODUCTION

According to a 2015 WHO report, globally one in 10 people (i.e., 600 million people) become ill after ingesting contaminated food with microorganisms with an estimated death of 4,20,000 every year. Assuming the US data to the other parts of the world would mean that up to 1/3rd of the inhabitants in developed nations are suffering from foodborne illness each year, whereas the problem is possible to be more extensive in developing nations [26]. Among the 31 recognized pathogenic foodborne microbes, Norovirus, *Salmonella, Campylobacter, Clostridium perfringens*, *Staphylococcus aureus*, *Escherichia coli* O157:H7, *Listeria monocytogenes*, *Clostridium botulinum,* and *Vibrio* have been identified as the foremost etiological agents accountable for food-associated diseases, hospitalizations, and deaths [78]. Moreover, the consequences of foodborne pathogens notably hamper economic development through the onerous expenses on health care, loss of productivity during illnesses, and restricted sales of food products [8, 78].

However, for effective prevention, early and efficient detection is critical. Traditional culture-based methods for detecting the pathogenic bacteria present in food are by culturing them on agar plates, which is the gold standard as it yields a number of information about pathogens (number, serotype, etc.). However, they are more laborious, require technical expertise, and are time-consuming with the possibility of false-negative outcomes [8, 35]. Preliminary identification itself may require 2–3 days and for confirmation, more than a week is required [34]. In addition, cultural methods may be limited by their low sensitivity, and detection incapability if pathogens are injured or assumed VBNC state [18]. Rapid methods like immunological (ELISA, LFA, etc.) and biosensor-based approaches are more time-competent and labor-saving. However, more false-positive results, the delicate nature of reagents (antibodies), the requirement of specialized equipment like plate readers, and high investment costs have kept them outside the reach of many resource-limited settings [21].

As far as nucleic acid-based detection methods such as end-point PCR, and real-time PCR are concerned, the necessitation of sophisticated instruments (thermocycler, gel documentation system, etc.) are areas of concern [21]. Failure in detecting pathogens in foods would augment the transmission risk and may lead to fatalities, distribution of contaminated meat products, and unnecessary accumulation of products [34, 45]. Therefore, there is a necessity for detection techniques that satisfies the "ASSURED" principles ("affordable, sensitive, specific, user-friendly, rapid, and robust, equipment-free, and deliverable to end-users") [65]. Adoption of isothermal technology is rising in the microbial food safety field as it helps in precise detection of pathogens in varied food matrices in limited time and facilities.

12.2 ISOTHERMAL NUCLEIC ACID AMPLIFICATION TECHNIQUES

Isothermal amplification techniques amplify the target nucleic acids with the aid of enzymes such as *Bst* DNA polymerase, RNase H, helicase, avian myeloblastosis virus reverse transcriptase (AMVRT), T7 RNA polymerase, etc., with few methods requiring a single and few entails multiple enzymes [22, 31, 46]. The amplification process can be achieved by incubating the reaction mix at a single temperature and there is no obligation of thermal cycling at several temperatures [52]. The methods usually offer high specificity, rapidity, sensitivity, cost-effectiveness, and sophisticated equipment-free testing [47]. The visualization of the amplified products is possible with the naked eye as several dyes such as HNB, SYBR Green, etc., can facilitate the result interpretation [47]. These isothermal-based techniques can also be attached with some microdevices (microfluidic chips, lateral flow, centrifugal fluid devices) which are appropriate for on-site detection under field conditions [31, 32]. There are a number of isothermal amplification methods such as loop-mediated isothermal amplification (LAMP), rolling circle amplification (RCA), saltatory rolling circle amplification (SRCA), nucleic acid sequence-based amplification (NASBA), single primer isothermal amplification (SPIA), recombinase polymerase amplification (RPA), polymerase spiral reaction (PSR), helicase dependent amplification (HDA), cross-priming amplification (CPA), competitive annealing mediated isothermal amplification (CAMP), strand displacement amplification (SDA) and strand exchange amplification (SEA), that are deployed for the detection of foodborne pathogens and other food quality assessment testing [31, 47, 48, 57]. Different techniques entail different kinds of enzymes, various number of primers, and incubation times (Figure 12.1). Some of the constraints of these methods are false-positive results, background amplification, the requirement of multiple primers (for some methods) and complex primer designing process, cross-reactivity, self-hybridization, pre-processing (extraction of nucleic acid), and non-quantitative nature (except for few) [31]. Despite these constraints, these techniques are promising alternative nucleic acid amplification tests (NAAT) that are not only economical but also resource- and time savvy.

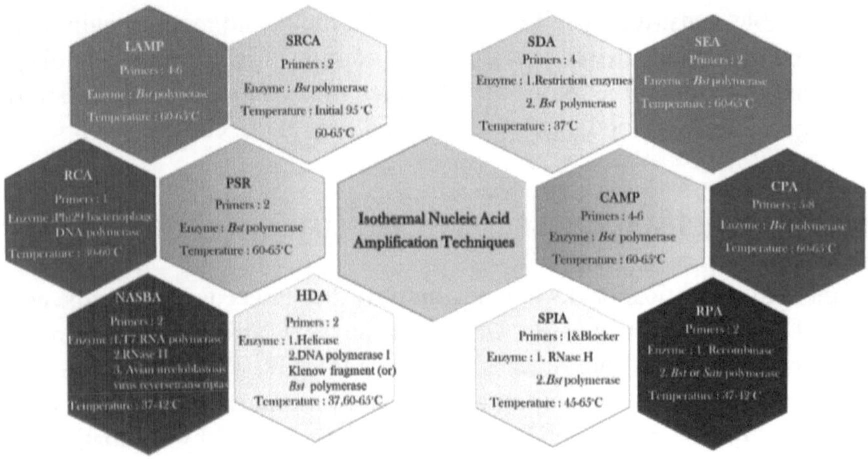

FIGURE 12.1 Requirements of enzymes, primer, and incubation time of various isothermal amplification techniques.

12.3 LOOP-MEDIATED ISOTHERMAL AMPLIFICATION (LAMP)

LAMP-based detection methods have been developed and used over the last decades [57, 84]. It requires 4–6 oligonucleotides and an enzyme (*Bst* DNA polymerase) that has an enhanced strand displacement action for isothermal amplification at 60–66°C. The primers for the LAMP assay are designed to spot 6 to 8 distinct zones of the target template [57]. During amplification, the pyrophosphate ions are generated that act with magnesium ions to produce magnesium pyrophosphate (white precipitates). This white precipitate is a signal for the positive reaction [54]. It is also feasible to visualize LAMP results with the naked eye by means of dyes such as calcein, HNB, and SYBR Green I (31). The result interpretation is also done by other real-time and end-point detection approaches such as real-time turbidimetry, probes, and gel electrophoresis. The LAMP assay can also be integrated into the sensor-based and lab-on-a-chip-type devices for point-of-care detection purposes [31, 32]. It is the most used and studied isothermal method for the analysis of microbiological food safety and a few assays are also commercially available. Eiken Chemical, Co., Ltd. (Tokyo, Japan) commercialized the Loopamp® *Campylobacter* detection kit, which requires specific equipment and pre-enrichment. The Loopamp® *Campylobacter* kit has been described as convenient for testing *C. jejuni*/*C. coli* contaminated chicken meat [84]. The analysis presented that the LAMP test achieved 98.5% sensitivity related to conventional culture methods, and the operation time was 23.5–25.5 h including the enrichment step, in contrast with the conventional method which requires 4–6 days. LAMP was deployed to screen foodborne *Salmonella, Staphylococcus aureus, Clostridium perfringens, Escherichia coli* O157:H7, etc. [42, 61–63]. In a recent study [44], an easy-to-perform LAMP assay was developed and evaluated. The method was capable of discriminating between test samples that are either below or above the industry (Australia) set the target for *Campylobacter*, i.e., 6000 CFU/carcass corresponding to 12 CFU/ml chicken rinse with >90% accuracy comparative to culture methods.

12.4 ROLLING CIRCLE AMPLIFICATION (RCA)

The rolling circle amplification (RCA) technique exploits the extremely processive polymerase activity of the Phi29 bacteriophage DNA polymerase and strand displacement activity. It operates at a constant temperature generally between 30 and 60°C [11, 17, 56]. The oligonucleotides anneal to the circular single-stranded DNA (target) and the replication begins with the aid of polymerase enzyme in a circular way. By elongation, it completes a loop of nucleotide sequence and reaches the initiation point. The strand displacement enzyme permits the newly synthesized strand to uninterruptedly displace the formerly formed strand as polymerization advances. The synthesis of the subsequent loop begins in an analogous manner and the product of amplification is a single-stranded long-chain DNA having several simultaneous complementary repeats to the circular template [11, 37, 39]. In linear/single primer RCA, the end product is linear for DNA/RNA targets [11]. The identification in the case of RCA method is accomplished with the aid of a padlock probe. The probes (padlock) comprise two specific sequences, one for complementary binding with the target sequence, the other for making a circular template and providing room for the hybridization of primers [81]. In addition, this technique can also be used at 30°C or room temperature and completed in 1–2 h [81]. RCA-based detection methods have been deployed for the detection of food-transmitted pathogens such as *E. coli* O157: H7, *Salmonella, Listeria monocytogenes, Staphylococcus aureus*, etc. [24, 30,

40, 43]. A chemiluminescent aptasensor-based RCA method was developed for screening *Salmonella typhimurium* contamination in meat samples. The limits of detection of the method were 10 CFU/mL and the method was also evaluated in real-world pork samples for the detection of *S. typhimurium* [24].

12.5 SALTATORY ROLLING CIRCLE AMPLIFICATION (SRCA)

SRCA is an innovative isothermal testing approach, which is a shortened form of the RCA initiated by Lizardi et al. [39], and it requires initial thermal denaturation. Since *Bst* DNA polymerase can generate template-independent DNA, it synthesizes at the connection of non-closed circular DNA and progresses to synthesize DNA at the principal primer binding site, simply like saltating a nick [97]. After that, SRCA exhibits an analogous amplification sequence to RCA. Contrasting RCA, SRCA does not need ligase and padlock probes to consummate the cyclization [90, 91]. SRCA is a blend of isothermal amplification and endpoint PCR (needing a couple of PCR primers). It does not demand a thermocycler, gel electrophoresis, or documentation system. The SRCA-based methods have been developed for a few pathogens of foodborne significance such as *Shigella* [77], *S. aureus* [90], *Clostridium perfringens* [48], *C. jejuni* [49] *Salmonella* [51]. A recently developed SRCA method for screening *C. Perfringens* contamination in meat samples displayed an analytical sensitivity and limit of detection of 80 fg, and 80 CFU/g of pork. The developed SRCA method was 10-fold more sensitive than endpoint and real-time PCR assays [49].

12.6 NUCLEIC ACID SEQUENCE-BASED AMPLIFICATION (NASBA)

NASBA is a type of isothermal amplification technique, which can restrictively do mRNA amplification even in the presence of genomic DNA [9, 69]. NASBA reaction mix requires one forward and reverse primer and three kinds of enzymes namely, T7 RNA polymerase, RNase H, and AVMRT. The 5′ end of the forward oligos has a T7 promoter sequence and complementary sequence to the RNA at the 3′ end. The forward oligos fix to the target sequence and extension occurs with the aid of AVMRT to produce a hybrid strand (RNA-DNA). RNase H degrades the RNA strand of the hybrid to generate an ssDNA copy. The reverse primer fixes to this ssDNA and with the aid of AVMRT, a dsDNA along with the T7 promoter sequence is synthesized. Subsequently, T7 RNA polymerase binds to the dsDNA to generate a copy of the opposite sense ssRNA of the initial target (complementary) RNA. Then, the reaction arrives in the cyclic phase to synthesize more complementary RNA strands [31, 86). As NASBA enzymes are heat labile, amplification can be accomplished at a comparatively low temperature, with ideal conditions of 41°C for 90–120 min [7]. The results can be interpreted using agarose gel electrophoresis, ELISA, enzyme-linked gel assay, electrochemiluminescence, and real-time detection with molecular beacons [3, 53]. NASBA also has the potential to detect viable pathogens in various food matrices [10]. NASBA-based detection methods have been deployed for the detection of foodborne pathogens such as *C. jejuni, L. monocytogenes, Salmonella*, etc. [55, 71, 92]. A molecular beacon-real-time NASBA assay was developed for testing *L. monocytogenes* contamination up to 40 cells/reaction [55]. Commercial NASBA kits are also available for the testing of *L. monocytogenes, Salmonella*, and *V. cholerae* [13, 19, 55].

12.7 SINGLE PRIMER ISOTHERMAL AMPLIFICATION (SPIA)

SPIA is an isothermal amplification-based detection method that requires a single chimeric primer (DNA fragment at the 3′ end and an RNA fragment at the 5′ end), RNase H, DNA polymerase, and a blocker oligo [33]. Initially, the chimeric primers bind to their target site on the DNA and are amplified by the DNA polymerase. RNase H cleaves the RNA fragment of the amplified product which exposes the primer-binding site for the new chimeric primer to bind. DNA polymerase with its strong strand-displacement activity produces a large amount of ssDNA from the previous product. The blocker oligo that possesses a sequence complementary to the target sequence of the primer binding site is to stop the extension of the primer. After binding to the target sequence, the blocker oligo cannot extend as its 3′ end is modified. The visualization of results can be done by gel electrophoresis, SYBR Green II dye and real-time monitoring [31, 33, 88]. SPIA-based methods have been used for the detection of food-associated pathogens such as *Salmonella, L. monocytogenes, S. aureus, Vibrio parahaemolyticus*, etc. [75, 87–89]. A recently established SPIA assay for screening *V. parahaemolyticus* contamination displayed the limit of detection of 5.7 CFU/mL and 42 CFU/g in pure culture and raw oysters, respectively. On applying to real-world samples, the SPIA approach revealed 98.21% specificity, 100% sensitivity, and 99.17% accuracy in detecting *V. parahaemolyticus* from contaminated seafood samples [88].

12.8 RECOMBINASE POLYMERASE AMPLIFICATION (RPA)

RPA is an isothermal amplification technique that principally depends on two enzymes comprising, recombinase, and DNA polymerase, and a single-stranded DNA binding protein (SSB). Initially, the recombinase closely combines with the primers of DNA to produce a complex (primer-recombinase). The complex enables strand exchange at similar sites on dsDNA and facilitates the primer insertion to the complementary dsDNA followed by PCR-like extension of the target sequence. The unwinding of the template occurs due to SSB, and a new complementary strand of DNA originates because of the polymerase. RPA amplification can be achieved at a persistent low temperature (37–42°C) and 10–20 min run time without any denaturation [6, 31, 59]. The visualization of the amplification can be accomplished either by endpoint or in real-time. The real-time visualization of amplification is done using sensor-based devices or exo-probes (fluorescence probes). An appreciated feature of the RPA method is the availability (TwistDX, Cambridge, UK) of a variety of ready-to-use kits in the market that contains the reagents for both endpoint (lateral flow/gel electrophoresis) and real-time assays. This makes the RPA system instrument-independent and increases the portability of the tests [11, 31, 70]. RPA-based detection methods are established for testing foodborne pathogens such as *Campylobacter, L. monocytogenes, S. aureus, Salmonella*, etc. [23, 28, 29]. A duplex RPA assay developed for concurrent detection of *L. monocytogenes* and *S. aureus* revealed a sensitivity of 10^2 copies/μL and after 4 h incubation, the limit of detection improved to 270 CFU/tube and 230 CFU/tube for *S. aureus* and *L. monocytogenes*, respectively [23]. Commercial RPA kit (TwistAmp™ Basic kit, TwistDX, UK) is available for testing of *L. monocytogenes, S. aureus,* and *Salmonella* contamination [1, 27].

12.9 POLYMERASE SPIRAL REACTION (PSR)

PSR is a simple isothermal amplification method that requires a single pair of primer and a DNA polymerase enzyme. Two primers with reverse sequences (exogenous/botanic origin) integrated at the 5′ end of the primers of the target gene are required for the PSR system [97]. When the temperature reaches 61–65°C, the ds DNA of the template unlocks due to betaine. The actual target gene's forward primer segment anneals to one ss DNA and extends and forms a structure A. After the melting of structure A, the target gene's reverse primer segment hybridizes with it and extends to form structure B. The double strand of structure B melts and forms a single chain. As the exogenous gene primer attached at the 5′ end of the actual target gene primers is simple reverse to each other, the structure B curls to form a structure C. The 3′ end of complementary to reverse exogenous sequence continues to extend and forms a spiral structure. The mechanism of PSR is clearly explained by Liu et al. [38]. Various formats of visualization including gel electrophoresis and dye based (SYBR Green and HNB dye) detection can be employed to interpret the results, making PSR suitable for field application [46]. PSR method has been deployed for the detection of foodborne pathogens such as *Salmonella, S. aureus, C. perfringens, Campylobacter, V. parahaemolyticus*, etc. [25, 45, 47, 48, 83]. A recently established PSR assay for screening pork for *Campylobacter coli* contamination revealed analytical sensitivity and detection limit of 62 fg DNA/tube and 100 CFU/g pork, respectively [45].

12.10 HELICASE DEPENDENT AMPLIFICATION (HDA)

HDA is another isothermal amplification technique closely resembling *in vivo* DNA replication. Helicase enzyme is used in place of heat for strand separation, which obviates the necessity of thermal denaturation and the thermal cycling process. The process begins with the unwinding of dsDNA by helicase followed by binding of forward and reverse primer and elongation mediated by a polymerase enzyme. After elongation, the newly produced dsDNA is unwinded by helicase and the exponential amplification takes place. SSBP is added to the reaction blend to stabilize the ssDNA by preventing the reassociation of separated complementary DNA strands. The HAD protocol usually operates in the temperature range of 60–65°C and runtime range of 60–120 min [11, 31]. There are two types of HDA, viz. thermophilic (tHDA) and mesophilic (mHDA), the former operates at 37°C with UvrD helicase (*E. coli*-derived) and DNA polymerase I Klenow fragment, and the latter operates at 60–65°C in the presence of Tte-UvrD (*Thermoanaerobacter tengcongenesis* derived) and *Bst*DNA polymerase. In the tHDA method, stabilization of ssDNA is not required [2, 5, 72]. Therefore, a pair of primers, two enzymes (helicase and polymerase), and a visualization system are sufficient to establish the HDA method. Any visualization system such as fluorescence, sensor-based, electrochemical-based, molecular beacons, hairpin probes, immunochromatography, LFA, microdevices, microchips, etc., can be integrated with the HDA system for the detection of amplified products [5, 31, 66]. The HDA technique was effectively used to detect a variety of foodborne pathogens such as *Listeria monocytogenes, Salmonella, Cronobacter*, etc. [14, 60, 80]. A tHDA-based detection method developed for detecting *Cronobacter* showed a detection limit of 10^0 and 10^1 CFU/ml *Cronobacter* in pure culture and artificially contaminated powdered infant formula, respectively [80].

12.11 CROSS-PRIMING AMPLIFICATION (CPA)

CPA is another isothermal amplification method that necessitates multiple primers and probes. DNA polymerase enzyme displaces the strand and facilitates the initiation of the reaction. There are two forms of CPA, namely single cross-priming amplification (sCPA) and double cross-priming amplification (dCPA) which require single and double-cross primers, respectively. The cross primer comprises complementary sequences to target nucleic acid on one side and non-complementary sequences on the other side. The primer, which is upstream of the cross primer, binds, and extends, thus displacing the cross-primer strand. The annealing spot for other oligos is added at the cross primer's non-complementary side. Accordingly, conjunctive annealing proceeds on this strand with a different set of primers. In the extension stage, these strands replace each other to synthesize numerous shorter fragments and amplification progresses exponentially. The CPA can be integrated with LFA and immunochromatographic strips for result interpretation [15, 31, 82, 96]. The CPA was deployed for screening contamination of a few foodborne pathogens including *L. monocytogenes, Yersinia enterocolitica, Bacillus cereus*, etc. [76, 93, 94]. The CPA method integrated with pre-enrichment (37°C for 24 h) and immunoblotting for detection of *Y. enterocolitica* achieved a detection limit of 10 CFU/ml and 100 CFU/100 g of pure bacterial culture and milk powder, respectively [93].

12.12 COMPETITIVE ANNEALING MEDIATED ISOTHERMAL AMPLIFICATION (CAMP)

Similar to LAMP, this technique banks on a DNA polymerase-mediated auto-cycling DNA synthesis and strand displacement. The CAMP reaction requires a minimum of two pairs of specially designed primers and loop primers can also be added for accelerating the reaction. The main pairs comprise an N-linked forward primer (NF) and the complementary one of an N-linked (Nc) reverse primer (NR). The NF and NR sequences can be obtained from PCR primers. The other primer pair (F1c and R1) are complementary to the forward primer (FP) and the reverse primer (RP) of a normal

PCR reaction, respectively. The CAMP reaction with *Bst* DNA polymerase is incubated at 60–65°C. The results of CAMP amplification can be interpreted with the help of both endpoint and real-time monitoring [4, 36, 41]. A few CAMP-based assays have been standardized for the detection of foodborne pathogens such as *L. monocytogenes* [36] and *B. cereus* [4]. The CAMP assay for screening *B. cereus* contamination in milk displayed a detection limit of 59 CFU/mL of artificially spiked milk [4]. The CAMP assay combined with the enrichment (37°C for 16 h) step could consistently detect up to 1 CFU/mL *L. monocytogenes* in artificially spiked milk samples [36].

12.13 STRAND DISPLACEMENT AMPLIFICATION (SDA)

SDA depends on bifunctional primers integrating both target identification and endonucleases target segments. This method works with initial heat denaturation followed by the addition of restriction enzymes and polymerase and incubation at 37°C. After separation of the strand, the bifunctional primers extend adding the restriction target into the amplicon. The amplicon is released by binding and extension of bumper primers. Consecutive rounds of primer binding produce dsDNA integrating restriction sites and the restriction endonucleases act on those sites to nick a single strand from the newly synthesized duplex strand. Following nicking, the polymerase displaces the present strand and incorporates a fresh amplicon. This nick and run mechanism repeat and results in an exponential amplification [11, 73, 74]. SDA-based detection methods have been deployed for the detection of foodborne pathogens such as *Salmonella, E. coli* O157:H7, *S. aureus*, etc. [15, 58, 79]. A lateral flow biosensor-based aptamer mediated SDA method for detection of *E. coli* O157:H7 could detect as low as 10 CFU per test [79].

12.14 STRAND EXCHANGE AMPLIFICATION (SEA)

SEA is another innovative isothermal method that entails only a pair of oligos and an enzyme (*Bst* DNA polymerase). Under the isothermal condition, the single-stranded denaturation bubbles of dsDNA are invaded by a pair of explicit primers, and extension of the original duplex happens due to the presence of DNA polymerase leading to the generation of ds DNA products. In the meantime, the previously formed dsDNA amplicons are separated, hybridized, and extended for several rounds for further synthesizing of multiple dsDNA amplicons. The SEA technique can also detect RNA without the requirement of reverse transcription which is usually required in other isothermal amplification techniques [20, 31, 95]. The SEA method has been applied for the detection of foodborne pathogens such as *L. monocytogenes* [95] and *Cronobacter* [20]. With SEA based method, *Cronobacter* spp. in the range of 10^1–10^6 CFU/mL can be precisely determined with the visual detection limit of 10 CFU/mL within 1 h time in the infant formula milk powder [20].

12.15 CONCLUSION

In this chapter, we discussed most of the isothermal amplification techniques that have got applications in testing foodborne pathogens in various food matrices. Each technique has its own merits and demerits. For example, the SRCA method has the advantage of using a simple PCR primer, instead of specially designed primers like in the case of CAMP. However, the same SRCA method demands a heat-mediated initial denaturation step for strand separation. Thus, picking any method for testing food samples should be made after analyzing all the advantages and disadvantages vis-à-vis the available resources and facilities of the laboratory. And as researchers throughout the globe are striving to bring out more simple and handy techniques, we can expect some more exciting addition of robust and powerful isothermal amplification techniques to the bunch of techniques available today.

KEYWORDS

- **cross-priming amplification**
- **detection**
- **foodborne pathogens**
- **helicase dependent amplification**
- **isothermal amplification**
- **molecular diagnostics**
- **nucleic acid**
- **single primer isothermal amplification**
- **strand displacement amplification**

REFERENCES

1. James, A., & Macdonald, J., (2015). Recombinase polymerase amplification: Emergence as a critical molecular technology for rapid, low-resource diagnostics. *Expert Rev. Mol. Diagn., 15*, 1475–1489.
2. An, L., Tang, W., Ranalli, T. A., Kim, H. J., Wytiaz, J., & Kong, H., (2005). Characterization of a thermostable UvrD helicase and its participation in helicase-dependent amplification. *J. Biol. Chem., 280*, 28952–28958.
3. Asadi, R., & Mollasalehi, H., (2021). The mechanism and improvements to the isothermal amplification of nucleic acids, at a glance. *Anal. Biochem., 631*, 114260.
4. Bao, H., Li, W., Yue, X., Wu, J., Qiao, Y., Peng, Q., Shi, B., et al., (2020). Rapid and simple detection of *Bacillus cereus* in milk by real-time competitive annealing mediated isothermal amplification. *The Analyst., 145*, 6677–6682.
5. Barreda-García, S., Miranda-Castro, R., Santos-Álvarez, N., Miranda-Ordieres, A. J., & Lobo-Castañón, M. J., (2018). Helicase-dependent isothermal amplification: A novel tool in the development of molecular-based analytical systems for rapid pathogen detection. *Anal. Bioanal. Chem., 410*, 679–693.
6. Boyle, D. S., Lehman, D. A., Lillis, L., Peterson, D., Singhal, M., Armes, N., & Overbaugh, J., (2013). Rapid detection of HIV-1 proviral DNA for early infant diagnosis using recombinase polymerase amplification. *MBio., 4*, e00135–13.
7. Brennan, D., Coughlan, H., Clancy, E., Dimov, N., Barry, T., Kinahan, D., & Galvin, P., (2017). Development of an on-disc isothermal *in vitro* amplification and detection of bacterial RNA. *Sens. Actuators B Chem., 239*, 235–242.
8. Campbell, V. R., Carson, M. S., Lao, A., Maran, K., Yang, E. J., & Kamei, D. T., (2021). Point-of-need diagnostics for foodborne pathogen screening. *SLAS Technol., 26*, 55–79.
9. Compton, J., (1991). Nucleic acid sequence-based amplification. *Nature, 350*, 91, 92.
10. Cook, N., (2003). The use of NASBA for the detection of microbial pathogens in food and environmental samples. *J. Microbiol. Methods., 53*, 165–174.
11. Craw, P., & Balachandran, W., (2012). Isothermal nucleic acid amplification technologies for point-of-care diagnostics: A critical review. *Lab on a Chip., 12*, 2469–2486.
12. Cusato, S., Tavolaro, P., & Oliveira, C. A. F. D., (2012). Implementation of hazard analysis and critical control points system in the food industry: Impact on safety and the environment. In: *Novel Technologies in Food Science* (pp. 21–37). Springer, New York, NY.
13. D'Souza, D. H., & Jaykus, L. A., (2003). Nucleic acid sequence based amplification for the rapid and sensitive detection of *Salmonella enterica* from foods. *J. Appl. Microbiol., 95*, 1343–1350.
14. Du, X. J., Zhou, T. J., Li, P., & Wang, S., (2017). A rapid *Salmonella* detection method involving thermophilic helicase-dependent amplification and a lateral flow assay. *Mol. Cell. Probes., 34*, 37–44.
15. Fang, Z., Wu, W., Lu, X., & Zeng, L., (2014). Lateral flow biosensor for DNA extraction-free detection of *Salmonella* based on aptamer mediated strand displacement amplification. *Biosens. Bioelectron., 56*, 192–197.
16. Feng, T., Li, S., Wang, S., & Pan, J., (2018). Cross priming amplification with nucleic acid test strip analysis of mutton in meat mixtures. *Food Chemistry, 245*, 641–645.
17. Fire, A., & Xu, S. Q., (1995). Rolling replication of short DNA circles. *Proc. Natl. Acad. Sci. Unit. States Am., 92*, 4641–4645.

18. Foddai, A. C., & Grant, I. R., (2020). Methods for detection of viable foodborne pathogens: Current state-of-art and future prospects. *Appl. Microbiol. Biotechnol., 104*, 4281–4288.
19. Fykse, E. M., Skogan, G., Davies, W., Olsen, J. S., & Blatny, J. M., (2007). Detection of *Vibrio cholerae* by real-time nucleic acid sequence-based amplification. *Appl Environ. Microbiol., 73*, 1457–1466.
20. Gao, Y., Ye, Y., Xu, J., Wu, Q., Yao, B., & Chen, W., (2021). Rapid and easy quantitative identification of *Cronobacter* spp. in infant formula milk powder by isothermal strand-exchange-amplification based molecular capturing lateral flow strip. *Food Control, 126*, 108048.
21. Ghatak, S., (2020). Strategies for elimination of foodborne pathogens, their influensive detection techniques and drawbacks. In: *Meat Quality Analysis* (pp. 267–286). Academic Press.
22. Gill, P., & Ghaemi, A., (2008). Nucleic acid isothermal amplification technologies-a review. *Nucleos. Nucleot. Nucl., 27*, 224–243.
23. Guo, Z., Chen, Z., Liu, X., Chen, J., Chen, G., Liu, Z., & Lan, Q., (2019). Detection of *Listeria monocytogenes* and *Staphylococcus aureus* via duplex recombinase polymerase amplification. *J. Food Saf., 39*, e12628.
24. Hao, L., Gu, H., Duan, N., Wu, S., Ma, X., Xia, Y., & Wang, Z., (2017). A chemiluminescent aptasensor based on rolling circle amplification and Co^{2+}/N-(aminobutyl)-N-(ethylisoluminol) functional flowerlike gold nanoparticles for *Salmonella typhimurium* detection. *Talanta., 164*, 275–282.
25. He, S., Jang, H., Xu, K., Wang, J., Pang, B., Si, X., Jin, M., et al., (2020). Rapid visualized isothermal nucleic acid testing of *Vibrio parahaemolyticus* by polymerase spiral reaction. *Anal. Bioanal. Chem., 412*, 93–101.
26. Kaferstein, F., & Abdussalam, M., (1999). Food safety in the 21st century. *Bull. WHO, 77*, 347–351.
27. Kersting, S., Rausch, V., Bier, F. F., & Nickisch-Rosenegk, M. V., (2014). Multiplex isothermal solid-phase recombinase polymerase amplification for the specific and fast DNA-based detection of three bacterial pathogens. *Microchim. Acta., 181*, 1715–1723.
28. Kim, J. Y., & Lee, J. L., (2016). Rapid detection of *Salmonella enterica* serovar enteritidis from eggs and chicken meat by real-time recombinase polymerase amplification in comparison with the two-step real-time PCR. *J. Food Saf., 36*, 402–411.
29. Kim, J. Y., & Lee, J. L., (2017). Development of a multiplex real-time recombinase polymerase amplification (RPA) assay for rapid quantitative detection of *Campylobacter coli* and *jejuni* from eggs and chicken products. *Food Control, 73*, 1247–1255.
30. Kordas, A., Papadakis, G., Milioni, D., Champ, J., Descroix, S., & Gizeli, E., (2016). Rapid *Salmonella* detection using an acoustic wave device combined with the RCA isothermal DNA amplification method. *Sens. Bio-Sens. Res., 11*, 121–127.
31. Kumar, Y., (2021). Isothermal amplification-based methods for assessment of microbiological safety and authenticity of meat and meat products. *Food Control., 121*, 107679.
32. Kumar, Y., Bansal, S., & Jaiswal, P., (2017). Loop-mediated isothermal amplification (LAMP): A rapid and sensitive tool for quality assessment of meat products. *Compr. Rev. Food Sci. Food Saf., 16*, 1359–1378.
33. Kurn, N., Chen, P., Heath, J. D., Kopf-Sill, A., Stephens, K. M., & Wang, S., (2005). Novel isothermal, linear nucleic acid amplification systems for highly multiplexed applications. *Clin. Chem., 51*, 1973–1981.

34. Law, J. W. F., Ab Mutalib, N. S., Chan, K. G., & Lee, L. H., (2015). Rapid methods for the detection of foodborne bacterial pathogens: Principles, applications, advantages and limitations. *Front. Microbiol.*, *5*, 770.
35. Lee, N., Kwon, K. Y., Oh, S. K., Chang, H. J., Chun, H. S., & Choi, S. W., (2014). A multiplex PCR assay for simultaneous detection of *Escherichia coli* O157: H7, *Bacillus cereus*, *Vibrio parahaemolyticus*, *Salmonella spp.*, *Listeria monocytogenes*, and *Staphylococcus aureus* in Korean ready-to-eat food. *Foodborne Pathog. Dis.*, *11*, 574–580.
36. Li, W., Mao, R., Yue, X., Wu, J., Wu, R., Qiao, Y., & Du, Y., (2020). Competitive annealing mediated isothermal amplification (CAMP) for rapid and simple detection of *Listeria monocytogenes* in milk. *Food Control*, *117*, 107347.
37. Liu, D., Daubendiek, S. L., Zillman, M. A., Ryan, K., & Kool, E. T., (1996). Rolling circle DNA synthesis: Small circular oligonucleotides as efficient templates for DNA polymerases. *J. Am. Chem. Soc.*, *118*, 1587–1594.
38. Liu, W., Dong, D., Yang, Z., Zou, D., Chen, Z., Yuan, J., & Huang, L., (2015). Polymerase spiral reaction (PSR): A novel isothermal nucleic acid amplification method. *Sci. Rep.*, *5*, 12723–12730.
39. Lizardi, P. M., Huang, X., Zhu, Z., Bray-Ward, P., Thomas, D. C., & Ward, D. C., (1998). Mutation detection and single-molecule counting using isothermal rolling-circle amplification. *Nat. Gent.*, *19*, 225–232.
40. Long, Y., Zhou, X., & Xing, D., (2011). Sensitive and isothermal electrochemiluminescence gene-sensing of *Listeria monocytogenes* with hyperbranching rolling circle amplification technology. *Biosens. Bioelectron.*, *26*, 2897–2904.
41. Mao, R., Qi, L., Li, J., Sun, M., Wang, Z., & Du, Y., (2018). Competitive annealing mediated isothermal amplification of nucleic acids. *Analyst*, *143*, 639–642.
42. Maruyama, F., Kenzaka, T., Yamaguchi, N., Tani, K., & Nasu, M., (2003). Detection of bacteria carrying the *stx 2* gene by in situ loop-mediated isothermal amplification. *Appl Environ. Microbiol.*, *69*, 5023–5028.
43. Maruyama, F., Kenzaka, T., Yamaguchi, N., Tani, K., & Nasu, M., (2005). Visualization and enumeration of bacteria carrying a specific gene sequence by in situ rolling circle amplification. *Appl. Environ. Microbio.*, *71*, 7933–7940.
44. Mason, M. G., Blackall, P. J., Botella, J. R., & Templeton, J. M., (2020). An easy-to-perform, culture-free *Campylobacter* point-of-management assay for processing plant applications. *J. Appl. Microbiol.*, *128*, 620–629.
45. Milton, A. A. P., Momin, K. M., Priya, G. B., Das, S., Angappan, M., Sen, A., Sinha, D. K., et al., (2022). Development and evaluation of a novel polymerase spiral reaction-based testing technique for same-day visual detection of *Campylobacter coli* in pork. *Food Microbiol.*, 104066.
46. Milton, A. A. P., Momin, K. M., Priya, G. B., Das, S., Angappan, M., Sen, A., et al., (2021a). Development of a novel visual detection technique for *Campylobacter jejuni* in chicken meat and caecum using polymerase spiral reaction (PSR) with pre-added dye. *Food Control.*, *126*, 108064.
47. Milton, A. A. P., Momin, K. M., Ghatak, S., Priya, G. B., Angappan, M., Das, S., Sanjukta, R. K., et al., (2021b). Development of a novel polymerase spiral reaction (PSR) assay for rapid and visual detection of *Clostridium perfringens* in meat. *Heliyon.*, *7*, e05941.
48. Milton, A. A. P., Momin, K. M., Ghatak, S., Thomas, S. C., Priya, G. B., Angappan, M., Das, S., et al., (2021c). Development of a novel polymerase spiral reaction (PSR) assay for rapid and visual detection of *Staphylococcus aureus* in meat. *LWT*, *139*, 110507.

49. Milton, A. A. P., Momin, K. M., Priya, G. B., Ghatak, S., Gandhale, P. N., Angappan, M., Das, S., & Sen, A., (2021d). A novel in situ methodology for visual detection of *Clostridium perfringens* in pork harnessing saltatory rolling circle amplification. *Anaerobe*, *69*, 102324.
50. Milton, A. A. P., Momin, K. M., Priya, G. B., Das, S., Angappan, M., Sen, A., & Ghatak, S., (2021e). Novel saltatory rolling circle amplification assay for rapid and visual detection of *Campylobacter jejuni* in chicken meat. *LWT*, *149*, 111807.
51. Milton, A. A. P., Momin, K. M., Priya, G. B., Ghatak, S., Das, S., Gandhale, P. N., & Sen, A., (2021f). Development of novel visual detection methodology for *Salmonella* in meat using saltatory rolling circle amplification. *J. Appl. Microbiol.*, *131*, 2361–2371.
52. Momin, K. M., Milton, A. A. P., Ghatak, S., Thomas, S. C., Priya, G. B., Das, S., & Sen, A., (2020). Development of a novel and rapid polymerase spiral reaction (PSR) assay to detect *Salmonella* in pork and pork products. *Mol. Cell. Probes.*, *50*, 101510.
53. Moon, Y. J., Lee, S. Y., & Oh, S. W., (2022). A review of isothermal amplification methods and food-origin inhibitors against detecting foodborne pathogens. *Foods.*, *11*, 322.
54. Mori, Y., Nagamine, K., Tomita, N., & Notomi, T., (2001). Detection of loop-mediated isothermal amplification reaction by turbidity derived from magnesium pyrophosphate formation. *Biochem. Biophys. Res. Commun.*, *289*, 150–154.
55. Nadal, A., Coll, A., Cook, N., & Pla, M., (2007). A molecular beacon-based real time NASBA assay for detection of *Listeria monocytogenes* in food products: Role of target mRNA secondary structure on NASBA design. *J Microbiol Methods.*, *68*, 623–632.
56. Nilsson, M., Malmgren, H., Samiotaki, M., Kwiatkowski, M., Chowdhary, B. P., & Landegren, U., (1994). Padlock probes: Circularizing oligonucleotides for localized DNA detection. *Science*, *265*, 2085–2088.
57. Notomi, T., Okayama, H., Masubuchi, H., Yonekawa, T., Watanabe, K., Amino, N., & Hase, T., (2000). Loop-mediated isothermal amplification of DNA. *Nucleic Acids Res.*, *28*, e63.
58. Pang, Y. H., Wang, Y. X., Chen, W. M., Shen, X. F., & Zhang, Y., (2021). Cyclic strand displacement polymerase reaction to turn-on molecular beacons for rapid detection of *Staphylococcus aureus*. *Food Biosci.*, *44*, 101405.
59. Piepenburg, O., Williams, C. H., Stemple, D. L., & Armes, N. A., (2006). DNA detection using recombination proteins. *PLoS Biol.*, *4*, e204.
60. Pisamayarom, K., & Chaumpluk, P., (2017). Rapid *Listeria monocytogenes* assay based on helicase dependent amplification (HDA) and nucleic acid hybridization in blue silver nanoplates. *Int. J. Res. Granthaalayah.*, *5*, 322–335.
61. Priya, G. B., Agrawal, R. K., Milton, A. A. P., Mendiratta, S. K., Singh, B. R., Kumar, D., & Gandham, R. K., (2021). Isothermal amplification assay for visual on-site detection of *Staphylococcus aureus* in chevon. *Food Biotechnol.*, *35*, 221–236.
62. Priya, G. B., Agrawal, R. K., Milton, A. A. P., Mishra, M., Mendiratta, S. K., Luke, A., & Rajkhowa, S., (2020). Rapid and visual detection of *Salmonella* in meat using invasin A (invA) gene-based loop-mediated isothermal amplification assay. *LWT*, *126*, 109262.
63. Priya, G. B., Agrawal, R. K., Milton, A. A. P., Mishra, M., Mendiratta, S. K., Agarwal, R. K., & Kumar, D., (2018). Development and evaluation of isothermal amplification assay for the rapid and sensitive detection of *Clostridium perfringens* from chevon. *Anaerobe.*, *54*, 178–187.
64. Priyanka, B., Patil, R. K., & Dwarakanath, S., (2016). A review on detection methods used for foodborne pathogens. *Indian J. Med. Res.*, *144*, 327–338.

65. Puiu, M., & Bala, C., (2020). Microfluidics-integrated biosensing platforms as emergency tools for on-site field detection of foodborne pathogens. *TrAC–Trends Anal. Chem., 125*, 115831.
66. Qi, H., Yue, S., Bi, S., Ding, C., & Song, W., (2018). Isothermal exponential amplification techniques: From basic principles to applications in electrochemical biosensors. *Biosens. Bioelectron., 110*, 207–217.
67. Mao, R., Qi, L., Li, J., Sun, M., Wang, Z., & Du, Y., (2018). Competitive annealing mediated isothermal amplification of nucleic acids. *Analyst., 143*, 639–642.
68. Schlundt, J., (2002). New directions in foodborne disease prevention. *Int. J. Food Microbiol., 78*, 3–17.
69. Simpkins, S. A., Chan, A. B., Hays, J., Öpping, B. P., & Cook, N., (2000). An RNA transcription-based amplification technique (NASBA) for the detection of viable *Salmonella enterica. Lett. Appl. Microbiol., 30*, 75–79.
70. TwistAmp DNA amplification kits, Combined instruction manual. In: T. Limited (Ed.), twistdx.co.UK 2018, (pp. 1–62).
71. Uyttendaele, M., Bastiaansen, A., & Debevere, J., (1997). Evaluation of the NASBA® nucleic acid amplification system for assessment of the viability of *Campylobacter jejuni. Int. J. Food Microbiol., 37*, 13–20.
72. Vincent, M., Xu, Y., & Kong, H., (2004). Helicase-dependent isothermal DNA amplification. *EMBO Reports, 5*, 795–800.
73. Walker, G. T., Fraiser, M. S., Schram, J. L., Little, M. C., Nadeau, J. G., & Malinowski, D. P., (1992a). Strand displacement amplification—An isothermal, *in vitro* DNA amplification technique. *Nucleic Acids Res., 20*, 1691–1696.
74. Walker, G. T., Little, M. C., Nadeau, J. G., & Shank, D. D., (1992b). Isothermal *in vitro* amplification of DNA by a restriction enzyme/DNA polymerase system. *Proc. Natl. Acad. Sci., 89*, 392–396.
75. Wang, J., Li, R., Hu, L., Sun, X., Wang, J., & Li, J., (2016). Development of a quantitative fluorescence single primer isothermal amplification-based method for the detection of *Salmonella. Int. J. Food Microbiol., 219*, 22–27.
76. Wang, Y., Wang, Y., Ma, A., Li, D., & Ye, C., (2014). Rapid and sensitive detection of *Listeria monocytogenes* by cross-priming amplification of *lmo0733* gene. *FEMS Microbiol. Lett., 361*, 43–51.
77. Wang, Z., Yang, Q., Zhang, Y., Meng, Z., Ma, X., & Zhang, W., (2018). Saltatory rolling circle amplification (SRCA): A novel nucleic acid isothermal amplification technique applied for rapid detection of *Shigella* spp.in vegetable salad. *Food Anal. Methods, 11*, 504–513.
78. World Health Organization, (2015). *WHO Estimates of the Global Burden of Foodborne Diseases: Foodborne Disease Burden Epidemiology Reference Group 2007–2015*. World Health Organization.
79. Wu, W., Zhao, S., Mao, Y., Fang, Z., Lu, X., & Zeng, L., (2015). A sensitive lateral flow biosensor for *Escherichia coli O157: H7* detection based on aptamer mediated strand displacement amplification. *Anal. Chim. Acta., 861*, 62–68.
80. Xu, D., Ming, X., Gan, M., Wu, X., Dong, Y., Wang, D., & Xu, F., (2018). Rapid detection of *Cronobacter* spp. in powdered infant formula by thermophilic helicase-dependent isothermal amplification combined with silica-coated magnetic particles separation. *J. Immunol. Methods., 462*, 54–58.

81. Xu, X., Su, Y., Zhang, Y., Wang, X., Tian, H., Ma, X., & Xu, W., (2021). Novel rolling circle amplification biosensors for foodborne microorganism detection. *Trac-Trends Anal. Chem., 141*, 116293.
82. Xu, G., Hu, L., Zhong, H., Wang, H., Yusa, S. I., Weiss, T. C., & You, Q., (2012). Cross priming amplification: Mechanism and optimization for isothermal DNA amplification. *Sci. Rep., 2*, 1–7.
83. Xu, W., Gao, J., Zheng, H., Yuan, C., Hou, J., Zhang, L., & Wang, G., (2019). Establishment and application of polymerase spiral reaction amplification for *Salmonella* detection in food. *J. Microbiol. Biotechnol., 29*, 1543–1552.
84. Yamazaki, W., Taguchi, M., Ishibashi, M., Kitazato, M., Nukina, M., & Misawa, N., (2008). Development and evaluation of a loop-mediated isothermal amplification assay for rapid and simple detection of *Campylobacter jejuni* and *Campylobacter coli*. *J. Med. Microbiol., 57*, 444–451.
85. Yamazaki, W., Taguchi, M., Kawai, T., Kawatsu, K., Sakata, J., Inoue, K., & Misawa, N., (2009). Comparison of loop-mediated isothermal amplification assay and conventional culture methods for detection of *Campylobacter jejuni* and *Campylobacter coli* in naturally contaminated chicken meat samples. *Appl. Environ. Microbiol., 75*, 1597–1603.
86. Yan, L., Zhou, J., Zheng, Y., Gamson, A. S., Roembke, B. T., Nakayama, S., & Sintim, H. O., (2014). Isothermal amplified detection of DNA and RNA. *Mol. Biosyst., 10*, 970–1003.
87. Yang, Q., Xu, H., Zhang, Y., Liu, Y., Lu, X., Feng, X., & Zhang, W., (2020). Single primer isothermal amplification coupled with SYBR green II: Real-time and rapid visual method for detection of *Listeria monocytogenes* in raw chicken. *LWT, 128*, 109453.
88. Yang, Q., Guo, W., Liu, Y., Zhang, Y., Ming, R., Yuan, Y., & Zhang, W., (2021). Novel single primer isothermal amplification method for the visual detection of *Vibrio parahaemolyticus*. *Food Anal. Methods, 14*, 1995–2002.
89. Yang, Q., Zhang, Y., Li, C., Zhao, Y., Ma, X., Zhang, W., & Zhang, S., (2018). A rapid and visual single primer isothermal amplification-based method for the detection of *Staphylococcus aureus* in raw pork products. *Food Anal. Methods, 11*, 3113–3120.
90. Yang, Q., Zhang, Y., Li, S., Lu, X., Yuan, Y., & Zhang, W., (2019). Saltatory rolling circle amplification for sensitive visual detection of *Staphylococcus aureus* in milk. *J. Dairy Sci., 102*, 9702–9971.
91. Yuan, N., Zhang, Y., Xu, H., Zhou, Z., Lu, X., Chen, T., & Zhang, W., (2020). Development of the saltatory rolling circle amplification assay for rapid and visual detection of *Alicyclobacillus acidoterrestris* in apple juice. *J. Agric. Food Chem. 68*, 4538–4545.
92. Zhai, L., Liu, H., Chen, Q., Lu, Z., Zhang, C., Lv, F., & Bie, X., (2019). Development of a real-time nucleic acid sequence–based amplification assay for the rapid detection of *Salmonella* spp. from food. *Braz. J. Microbiol., 50*, 255–261.
93. Zhang, H., Feng, S., Zhao, Y., Wang, S., & Lu, X., (2015). Detection of *Yersinia enterocolitica* in milk powders by cross-priming amplification combined with immunoblotting analysis. *Int. J. Food Microbiol., 214*, 77–82.
94. Zhang, J., Di, B., Shan, H., Liu, J., Zhou, Y., Chen, H., & Bai, Z., (2019). Rapid detection of *Bacillus cereus* using cross-priming amplification. *J. Food Prot., 82*, 1744–1750.
95. Zhang, M., Wang, X., Han, L., Niu, S., Shi, C., & Ma, C., (2018). Rapid detection of foodborne pathogen *Listeria monocytogenes* by strand exchange amplification. *Anal. Biochem., 545*, 38–42.

96. Zhong, J., & Zhao, X., (2018). Isothermal amplification technologies for the detection of foodborne pathogens. *Food Anal. Methods, 11*, 1543–1560.
97. Zyrina, N. V., Zheleznaya, L. A., Dvoretsky, E. V., Vasiliev, V. D., Chernov, A., & Matvienko, N. I. N., (2007). BspD6I DNA nickase strongly stimulates template-independent synthesis of non-palindromic repetitive DNA by *Bst* DNA polymerase. *J. Biol. Chem., 388*, 367–372.

CHAPTER 13

PGPR Biotechnology for Management of Biotic and Abiotic Stresses in Agricultural Plants: Recent Developments

C. JIMTHA JOHN

Assistant Professor, Department of Integrated MSc Biology, St. Joseph's College, Irinjalakuda, Thrissur, Kerala, India

ABSTRACT

This chapter aims to investigate the plant growth promotion and biocontrol potential of rhizobacteria isolated from the rhizosphere of selected medicinal plants. *In vivo* plant growth enhancement effect of rhizobacteria on the model plant system was checked. The carotenoid content modulation effect of selected PGPR was also examined. The study also focused on identifying the ability of selected PGPR on enhancing soil nutrient quality. Selected PGPR were identified based on 16S rDNA gene sequencing as *Bacillus niacini* and *Chitinophaga eiseniae*. The isolated strain of *Bacillus niacini* was found to possess four plant growth promoting traits (ACC deaminase production, N fixation, IAA production, ammonia production) out of five tested and was found to increase growth of the treated plants considerably. The strain also caused a significant increase in the carotenoid content of the treated plants (6.0171 μmol ml^{-1}) compared to that of control (0.0028 μmol ml^{-1}). It was also found to improve the carbon, nitrogen, and hydrogen content of the treated soil, thus improving soil quality. The biocontrol possessing strain

Bioresources and Bioprocess in Biotechnology for a Sustainable Future. Leonardo Sepúlveda Torre, Juan Carlos Contreras-Esquivel, Ann Rose Abraham, & A. K. Haghi (Eds.)

Chitinophaga eiseniae caused inhibition of growth of *Fusarium oxysporum* (70%) and *Rhizoctonia solani* (71%). The crude extract of the isolate was found very effective against *Fusarium oxysporum* and was able to prevent the development of symptoms of vascular wilt on ginger rhizome. The two strains were also found to be compatible after perpendicular streaking method, thus paving the way to be developed into a consortium for promoting plant growth and biocontrol.

13.1 INTRODUCTION

The challenges faced by modern agriculture include loss of soil fertility, increasing pest and pathogen attacks, fluctuating climatic factors, etc. Eco-friendly approaches like biofertilizers, biopesticides, and crop residue return are needed to enable sustainability and environmental safety of agricultural production [17]. Newer methods should be formulated to increase the food crop production yield without harming the environment to meet the requirement of the growing world population. The application of plant growth-promoting rhizobacteria (PGPR) to increase plant growth can be a potential solution for this issue. PGPR refers to soil bacteria that colonize the rhizosphere of plants. The PGPR also refers to stimulate plant growth by several mechanisms by growing in, on or around plant tissues [27]. It worth mentioning that the plant growth promoting rhizobacteria can promote plant growth either directly or indirectly. Direct methods of plant growth promotion involve N fixation, ACC deaminase production, phytohormone production, etc., and indirect methods include lytic enzyme production, antibiotic production, siderophore production, etc. [17].

PGPR can be also used for improving the carotenoid content of plants [21]. Carotenoids in general and beta carotene in particular have substantial effects in human health. Their biological activities can play important roles to avoid and cure several types of diseases and so, human health is greatly affected by the presence of carotenoid pigments [12].

Attacks by various plant pathogens cause reduction in the yield of crops worldwide. *Rhizoctonia solani* and *Fusarium oxysporum* are important plant pathogens. *Rhizoctonia solani* causes sheath blight disease in rice [30], damping off of tomato seedlings [2], etc. *Fusarium oxysporum* is the causative agent of fusarium wilt of banana [35], stem decay and root rot of *Amaranthus hybridus* [8], fusarium yellows in ginger [34], etc. Ecofriendly measures should be adopted to suppress these pathogens in order to protect the crops. The use of PGPR as biocontrol agents against these pathogens stands as an excellent way to control these pathogens.

Modern agriculture depends a lot on chemical fertilizers for providing essential plant nutrients. But their overuse has led to serious environmental impacts. The overuse of chemical N fertilizers has led to the contamination of ground water with nitrate. The gaseous oxides of nitrogen, which are derived from N fertilizers, are highly reactive and they pose a great threat to the stability of the ozone layer. Nitrogen runoff from the fields to surface waters has led to extensive pollution and eutrophication of lakes and rivers [7]. The rising cost of chemical pesticides and increasing consumer demand for pesticide-free food now encourages the development of good PGPR inoculants to increase the crop yield [10].

The objective of this research was to isolate efficient PGPR strains from the rhizosphere of medicinal plants, which can promote plant growth together with providing biocontrol against prominent plant pathogens, *Rhizoctonia solani,* and *Fusarium oxysporum*. The ability of selected PGPR isolates to increase the carotenoid content of plants and their capability to enhance soil nutrient content was also accessed through the study.

13.2 MATERIALS AND METHODS

13.2.1 ISOLATION OF PLANT GROWTH PROMOTING BACTERIA FROM THE RHIZOSPHERE OF MEDICINAL PLANTS

Soil adhering to the roots of medicinal plants (*Plectranthus amboinicus, Azadirachta indica, Bacopa monnieri,* and *Aloe barbadensis*) was separated by gentle tapping using sterilized forceps and was collected into separate sterilized plastic bags. Each of these samples was processed separately on the same day. About 10.0 g of rhizosphere soil was placed in 95 ml dilution blank containing sterile distilled water (10^{-1} dilution) and was shaken for 10 min in a rotary shaker. Then, 1.0 ml of this suspension was transferred into a 9 ml blank giving 10^{-2} dilution. This serial dilution was continued up to 10^{-10}. Aliquots from 10^{-4}, 10^{-5}, 10^{-6} dilutions were pour plated on nutrient agar medium [11]. After incubation, bacterial colonies formed were observed and morphologically different colonies were subcultured on nutrient agar plates to obtain pure cultures of the respective strains.

13.2.2 SCREENING FOR PLANT GROWTH PROMOTING TRAITS

13.2.2.1 INDOLE ACETIC ACID PRODUCTION DETECTION AND ESTIMATION

About 50 milliliter of nutrient broth containing 0.1% L-tryptophan was inoculated with 500 μl of 24 hrs. old bacterial cultures and incubated in an incubator shaker at 30 ± 0.1°C and 180 rpm for 48 hours in the dark. After incubation, the bacterial cultures were centrifuged at 10,000 rpm for 10 minutes at 4°C. Estimation of indole-3-acetic acid (IAA) in the supernatants was done using a colorimetric assay [12]. One ml of supernatant was mixed with 4 ml of Salkowski's reagent and absorbance of the resulting pink color was read after 30 minutes at 535 nm in a UV/Visible spectrophotometer. Appearance of pink color in test tubes indicated IAA production [18]. The tube showing maximum IAA production was chosen and the amount of IAA produced by the specific isolate was determined from the regression equation of the standard curve. The result was expressed as μg mL^{-1} over control.

13.2.2.2 AMMONIA PRODUCTION DETECTION

Ammonia production by the isolates was obtained by the method explained by Ajay Kumar [28]. Overnight grown bacterial cultures were inoculated into 10 ml of peptone broth and incubated at 30 ± 0.1°C for 48 hours. After incubation, 0.5 ml of Nessler's reagent was added to the tubes. The formation of yellow to brown precipitate indicated the presence of ammonia.

13.2.2.3 NITROGEN FIXING CAPABILITY DETECTION

Overnight grown bacterial cultures were streaked onto Jensen's medium. Inoculated plates were incubated for 4 days at 28°C [25].

13.2.2.4 ACC DEAMINASE PRODUCTION DETECTION

The presence of ACC deaminase activity in bacterial isolates was detected by streaking the isolates on plates with DF salts minimal medium [25].

13.2.2.5 HYDROGEN CYANIDE PRODUCTION

Bacterial cultures were streaked on nutrient agar medium containing 4.4 g per liter of glycine. A Whatman filter paper No. 1 soaked in 0.5% picric acid solution (in 2% sodium carbonate) was placed inside the lid of the plate. Plates were sealed with parafilm and incubated at 30 ± 0.1°C for 4 days [28].

13.2.2.6 PLANT STUDY

After analyzing the plant growth promoting properties, the isolates which gave positive results for at least 4 plant growth promoting properties were selected for further studies.

13.2.2.7 INOCULUM PREPARATION AND GREEN HOUSE EXPERIMENT

For plant study, method of Akansha Jain [23] was adopted with slight modification. The selected bacterial strains NM2, PK11, AV1, AV2, and NM3 were grown on nutrient agar plates. Single colonies of each bacterial isolates were transferred to 500 ml flasks containing 250 ml of nutrient broth and were placed on a rotary shaker for 48 hours at 27°C. One flask containing sterile nutrient broth was kept as control.

Seeds of Amaranthus were surface sterilized with 1% sodium hypochlorite for 30 s, rinsed twice with sterile distilled water and dried under a sterile air stream. The seeds were treated with suspensions of NM2, PK11, AV1, AV2, and NM3 separately. For seed coating, microbial suspensions were prepared in 1% carboxymethyl-cellulose, as the adhesive and the seeds were soaked in their respective suspensions for 10 hours at room temperature. After incubation, the microbial suspensions were drained off and the seeds were dried overnight in sterile Petri dishes. The same procedure was followed for the treatment of seeds to be used as a control except, they were soaked in sterile nutrient broth.

Polypropylene bags were filled with fertile soil (500 gm soil per bag). For each treatment, three polypropylene bags were used, and six seeds were sown in each bag. The pots were placed in the greenhouse, and irrigation was provided as and when required or at 2 days intervals. Observations were recorded after 15 days of sowing for the treatments which showed maximum growth. Growth promoting effects of the selected isolates on Amaranthus were evaluated by determining the root length (cm), shoot length (cm), root and shoot fresh and dry biomass (g) and number of lateral roots.

13.3 STATISTICAL ANALYSIS

All the data collected in the study were subjected to analysis of variance (ANOVA). The treatment mean values were compared by Duncan's multiple range test at $P \leq 0.05$ significance level. The software used for analysis was SPSS ver. 16 (SPSS Inc., Chicago, IL).

13.3.1 CHNS ANALYSIS OF THE SOIL

To know whether the growth of isolate improved the CHNS content of the soil, CHNS analysis was carried out. For CHNS analysis of the soil, 1 gram of soil was stored (sample 1) from the soil used to plant the seeds. After harvesting the plants, 1 gram of soil was again collected (sample 2). Both soil samples were analyzed using Euro Vector CHNS Analyzer Model No: EA 3000 at KFRI, Thrissur. Sulfanilamide was used as a standard for the analysis. CHNS analyzer instrument parameters: carrier pressure –100 kPa, purge flow –80 ml/min, oxygen flow –20 ml, $\Delta P\ O_2$ –35 kPa, sampling delay –7 seconds, run time –320 seconds, front furnace –980°C, GC oven –105°C.

13.3.2 CAROTENOID CONTENT ESTIMATION

Amaranthus is rich in carotenoids and several studies proved that the carotenoid content of many plants can be increased by the inoculation of efficient plant growth promoting strains. The treatment containing the set of plants which showed maximum growth due to the presence of the particular PGPR isolate and those in the control were selected for estimating their carotenoid content.

13.3.3 EXTRACTION OF CAROTENOIDS USING SOLVENTS

- 0.5 g of Amaranthus was at first step weighed and placed in a mortar.
- It was then crushed with a pestle.
- A mixture of hexane and acetone in the ratio of 1:1 was added into the mortar.
- The sample was crushed very carefully.
- 1.25 ml of acetone was added gradually at regular intervals.
- The solvents were collected separately and for double extraction, the process was repeated with the sample.

- The solvents containing carotenoids were filtered through a filter paper.
- The solvents were then transferred into a separating funnel.
- 12.5 ml of distilled water was added along with 12.5 ml of 10% NaCl solution.
- The mixture was shaken firmly and kept aside for the layers to separate.
- The upper layer contained carotenoids and it was collected separately after the removal of the water and NaCl solution.
- The extract was collected in tubes [24].

13.3.4 ESTIMATION OF CAROTENOID CONTENT

The carotenoid content of the samples was analyzed using the following formulas [40]:

$$\text{Anthocyanin} = 0.08173A_{537} - 0.00697A_{647} - 0.002228A_{663}$$

$$\text{Chl}_a = 0.01373A_{663} - 0.000897\,A_{537} - 0.003046\,A_{647}$$

$$\text{Chl}_b = 0.02405\,A_{647} - 0.004305\,A_{537} - 0.005507\,A_{663}$$

where; A_{537} is the absorbance of the extract solution in a 1-cm path length cuvette at a wavelength 537 nm; A_{647} is the absorbance of the extract solution in a 1-cm path length cuvette at a wavelength 647 nm; A_{663} is the absorbance of the extract solution in a 1-cm path length cuvette at a wavelength 663 nm; Chl_a is the absorption coefficient of chlorophyll a; Chl_b is the absorption coefficient of chlorophyll b.

$$\text{Carotenoids} = \frac{(A470 - (17.1 \times (\text{Chla} + \text{Chlb}) - 9.479 \times \text{Anthocyanin}))}{119.26}$$

A_{470} is the absorbance of the extract solution in a 1-cm path length cuvette at a wavelength 470 nm. The units in all the equations are micromoles per milliliter (μmol ml^{-1}) [40].

13.3.5 DETECTION OF BIOCONTROL PROPERTIES OF THE ISOLATES AGAINST PLANT PATHOGENS F. OXYSPORUM AND RHIZOCTONIA SOLANI

For detecting the biocontrol activity of the isolates against pathogens *F. oxysporum* and *Rhizoctonia solani,* method of Harish Kumar [29] was adopted with slight modification. For performing the dual culture plate assay, the rhizobacteria was streaked on the edge of the PDA in a Petri plate and a mycelial plug (5 mm diameter) of the respective pathogen was inoculated at the opposite edge of the plate. The plates were incubated for 5 days at 28 ± 1°C. The percentage of inhibition was calculated by using the formula:

$$(C - T)/C \times 100$$

where; C is the radial growth of fungal pathogen in control; T is the radial growth of fungal pathogen in dual culture [29].

13.3.6 EXTRACTION OF CULTURE SUPERNATANT AND ANALYZING ITS EFFICIENCY

For assay of antifungal activity in culture filtrates, method of Zacky FA and Ting AY [43] was adopted with slight modification. Bacterial cultures were grown 10 days at 28°C with aeration. Cells were removed by centrifugation for 10 min at 10,000 rpm. The culture supernatant was then extracted twice with ethyl acetate and evaporated. The resulting crude extract was dissolved in methanol and stored at 4°C. Small Petri plates (35 mm in diameter) were filled with 200 µl of crude extract and molten SDA was incorporated into it. After the plates were cooled, *Fusarium oxysporum* and *Rhizoctonia solani* inoculums were placed on the agar surface and the plates were incubated for 20 to 36 hours at 28°C. Control plates were prepared and incubated similar, substituting the extracts with methanol.

13.3.7 COMPATIBILITY TESTING

The isolates that were decided to be formulated into a consortium was tested for their compatibility among each other. Compatibility between the isolates was tested by the perpendicular streaking method in nutrient agar plates [22]. The plates were incubated at 28 ± 2°C for 48 hours and observed for inhibition among the isolates. The ability of the isolates to grow simultaneously in the medium without inhibiting each other indicates that the strains are compatible.

13.3.8 MORPHOLOGICAL, BIOCHEMICAL, AND MOLECULAR IDENTIFICATION OF THE ISOLATES

The preliminary identification of the selected isolates was done by analyzing the colony morphology, gram staining and biochemical tests. Isolates were tested for catalase, VP, urea hydrolysis, indole production and citrate utilization tests [3].

The identity of the selected isolates was confirmed by performing 16S rDNA sequence analysis. The genomic DNA of the isolates was isolated using a DNA isolation kit from Chromous Biotech Bacterial Genomic DNA Mini Spin Kit (RKT 17). The purity of the isolated DNA was confirmed by agarose gel electrophoresis. Genomic DNA isolated was used as the template for PCR amplification of 16S rDNA using the primers 16S F (5′GAG TTT GAT CCT GGC TCA G-3′) and 16SR (5′GAT ATT ACC GCG GCG CCT G3′). PCR was carried out in a 50 μL reaction volume containing 50 μg of genomic DNA, 20 pico moles of each primer (both forward and reverse), 1x Emerald Amp GT PCR Master Mix (Takara Bio Inc.). PCR was carried out for 35 cycles in a Mycycler™ (Bio-Rad, USA) with the initial denaturation at 94°C for 3 min, cyclic denaturation at 94°C for 30 seconds, annealing at 58°C for 30 seconds and extension at 72°C for 2 min with a final extension of 7 min at 72°C. The PCR product formation was confirmed by agarose gel electrophoresis for which 8 μl of PCR product was loaded on 1.5% agarose gel and 1xTBE gel running buffer. After running, the gel was removed, and the bands were visualized under a UV transilluminator. The product was more refined for its use as the template for sequencing PCR using Big Dye Terminator Sequence Reaction Ready Mix (Applied Biosystem) [25]. The sequencing of the PCR product was done in Scigenome, Ernakulam, Kerala (India). The sequence data was then subjected to BLAST analysis. The phylogenetic analysis of the 16S rDNA sequence was conducted by MEGA 5 using the neighbor-joining method with 1,000 bootstrap replicates.

13.4 RESULTS

13.4.1 ISOLATION OF BACTERIAL STRAINS

Isolation of rhizobacteria from the rhizosphere soil samples of four medicinal plants has resulted in the purification of 34 strains, which were screened for various plant growth promoting properties. Around 12 bacterial isolates were obtained from the rhizosphere of *Plectranthus amboinicus*, 7 from the rhizosphere of *Azadirachta indica,* 8 from the rhizosphere of *Bacopa monnieri* and 7 from the rhizosphere of *Aloe Barbadensis.*

13.4.2 SCREENING OF PLANT GROWTH PROMOTING CHARACTERISTICS

Plant growth promoting features of the isolates were analyzed. While screening for IAA production, 16 isolates were found to be positive (Figure 13.1). Around 14 isolates were found to be positive for nitrogen fixing property (Figure 13.1) and 27 isolates were found to be positive for ammonia production (Figure 13.1). ACC deaminase production was shown by 15 isolates (Figure 13.1). HCN production was detected in only 2 isolates.

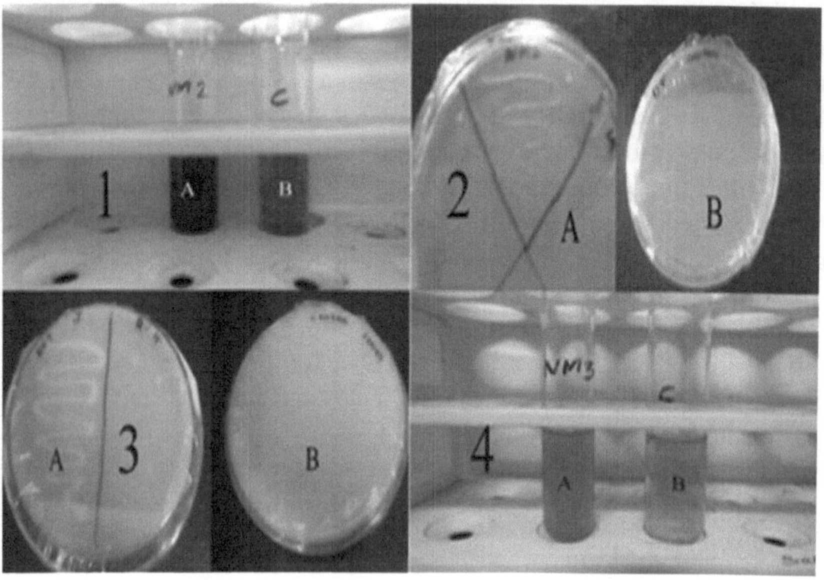

FIGURE 13.1 Plant growth promoting traits of isolated PGPR.

Note: 1: IAA production; 2: ACC deaminase production; 3: Nitrogen fixation; 4: Ammonia production; A: Positive; B: Negative.

After quantitative estimation, the isolate (NM2) which showed the good IAA production was found to produce 40 μg per ml of IAA.

13.4.3 STUDIES ON THE EFFECT OF PGPR ON GROWTH PARAMETERS OF AMARANTHUS SPECIES

After analyzing the plant growth promoting properties, the five isolates selected for plant study were NM2, PK11, AV1, AV2, and NM3. After plant study it was found that the seeds of the plants treated with NM2 isolate showed maximum growth compared to others followed by NM3 isolate (Figure 13.2).

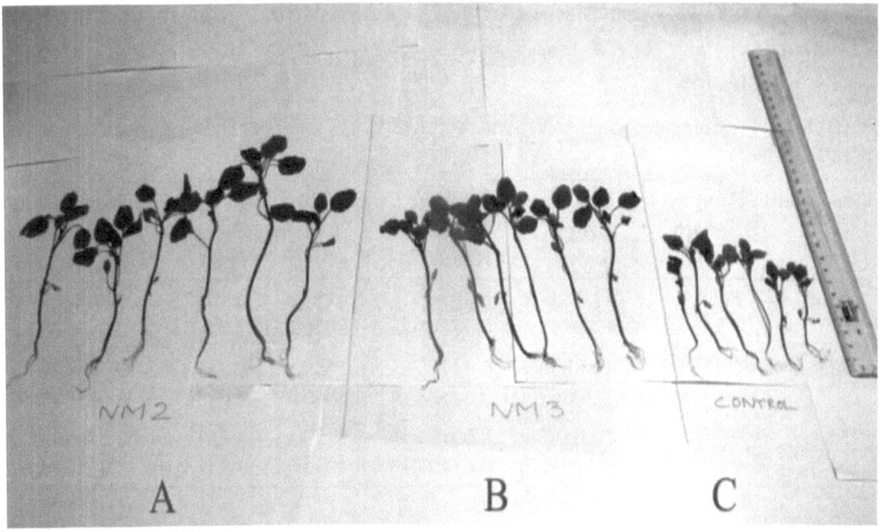

FIGURE 13.2 Seedling establishment and growth promotion analysis of the selected isolates using Amaranthus seedlings (pot experiment).

Note: A: Plants inoculated with NM2; B: Plants inoculated with NM3; and C: Untreated control plants.

From statistical analysis it was understood that the plants whose seeds were treated with PGPR isolates showed significant increase in growth parameters compared to that of control. Even though plants treated with NM2 isolate showed an increase in all growth parameters compared to that of control, the number of leaves and dry weight of plants treated with NM3 isolate does not show a significant increase compared to that of control.

NM2 isolate treatment increased root length of the plants by 1.6-fold (2.8133 cm) compared to that of control plants (1.7017 cm) (Table 13.1). Likewise, high value for the shoot length of plants were also obtained by NM2 treatment which was about 1.7-fold (20.0467 cm) higher than that of control plants (11.1483 cm) (Table 13.1). Significant increase in fresh weight of plants, about 3.4-fold (0.3517 gm) compared to control (0.1017 gm) was also obtained and a notable increase in the dry weight of plants, about 4.7-fold (0.04050 gm) compared to that of control (0.00850 gm) were also obtained with NM2 treatment (Table 13.1). About 1.2-fold (6.8850) rise in the number of leaves compared to that of control (5.4950) and 1.4-fold (8.0517) increase in the number of lateral roots compared to control (5.4433) were also observed for NM2 treatment received plants (Table 13.1).

TABLE 13.1 Enhanced Growth of Amaranthus Plants Obtained Through Inoculation with the PGPR Isolates

Treatment	Root Length (cm)	Shoot Length (cm)	Number of Lateral Roots	Number of Leaves	Fresh Weight (gm)	Dry Weight (gm)
Control	1.7017 ± 0.15342	11.1483 ± 1.10892	5.4433 ± 0.58284	5.4950 ± 0.86269(a)	0.1017 ± 0.02401	0.00850 ± 0.000548(a)
NM3	2.4367 ± 0.15449(a)	14.8850 ± 1.20669(a)	6.4417 ± 0.54584(a)	6.1650 ± 0.58671(a)	0.2550 ± 0.03507(a)	0.01467 ± 0.001033(a)
NM2	2.8133 ± 0.09585(b)	20.0467 ± 0.67911(b)	8.0517 ± 0.64493(b)	6.8850 ± 0.40461(b)	0.3517 ± 0.04956(b)	0.04050 ± 0.038615(b)

Values are means of three replicates. Mean values (mean ± SD) followed by the same alphabet are not significantly different in the column by Duncan's multiple range test ($p \leq 0.05$).

13.4.4 CHNS ANALYSIS OF SOIL

The result of the CHNS analysis of soil samples is given in Table 13.2.

TABLE 13.2 CHNS Analysis of Soil

SL. No.	Sample Name	Sample Weight (mg)	N (%)	C (%)	H (%)	S (%)
1.	Sample 1	9.6	1.613	1.985	0.534	–
2.	Sample 2	8.5	1.859	2.449	0.677	–

Note: Sample 1: Soil collected before the introduction of NM2 isolate; Sample 2: Soil collected after harvesting the plants whose seeds were treated with NM2 isolate.

From CHNS analysis, it was found that after the growth of Amaranthus plants which received NM2 isolate treatment, the carbon, nitrogen, and hydrogen contents of the soil were increased.

13.4.5 CAROTENOID CONTENT ESTIMATION

The carotenoid content estimation of the control plants and plants which received NM2 treatment was done. The absorbance shown by the plant extracts at various optical densities are given in Table 13.3.

TABLE 13.3 Absorbance Shown by the Extracted Pigments of Plants at Various Optical Densities

Plant Extract Sample	Absorbance			
	At 537 nm	At 647 nm	At 663 nm	At 470 nm
Extract of plants treated with NM2 isolate	0.3418	0.29	0.2526	0.588
Control	0.2007	0.1869	0.1561	0.2733

After calculation, the carotenoid content of the plant extract sample taken from NM2 isolate treated plant was found to be 6.0171 μmol ml^{-1} and the carotenoid content of plant extract sample taken from control plant was 0.0028 μmol ml^{-1}. The results showed that the carotenoid content of plants treated with NM2 isolate showed a large increase compared to that of control.

13.4.6 SCREENING OF PGPR FOR ANTI-PHYTOPATHOGENIC ACTIVITY

All the 34 isolated PGPR were screened for antagonistic activity against various phytopathogens by dual culture. PK3 isolate showed antagonism towards both *Fusarium oxysporum* and *Rhizoctonia solani*. NM2, AV2, PK5, PK9, AV1, and PK1 showed antagonistic activity towards only *Fusarium oxysporum*. The results of dual culture plate assays of the isolates with the plant pathogens *Fusarium oxysporum* and *Rhizoctonia solani* is given in Table 13.4.

13.4.7 ANTIFUNGAL ACTIVITY OF THE CRUDE EXTRACT

Crude extract of the bacterial isolate PK3 showed the maximum inhibition of the growth of the pathogen *Fusarium oxysporum.* The pathogen was not able to grow on the medium supplemented with 200 μl of the crude extract while, it showed appreciable growth on control plate supplemented with 200 μl of methanol (Figure 13.3). The crude extract was not able to inhibit the growth of *Rhizoctonia solani.*

TABLE 13.4 Percentage of Inhibition of Growth of Fungal Plant Pathogens Due to the Presence of PGPR Isolates

Isolate	Fungal Pathogen	C (cm)	T (cm)	C-T	C-T/C	Percentage of Inhibition (C-T/C×100)
NM2	*Fusarium oxysporum*	5	2	3	0.6	60%
AV2	*Fusarium oxysporum*	5	3	2	0.4	40%
PK5	*Fusarium oxysporum*	5	2	3	0.6	60%
PK9	*Fusarium oxysporum*	5	2.5	2.5	0.5	50%
AV1	*Fusarium oxysporum*	5	2	3	0.6	60%
PK1	*Fusarium oxysporum*	5	2	3	0.6	60%
PK3	*Rhizoctonia solani*	7	2	5	0.7142	71%
PK3	*Fusarium oxysporum*	5	1.5	3.5	0.7	70%

Note: C: Radial growth of fungal pathogen in control; and T: Radial growth of fungal pathogen in dual culture.

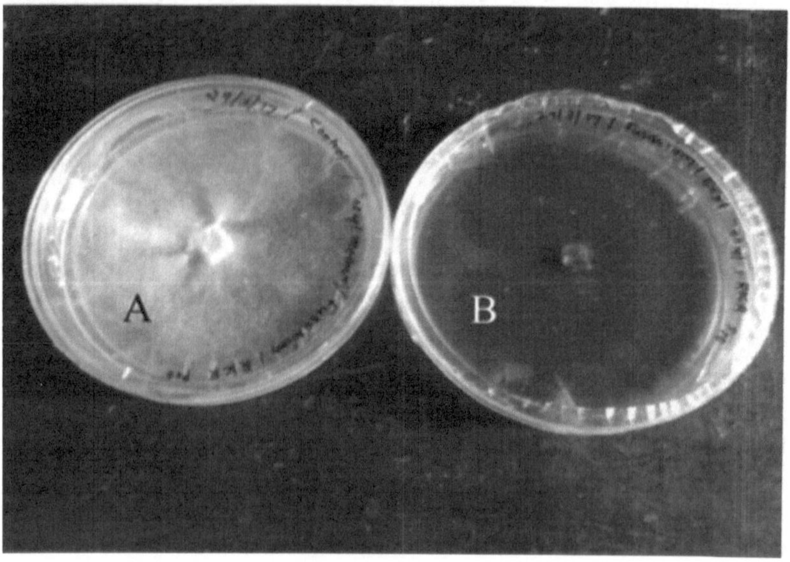

FIGURE 13.3 Effect of PK3 culture extract on growth of *Fusarium oxysporum.*

Note: A: Control plate showing growth of *Fusarium oxysporum* on SDA plate containing 200 µl of methanol; and B: Growth of *Fusarium oxysporum* on SDA plate containing 200 µl of the culture extract of PK3 isolate.

13.4.8 COMPATIBILITY TESTING OF THE ISOLATES

The compatibility of the isolates NM2 and PK3 was tested by the perpendicular streaking method on nutrient agar plates and the isolates were found to be compatible with each other and were able to grow simultaneously on the culture plate without any inhibition in growth. So if these two isolates are used to make a consortium to increase plant growth, they can increase plant growth and provide resistance against the pathogens without inhibiting each other.

13.4.9 ANALYZING THE EFFICIENCY OF THE CRUDE EXTRACT FOR THE BIOCONTROL OF FUSARIUM OXYSPORUM INFECTION ON ZINGIBER OFFICINALE

The use of crude extract of PK3 isolate for the biocontrol of *Fusarium oxysporum* infection on *Zingiber officinale* yielded good results. After the incubation period, the slice of the rhizome inoculated with the pathogen showed surface depression, wilting, softening, and browning of the tissues. The slice of the rhizome sprayed with the extract was intact and didn't show any disease symptoms as shown by the slice which was not treated with the extract (Figure 13.4).

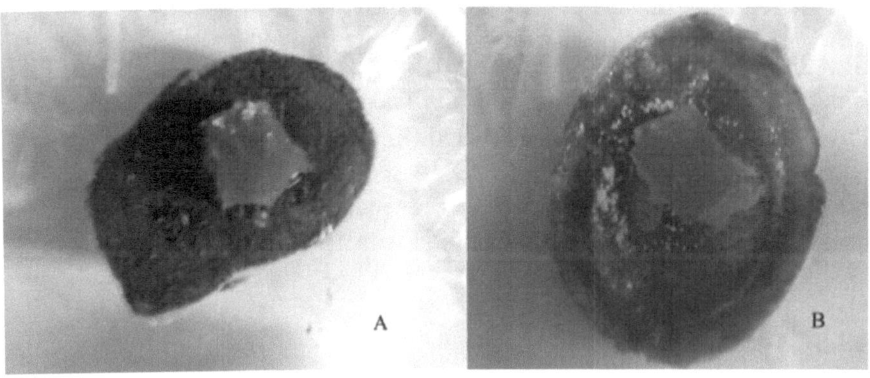

FIGURE 13.4 Effect of crude extract of PK3 on the growth of *Fusarium oxysporum*.

Note: A: Slice of ginger rhizome sprayed with crude extract and inoculated with *Fusarium oxysporum*; and B: Slice of ginger rhizome inoculated with *Fusarium oxysporum*.

13.4.10 IDENTIFICATION OF THE ISOLATES

13.4.10.1 MORPHOLOGICAL AND BIOCHEMICAL CHARACTERIZATION OF THE SELECTED PGPR ISOLATES

Upon morphological examination, the colonies of NM2 were flat, cream colored and fimbriate. The colonies of PK3 were convex, smooth, and yellow colored. After Gram's staining, NM2 isolate appeared as Gram positive long rods and PK3 appeared as Gram negative rods of medium size. NM2 isolate was found positive for catalase, citrate utilization tests and negative for urea hydrolysis, Voges Proskauer and indole production tests. PK3 isolate gave the same results for biochemical tests as that of NM2 isolate.

13.4.10.2 MOLECULAR IDENTIFICATION OF SELECTED PGPR

The DNA of the selected PGPRs was successfully isolated using Chromous Biotech Bacterial Genomic DNA Mini Spin Kit (RKT 17). The concentration and purity of isolated DNA were checked using agarose gel electrophoresis. When viewed under UV transilluminator, thick DNA bands were observed for both isolated NM2 and PK3 DNA samples. PCR amplification of 16S rDNA of selected rhizobacteria resulted in the formation of 1500 base pair product as confirmed by agarose gel electrophoresis.

BLAST analysis of the 16S rDNA 931 base pairs trimmed sequence of NM2 isolate showed 99% identity with *Bacillus niacini* strain GYR1 and BLAST analysis of the 16S rDNA 938 base pairs trimmed sequence of PK3 isolate showed 98% identity with *Chitinophaga eiseniae* strain YC6729. Phylogenetic tree was also constructed (Figure 13.5).

Phylogenetic analysis of 16S rDNA sequences of the bacterial isolates selected in the study along with other sequences. The analysis was conducted using MEGA5 neighbor-joining method with 1,000 boot-strap replicates.

13.5 DISCUSSION

As the world's population is rising steadily, food production should be also increased simultaneously to meet the requirements of the people. Heavy dependence on chemical fertilizers and pesticides to increase the crop yield have led to serious environmental impacts. Excessive use of fertilizers to increase plant growth and the application of the high amount of chemicals to destroy various plant pathogens and pests has resulted in the deterioration of the soil quality worldwide. The best way to increase growth and disease resistance among crops without harming the environment can be achieved through the use of plant growth promoting rhizobacteria (PGPR).

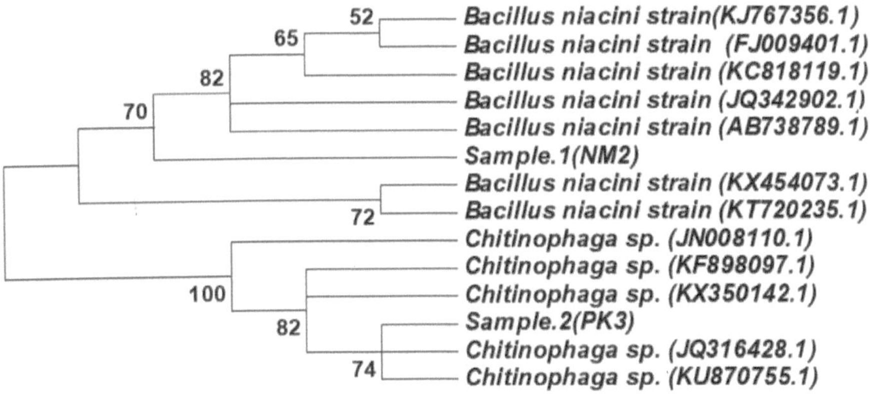

FIGURE 13.5 Phylogenetic tree.

The study was conducted to isolate efficient PGPR strains from the rhizosphere of medicinal plants to increase crop yield to a good extent, together with providing resistance against plant pathogens. Around 34 isolates obtained from the rhizosphere of various medicinal plants were screened for their plant growth promoting potentials like IAA production, nitrogen fixation, HCN production, ammonia production, ACC deaminase production and for biocontrol activity against *Fusarium oxysporum* and *Rhizoctonia solani* plant pathogens.

Among 34 isolates obtained, 16 were able to produce IAA in varying amounts. NM2 isolate obtained from the rhizosphere of *Azadirachta indica* showed maximum IAA production compared to other isolates. It was able to produce 40 μg of IAA per ml. PGPR isolates which produced 40 (±2.2) μg per ml IAA can promote better growth of plants compared to that of control [37]. Many studies have shown that optimum production of IAA by PGPR can promote plant growth in a variety of ways. For example, significant increases in rooting and root dry matter of cuttings of eucalyptus were observed when grown on IAA producing rhizobacteria inoculated substrate [41].

The capability to produce ACC deaminase enzyme was possessed by 15 isolates. Production of this enzyme helps to reduce the ethylene level in plants and various studies have revealed that the production of ACC deaminase by

PGPR helps in better plant growth [4]. For example, studies conducted by Ghosh et al., [13]; Grichko & Glick [19], etc. Plant growth promotion by the synergistic action of bacterial IAA and ACC deaminase has been suggested in numerous studies [14, 15]. NM2 isolate was found weakly positive for ACC deaminase production. Strains having a low ACC deaminase activity and capable of producing 30 to 47 μg ml^{-1} IAA has been found to increase seedling growth of *M. truncatula* [26].

The capability to fix atmospheric nitrogen was observed in 14 isolates. Several studies reported that atmospheric nitrogen fixing PGPR possesses the capability to fertilize several important agronomic plants such as rice [31], wheat [5], sugarcane [6], etc. Around 27 isolates were positive for ammonia production. The ammonia produced by rhizobacteria can act as an additional source of nitrogen in the soil [32]. The extensive use of chemical nitrogen fertilizers possesses several threats such as, pollution of the atmosphere due to evolution of greenhouse gasses like N_2O, NO, NH_3 from volatilization and denitrification of the applied fertilizers, groundwater toxicity due to leaching of NO_3^-–N [39]. The isolated strain NM2 was found positive for both N fixation and ammonia production. As we got good results in plant study with NM2 without using chemical fertilizers and as the strain was found to improve the nutrient quality of the soil, the use of diazotrophs such as NM2 can effectively reduce the dependence on chemical fertilizers.

After analyzing the plant growth promoting properties, the five isolates NM2, PK11, AV1, AV2, and NM3 were selected for plant study. Amaranthus plant was selected for the study. NM2 isolate treatment increased root length of the plants by 1.6-fold, shoot length by 1.7-fold, number of lateral roots by 1.4-fold, number of leaves by 1.2-fold, fresh weight by 3.4-fold and dry weight by 4.7-fold. NM3 isolate treatment increased root length of the plants by 1.4-fold, shoot length by 1.3-fold, number of lateral roots by 1.1 folds and fresh weight by 2.5-fold.

The CHNS analysis of soil showed an increase in nitrogen, hydrogen, and carbon content of soil after the growth of Amaranthus plants, whose seeds were treated with NM2 isolate. The ammonia produced by rhizobacteria can act as an additional source of nitrogen in the soil [32] and there will be a net release of protons into the soil for NH_4^+ uptake by the plant roots [38]. NM2 isolate was positive for ammonia production. The increase in hydrogen content of soil obtained may be due to an increased release of protons into the soil by the plant roots in order to uptake ammonia produced by the isolate. Various interactions between plant and rhizosphere microbial communities have been established through several studies. For example, following the application of 15N labeled fertilizer the excess of 15N in the microbial biomass increased significantly in both planted and control soils at up to 8 weeks after plant emergence, but then declined in control soils only. Retention of 15N in the microbial biomass in planted soil was attributed to the release of root-derived C from maize as estimated using the 13C abundance methodology. This release of exudates was suggested to promote microbial immobilization of the N and is consistent with greater microbial biomass in the rhizosphere [36]. The increase in carbon content observed after CHNS analysis of the planted soil may be due to some complex mechanisms like this, which occurred between the NM2 isolate and the Amaranthus plants.

Analysis of carotenoid content of plants which received NM2 treatment showed a significant increase (6.0171 μmol ml^{-1}) compared to that of control (0.0028 μmol ml^{-1}). Several studies have shown that inoculation with PGPR can increase carotenoid content of plants. For example, an increase in carotenoid content of *S. tuberosum* was seen when treated with PGPR species belonging to the genus *Bacillus* [1]. Biological activities of various carotenoids play important roles in the prevention and treatment of many kinds of diseases and so, human health is greatly affected by the presence of carotenoid pigments [12]. Isolate NM2 can be used as an efficient strain for increasing the carotenoid content of plants as evident from our study.

After biochemical, morphological, and molecular studies NM2 isolate, which was proved to be an effective plant growth promoting bacteria was identified as *Bacillus niacini.* Root-colonizing *Bacillus* mediated enhancement of plant growth is well known [42]. Significant increase in the seedling growth of *M. truncatula* was seen when inoculated with *B. niacini* strain. Four weeks after inoculation with *B. niacini*, the whole seedling biomass showed significant increases. About 2.2–2.5-fold increase in fresh weight of the roots and 3–4.7-fold increase in fresh weight of shoots were observed. Shoot dry weight increased by 4.6-fold. Seedling dry weight was increased by 4.6 times when compared to that of the non-inoculated plants. Stimulation of root branching was also observed [26]. *Bacillus* sp. is also reported to cause enhancement of biomass, nitrogen uptake and crop yield in groundnut [20].

PK3 isolate which proved to be an effective biocontrol agent was identified as *Chitinophaga eiseniae* after biochemical, morphological, and molecular studies. Several species of bacteria possess the capability to synthesize chitinase enzyme and this enzyme enable the bacteria to use chitin as a sole source of carbon and nitrogen. Some of the efficient chitinase producing bacteria have been shown to be potential agents for the biocontrol of plant pathogens including phytopathogenic fungi. Some of them are also shown to be effective against insect pests [9]. As chitin forms the structural component of both fungal cell walls and insect exoskeletons [9], chitinolytic bacteria can be effectively used against them and this helps in preventing diseases to the plants in an environmentally friendly way. *Chitinophaga* is one of the most notable chitinase producing Gram-negative bacteria [16]. Many works on establishing the potential of *Chitinophaga* as a biocontrol has not been done. However, it has been reported that some *Chitinophaga* species possess the ability to produce lantibiotics (lanthionine containing antibiotics). These secondary metabolites are known to possess antimicrobial activity. These compounds are highly active against filamentous fungi and yeasts [33]. From our study efficiency of *Chitinophaga eiseniae* as a biocontrol agent against the plant pathogens *Fusarium oxysporum* and *Rhizoctonia solani* has been proved. The results of growth inhibition of the plant pathogens in the presence of the isolate may be due to chitinase or lantibiotic produced by the isolate.

13.6 CONCLUSION

The present study has resulted in the identification of effective microbial strains which can be used in agriculture for promoting plant growth as well as for antifungal properties. The isolated strains were identified as *Bacillus niacini* and *Chitinophaga eiseniae*, respectively. The isolated strain of *Bacillus niacini* possessed four plant growth promoting properties out of five tested. The results of the plant study indicated that the treatment with *Bacillus niacini* increased carotenoid content of plants. It also helped in increasing, shoot length, root length, the number of leaves, number of lateral roots, fresh weight and dry weight of plants. The strain was also found to improve the nitrogen, carbon, and hydrogen content of the soil after inoculation. The isolated strain of *Chitinophaga eiseniae* was found to be a potential biocontrol agent against both *Fusarium oxysporum* and *Rhizoctonia solani* plant pathogens. The strain showed 70% or more inhibition towards both the pathogens in the dual culture plate assay. The crude extract was effective very against *Fusarium oxysporum.* The potency of the biocontrol activity of the crude extract was established through plant study with ginger rhizome. The pathogen was not able to produce the disease symptoms in the slice of ginger rhizome treated with the extract. The compatibility of the two isolates was confirmed by the perpendicular streaking method in order to avoid the chance of failure when they are applied as a consortium to the field.

The application of the identified strains is a potential solution for increasing the plant growth together with providing resistance to various plant pathogens. As these strains are found compatible, they can be formulated into a consortium if needed. The methods now adopted for increasing the crop yield heavily depend upon chemical fertilizers and pesticides and this has resulted in serious environmental problems. The usage of these strains to increase the plant growth can certainly reduce the dependence on the chemical fertilizers and pesticides and it also helps to restore the soil quality which is lost due to the excessive use of these chemical compounds. Besides these facts, these isolated strains can be also used to improve the carotenoid content of plants which in turn have a major impact on human health.

ACKNOWLEDGMENT

We are thankful to Dr. Radhakrishnan E K, Assistant professor, School of Biosciences, M G University, Kerala and Jishma, PhD Scholar M G University, Kerala for supporting us to do the molecular works for this research work.

KEYWORDS

- **ACC deaminase**
- **biochemical characterization**
- **biocontrol**
- **carotenoids**
- **IAA**
- **nitrogen fixation**
- **PGPR**
- **plant growth**

REFERENCES

1. Ahmed, A., & Hasnain, S., (2010). Auxin-producing *Bacillus sp.*: Auxin quantification and effect on the growth of *Solanum tuberosum*. *Pure Appl. Chem.*, *82*, 313–319. doi: 10.1351/PAC-CON-09-02-06.

2. Asaka, O., & Shoda, M., (1996). Biocontrol of *Rizhoctonia solani* damping-off of tomato with *Bacillus subtilis* RB14. *Appl. Environ. Microbiol., 62*, 4081–4085.
3. Bhattacharya, C., Harsha, P., Gupta, S., & Roy, S., (2014). *Isolation and Characterization of Bacterial Isolates from Agricultural Soil at Durg District, 4*, 221–226.
4. Bhattacharyya, P. N., & Jha, D. K., (2012). Plant growth-promoting rhizobacteria (PGPR): Emergence in agriculture. *World J. Microbiol. Biotechnol., 28*, 1327–1350. doi: 10.1007/s11274-011-0979-9.
5. Boddey, R. M., Baldani, V. L. D., Baldani, J. I., & Döbereiner, J., (1986). Effect of inoculation of *Azospirillum spp*. on nitrogen accumulation by field-grown wheat. *Plant Soil, 95,* 109–121. doi: 10.1007/BF02378857.
6. Boddey, R. M., Urquiaga, S., Alves, B. J. R., & Reis, V., (2003). Endophytic nitrogen fixation in sugarcane: Present knowledge and future applications. *Plant Soil, 252,* 139–149. doi: 10.1023/A:1024152126541.
7. Bohlool, B. B., Ladha, J. K., Garrity, D. P., & George, T., (1992). Biological nitrogen fixation for sustainable agriculture: A perspective. *Plant Soil Sci., 141*, 1–11. doi: 10.1017/CBO9781107415324.004.
8. Chen, W. Q., & Swart, W. J., (2002). The *in vitro* phytotoxicity of culture filtrates of *Fusarium oxysporum* to five genotypes of *Amaranthus hybridus*. *Euphytica, 127,* 61–67. doi: 10.1023/A:1019980600219.
9. Chernin, L. S., DelaFuente, L., Sobolev, V., Haran, S., Vorgias, C. E., Oppenheim, A. B., & Chet, I., (1997). Molecular cloning, structural analysis, and expression in *Escherichia coli* of a chitinase gene from *Enterobacter agglomerans*. *Appl. Environ. Microbiol., 63*, 834–839.
10. Compant, S., Duffy, B., Nowak, J., Cle, C., & Barka, E. A., (2005). Use of plant growth-promoting bacteria for biocontrol of plant diseases: Principles, mechanisms of action, and future prospects minireview use of plant growth-promoting bacteria for biocontrol of plant diseases: Principles, mechanisms of action, and future prospects. *Appl. Environ. Microbiol., 71*, 4951–4959. doi: 10.1128/AEM.71.9.4951.
11. Dinesh, R., Anandaraj, M., Kumar, A., Bini, Y. K., Subila, K. P., & Aravind, R., (2015). Isolation, characterization, and evaluation of multi-trait plant growth promoting rhizobacteria for their growth promoting and disease suppressing effects on ginger. *Microbiol. Res., 173*, 34–43. doi: 10.1016/j.micres.2015.01.014.
12. Eldahshan, O. A., Nasser, A., & Singab, B., (2013). Carotenoids. *J. Pharmacogn. Phytochem., 2,* 225–234.
13. Ghosh, S., Penterman, J. N., Little, R. D., Chavez, R., & Glick, B. R., (2003). Three newly isolated plant growth-promoting bacilli facilitate the seedling growth of canola, *Brassica campestris*. *Plant Physiol. Biochem., 41*, 277–281. doi: 10.1016/S0981-9428(03)00019-6.
14. Glick, B. R., & Glick, B. R., (2012). Plant growth-promoting bacteria: Mechanisms and applications. *Scientifica (Cairo). 2012*, 1–15. doi: 10.6064/2012/963401.
15. Glick, B. R., Cheng, Z., Czarny, J., & Duan, J., (2007). Promotion of plant growth by ACC deaminase-producing soil bacteria. *Eur. J. Plant Pathol., 119*, 329–339. doi: 10.1007/s10658-007-9162-4.
16. Gooday, G. W., (1990). The ecology of chitin degradation. In: Marshall, K. C., (ed.), *Advances in Microbial Ecology* (pp. 387–430). Springer US, Boston, MA. doi: 10.1007/978-1-4684-7612-5_10.
17. Gopalakrishnan, S., Sathya, A., Vijayabharathi, R., Varshney, R. K., Gowda, C. L. L., & Krishnamurthy, L., (2015). Plant growth promoting rhizobia: Challenges and opportunities. *3 Biotech, 5*, 355–377. doi: 10.1007/s13205-014-0241-x.

18. Gordon, S. A., & Weber, R. P., (1951). Colorimetric estimation of indole acetic acid. *Plant Physiol., 26*, 192–195.
19. Grichko, V. P., & Glick, B. R., (2001). Amelioration of flooding stress by ACC deaminase-containing plant growth-promoting bacteria. *Plant Physiol. Biochem., 39*, 11–17. doi: 10.1016/S0981-9428(00)01212-2.
20. Haldar, S., Roy, C. S., & Sengupta, S., (2011). Genetic and functional diversities of bacterial communities in the rhizosphere of *Arachis hypogaea*. *Antonie van Leeuwenhoek, Int. J. Gen. Mol. Microbiol., 100*, 161–170. doi: 10.1007/s10482-011-9570-5.
21. Hassan, W., David, J., & Bashir, F., (2014). ACC-deaminase and/or nitrogen-fixing rhizobacteria and growth response of tomato (*Lycopersicon pimpinellfolium* Mill.). *J. Plant Interact., 9*, 869–882. doi: 10.1080/17429145.2014.964785.
22. Ismet, A., (2012). Antiviral activities of streptomycetes against tobacco mosaic virus (TMV) in datura plant: Evaluation of different organic compounds in their metabolites. *African J. Biotechnol., 11*, 2130–2138. doi: 10.5897/AJB11.3388.
23. Jain, A., Singh, A., Singh, S., & Singh, H. B., (2015). Biological management of *Sclerotinia sclerotiorum* in pea using plant growth promoting microbial consortium. *J. Basic Microbiol., 55*, 961–972. doi: 10.1002/jobm.201400628.
24. Jeyanthi, R., L., Sharmila, S., Das, M. P., & Seshiah, C., (2014). Extraction and purification of carotenoids from vegetables. *J. Chem. Pharm. Res., 6*, 594–598.
25. Jimtha, J. C., Jishma, P., Arathy, G. B., Anisha, C., & Radhakrishnan, E., (2016). Identification of plant growth promoting rhizosphere *Bacillus sp.* WG4 antagonistic to *Pythium myriotylum* and its enhanced antifungal effect in association with *Trichoderma*. *J. Soil Sci. plant Nutr., 16*, 578–590. doi: 10.4067/S0718-95162016005000026.
26. Kisiel, A., & Kępczyńska, E., (2016). *Medicago truncatula* Gaertn. as a model for understanding the mechanism of growth promotion by bacteria from rhizosphere and nodules of alfalfa. *Planta, 243*, 1169–1189. doi: 10.1007/s00425-016-2469-7.
27. Kloepper, J. W., & Schroth, M., (1978). Plant growth-promoting rhizobacteria on radishes. *Proc. 4th Int. Conf. Plant Pathog. Bact.*
28. Kumar, A., Kumar, A., Devi, S., Patil, S., Payal, C., & Negi, S., (2012). Isolation, screening and characterization of bacteria from rhizospheric soils for different plant growth promotion (PGP) activities: An *in vitro* study. *Recent Res. Sci. Technol., 4*, 1–5.
29. Kumar, H., Bajpai, V. K., Dubey, R. C., Maheshwari, D. K., & Kang, S. C., (2010). Wilt disease management and enhancement of growth and yield of *Cajanus cajan* (L) var. Manak by bacterial combinations amended with chemical fertilizer. *Crop Prot., 29*, 591–598. doi: 10.1016/j.cropro.2010.01.002.
30. Malacca, M., Datta, K., Oliva, N., Velazhahan, R., Mew, T., Muthukrishnan, S., & Datta, S. K., (1999). Overexpression of thaumatin-like protein (PR5) gene in transgenic rice plants enhances resistance to sheath blight disease. *Theor. Appl. Genet., 98*, 1138–1145.
31. Malik, K. A., Bilal, R., Mehnaz, S., Rasul, G., Mirza, M. S., & Ali, S., (1997). Association of nitrogen-fixing, plant-growth-promoting rhizobacteria (PGPR) with kallar grass and rice. *Plant Soil, 194*, 37–44. doi: 10.1023/A:1004295714181.
32. McNeill, A., & Unkovich, M., (2007). *The Nitrogen Cycle in Terrestrial Ecosystems, Nutrient Cycling in Terrestrial Ecosystems.* Springer-Verlag, Berlin Heidelberg. doi: 10.1007/978-3-540-68027-7_2.
33. Mohr, K. I., Volz, C., Jansen, R., Wray, V., Hoffmann, J., Bernecker, S., Wink, J., Gerth, K., Stadler, M., & Müller, R., (2015). Pinensine: Die ersten antimykotischen lantibiotika. *Angew. Chemie, 127*, 11406–11410. doi: 10.1002/ANGE.201500927.

34. Moreira, S., Dutra, D., Rodrigues, A., Oliveira, J., Dhingra, O., & Pereira, O., (2013). Fungi and bacteria associated with post-harvest rot of ginger rhizomes in Espírito Santo, Brazil. *Trop. Plant Pathol., 38*, 218–226. doi: 10.1590/S1982-56762013000300006.
35. Ploetz, R. C., (2006). Fusarium wilt of banana is caused by several pathogens referred to as *Fusarium oxysporum* f. sp. *cubense*. *Phytopathology, 96*, 653–656. doi: 10.1094/PHYTO-96-0653.
36. Qian, J. H., Doran, J. W., & Walters, D. T., (1997). Maize plant contributions to root zone available carbon and microbial transformations of nitrogen. *Soil Biol. Biochem., 29*, 1451–1462. doi: 10.1016/S0038-0717(97)00043-6.
37. Raut, V., Shaikh, I., Naphade, B., Prashar, K., & Adhapure, N., (2017). Plant growth promotion using microbial IAA producers in conjunction with azolla: A novel approach. *Chem. Biol. Technol. Agric., 4*, 1. doi: 10.1186/s40538-016-0083-3.
38. Richardson, A. E., Barea, J. M., McNeill, A. M., & Prigent-Combaret, C., (2009). Acquisition of phosphorus and nitrogen in the rhizosphere and plant growth promotion by microorganisms. *Plant Soil, 321*, 305–339. doi: 10.1007/s11104-009-9895-2.
39. Shrestha, R. K., & Ladha, J. K., (1998). Nitrate in groundwater and integration of nitrogen-catch crop in rice-sweet pepper cropping system. *Soil Sci. Soc. Am. J., 62*, 1610–1619. doi: 10.2136/sssaj1998.03615995006200060019x.
40. Sims, D. A., & Gamon, J. A., (2002). Relationships between leaf pigment content and spectral reflectance across a wide range of species, leaf structures and developmental stages. *Remote Sens. Environ., 81*, 337–354. doi: 10.1016/S0034-4257(02)00010-X.
41. Teixeira, D. A., Alfenas, A. C., Mafia, R. G., Ferreira, E. M., De Siqueira, L., Maffia, L. A., & Mounteer, A. H., (2007). Rhizobacterial promotion of eucalypt rooting and growth. *Brazilian J. Microbiol., 38*, 118–123. doi: 10.1590/S1517-83822007000100025.
42. Wahyudi, A., Astuti, R., Widyawati, A., Meryandini, A., & Nawangsih, A., (2011). Characterization of *Bacillus sp.* strains isolated from rhizosphere of soybean plants for their use as potential plant growth for promoting rhizobacteria. *J. Microbiol. Antimicrob, 3*, 34–40.
43. Zacky, F. A., & Ting, A. S. Y., (2013). Investigating the bioactivity of cells and cell-free extracts of *Streptomyces griseus* towards *Fusarium oxysporum* f. sp. *cubense* race 4. *Biol. Control., 66*, 204–208. doi: 10.1016/j.biocontrol.2013.06.001.

CHAPTER 14

Infrared Spectroscopic Characterization of Some Synthetic Glycosides with Chromogenic Groups for Detection of Exo-Glycosidase Activity

ANA LAURA GARCÍA-PÉREZ,[1] OSCAR FERNANDO VÁZQUEZ-VUELVAS,[2] RODRIGO GUZMÁN-PEDRAZA,[3] AIDÉ SAENZ-GALINDO,[1] RODRIGO MACIAS-GARBETT,[1] and JUAN CARLOS CONTRERAS-ESQUIVEL[1]

[1]*School of Chemistry, Universidad Autonoma de Coahuila, Saltillo, Coahuila, Mexico*

[2]*Chemical Science Faculty, Universidad de Colima, Colima, Colima, Mexico*

[3]*School of Chemical Engineering, Autonomous University of Yucatan, Merida, Yucatan, Mexico*

ABSTRACT

A relevant aspect of carbohydrate analysis is the characterization of a helpful group of organic substrates widely used as target molecules to evaluate exo-glycosidase activities. Synthetic glycosides with chromogenic groups are relevant due to the facility and quickness to use them as a source of substrates. These can then be detected by rapid and straightforward colorimetric methods, employing just a piece of spectrophotometer equipment. The main characteristic of these organic substrates is the presence of an enzyme-hydrolyzable bond between a monosaccharide and aromatic heterocycle rings or phenyl rings with an *o*- or *p*-nitro activating substitution. These organic substrates are analytically convenient for the

Bioresources and Bioprocess in Biotechnology for a Sustainable Future. Leonardo Sepúlveda Torre, Juan Carlos Contreras-Esquivel, Ann Rose Abraham, & A. K. Haghi (Eds.)

evaluation of the hydrolytic enzymes such as α-glucosidase, β-glucosidase, β-galactosidase, and β-arabinofuranosidase. Given the importance of carbohydrate chemistry, these organic substrates have been analyzed using the vibrational technique attenuated total reflection (ATR)-Fourier transform infrared (FTIR) spectroscopy.

14.1 INTRODUCTION

Carbohydrates are a molecular group characterized by their biological origin as a predominant component of the vegetal and animal kingdoms as biopolymers or bonded to other organic structures. These compounds are also essential to life due to the numerous bioreactions in metabolic processes where they are involved [1, 2]. Carbohydrates are formed by small molecular units named monosaccharides, which may then be bonded to other equal or different structure monosaccharides to form polysaccharides, representing the most disseminated form of carbohydrates in nature [3]. Mono- or oligo saccharides are also present in the biological kingdom as intermediate entities during the metabolism through the biochemistry of polymerizing or degrading reactions, usually assisted and catalyzed by synthase or hydrolase enzymes [4].

For analytical purposes, the use of molecules to evaluate the catalytic capacity of hydrolytic enzymes has been instrumental to biotechnological methods of enzyme production. The analysis of the catalytical reactions is usually performed through sophisticated processes involving chromatography, specifically high-performance liquid chromatography (HPLC) [5]. Nonetheless, colorimetry presents an alternate technique widely used to evaluate or measure the catalytic products of enzymatic reactions. Colorimetric methodologies commonly employ organic compounds as enzymatic substrates. Such organic compounds feature bonds between monosaccharides and organic structures that are hydrolyzable at the oxygen atom linked to the anomeric carbon of the monosaccharide. The most interesting characteristic of these organic substrates is the enzymatic release of chromogenic molecules as products of the enzymatic reaction, which are ready to be spectrophotometrically measured [6–8]. As such, these molecules (Table 14.1) represent a valuable analytical tool to assess the production of enzymes for a wide variety of academic applications. The main advantage of using organic-carbohydrate compounds as enzymatic substrates lies in their application to the enzymology field of biotechnology and the evaluation of research- or industry-relevant enzymes, such as α-glucosidase and β-glucosidase, β-galactosidase, acetylation in protein-substrate enzymes (acetyl synthase) and β-arabinofuranosidase [9–11].

TABLE 14.1 Information of Organic Substrate and its Enzymes

Organic Substrates	Enzyme	E.C. Number
5-Bromo-4-Chloro-3-Indolyl-β-D-Galactopyranoside (X-Gal)	β-galactosidase	EC. 3.2.1.23
p-Nitrophenyl-β-D-galactopyranoside (*p*NPβGal)	β-galactosidase	EC. 3.2.1.23
2-Nitrophenyl-β-D-Gglucopyranoside (*o*NPβG)	β-glucosidase	EC. 3.2.1.21
p-Nitrophenyl-α-D-glucopyranoside (*p*NPαG)	α-glucosidase	EC. 3.2.1.20
p-Nitrophenyl-acetate (*p*NPA)	Acetylesterase	EC. 3.1.1.6
p-Nitrophenyl-β-L-arabinofuranoside (*p*NPβAra)	β-arabinofuranosidase	EC. 3.2.1.55

The analytical identification of these types of molecules is critical when organic synthesis is applied to produce these organic substrates. An important aspect is these compounds' relatively simple or small molecular structures. Usually, Nuclear magnetic resonance (NMR) techniques are the most used to elucidate and identify these molecules; however, this analytical equipment is a very sophisticated and expensive manner of analyzing organic substrates for most biochemical or biotechnological laboratories. Consequently, different, accessible equipment, and rapid, straightforward methods, such as Fourier transform infrared vibrational spectroscopy, may help identify and detect this kind of enzymatic organic substrates and composition samples [12].

The enzymatic organic substrates are constituted by a monosaccharide bonded to a hydrolyzable organic structure, which usually is a substituted phenyl ortho- or para-nitro group, and other entities such as an acetate group or aromatic heterocycle halogen-substituted rings. Given these molecular properties, Fourier transform infrared vibrational spectroscopy possesses a valuable potential to be used as an instrument to analyze specific signals or absorption bands to detect or identify molecules of organic substrates. Fourier transform infrared spectra display different zones along the spectral windows, which represent associated regions to specific structural characteristics of the analyzed samples; in fact, there is a region denominated as a functional group zone, which is the most pertinent aspect to be taken into account to utilize this Fourier transform infrared technology. Nonetheless, Fourier transform infrared spectroscopy may have some limitations in accurately assigning molecular structures to the spectrum of a sample; this technical restriction may be minimized with the continuous or repetitive use of this analytical tool, which statistically may contribute to identifying absorption patterns of a specific compound using a standard of the target organic substrate [12].

The adequate substrates to evaluate the relevant catalytic activities previously mentioned (α-glucosidase and β-glucosidase, β-galactosidase, acetylation in proteins enzymes (acetyl synthase), and β-arabinofuranosidase) are described as follows below. The molecular structure and the Attenuated total reflection (ATR)-Fourier transform infrared (FTIR) spectrum of each one is displayed and discussed to expose the advantages and benefits of vibrational spectroscopy to identify the main absorption peaks that describe the band pattern of every organic substrate. These compounds are 5-bromo-4-chloro-3-indolyl-β-D-galactopyranoside (X-Gal), p-nitrophenyl-β-D-galactopyranoside (pNPβGal), 2-nitrophenyl-β-D-glucopyranoside (oNPβG), p-nitrophenyl-α-D-glucopyranoside (pNPαG), p-nitrophenyl acetate (pNPA), and p-nitrophenyl-β-L-arabinofuranoside (p-NPβAra).

14.1.1 5-BROMO-4-CHLORO-3-INDOLYL-B-D-GALACTOPYRANOSIDE (X-GAL)

The 5-bromo-4-chloro-3-indolyl-β-D-galactopyranoside in Figure 14.1, is an organic compound with the heterocycle structure indole and substituents with halogens groups: chloro and bromo. The indole group is attached covalently to the galactoside group. 5-bromo-4-chloro-3-indolyl-β-D-galactopyranoside has been used in molecular biology and genetics to detect the gene LacZ [3, 13].

In the hydrolysis of the monosaccharides in Figure 14.2, the β-D-galactosidase acts breaking the union catalyzed the separation of 5-bromo-4-chloro-3-indolyl-β-D-galactopyranoside in galactose and 5-bromo-4-chloro-3-hydroxyindole. Subsequently, the 5-bromo-4-chloro-3-hydroxyindole obtained during the reaction is hydrolyzed, obtaining 5,5′-dibromo-4,4′-dichloro-indigo, a blue organic compound [13–16].

5-bromo-4-chloro-3-indolyl-β-D-galactopyranoside has been used successfully in medical applications to detect β-galactosidase in canine skin tissues, correlated to the production of large amounts of such enzyme in the senescent cells. The staining of tissues with 5-bromo-4-chloro-3-indolyl-β-D-galactopyranoside has been considered a simple and rapid method to detect β-galactosidase through the staining of skin tissues [14].

In addition, its use in the detection of the gene LacZ with a chromogenic method using 5-bromo-4-chloro-3-indolyl-β-D-galactopyranoside has good results and is widely reported in the literature (Table 14.2) for tissue staining applications and detecting of β-galactosidase activity specific for gen LacZ overexpression in tumor cells in animals [17, 18].

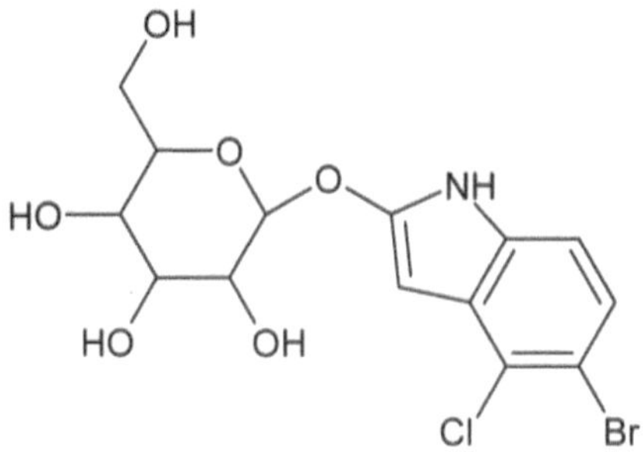

FIGURE 14.1 Chemical structure of the 5-bromo-4-chloro-3-indolyl-β-D-galactopyranoside (X-Gal).

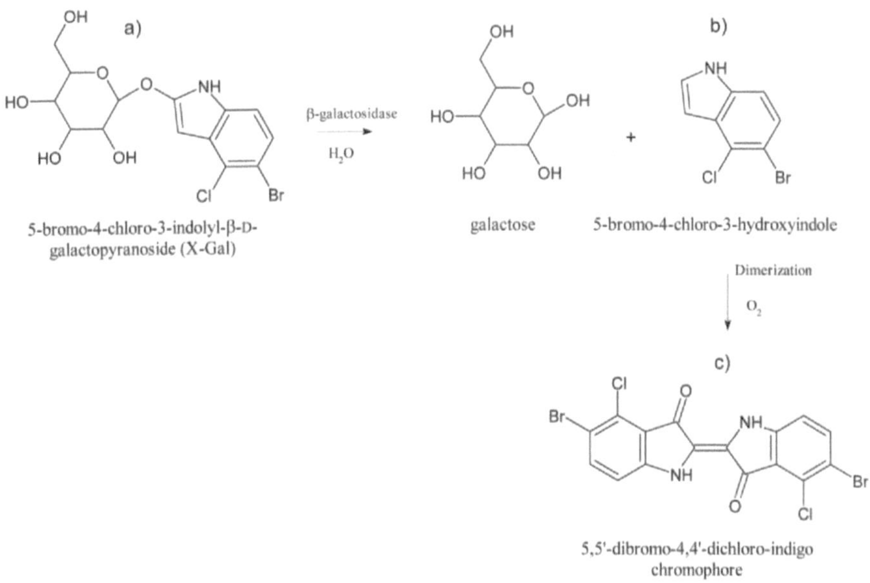

FIGURE 14.2 Scheme of the reaction decomposition (a) in presence of the β-galactosidase, produces the hydrolyzed product (b); finally, the dimerization and oxidization to form a chromophore blue (c) [13,15].

Source: Adapted from Ref. [13,15]

14.1.2 p-NITROPHENYL-B-D-GALACTOPYRANOSIDE (PNPBGAL)

The p-nitrophenyl-β-D-galactopyranoside in Figure 14.3, is a chromogenic monosaccharide-derivative substrate featuring a 4-nitrophenyl substituent in the 6th carbon of β-D-galactosidase. This substrate is widely employed for glycosidase hydrolysis [1].

TABLE 14.2 Some of the Most Recent Works in the Determination Gen LacZ Used Substrate 5-Bromo-4-Chloro-3-Indolyl-β-D-Galactopyranoside

Title	Authors	Year	References
BMP3 affects cortical and trabecular long bone development in mice.	Ivan Banovac, Lovorka Grgurevic, Viktorija Rumenovic, Slobodan Vukicevic and Igor Erjavec	2022	[19]
Developing a far-red fluorogenic beta-galactosidase probe for senescent cell imaging and photoablation.	Seung Koo Lee, Zhenhua Shen, Myung Shin Han and Ching-Hsuan Tung	2022	[20]
CRISPR/Cas9 gene editing in bacteria: Leading to Hematopoietic Stem Cell Editing in Pakistan.	Mohamed Abdul Naeem	2022	[21]
Protocol for whole-mount X-Gal staining combined with tissue clearing in embryo and adult mouse using CUBIC.	Haruko Watanabe-Takano, Moe Fukumoto, Shigetomo Fukuhara and Naoki Mochizuki	2022	[22]
A LacZ reporter with high activity in the human fungal pathogen *Candida albicans*.	Natalia Klimova, Siwei Chu and Bernard Turcotte	2021	[23]
Analysis of gene expression using LacZ reporter mouse lines.	Michael-Simon Krämer, Robert Feil and Hannes Schmidt	2021	[24]
Analysis of Dip2B expression in adult mouse tissues using the LacZ reporter gene.	Rajiv Kuman Sah, Noor Bahadar, Fatoumata Binta Bah, Salah Adlat, Zin Mar Oo, Luqing Zhang, Fawad Ali, M S Zobaer, Xuechao Feng and Yaowu Zheng	2021	[25]
Controlled release of rAAV vectors from APMA-functionalized contact lenses for corneal gene therapy.	Fernando Álvarez-Rivera, Ana Rey Rico, Jagadeesh K Venkatesan, Luis Díaz-Gómez, Magali Cucchiarini, Ángel Concheiro and Carmen Álvarez-Lorenzo	2020	[26]
Single-stranded adeno-associated virus serotype 9 mediated gene expression in astrocyte via intravenous delivery.	Xiao-Dan Wang	2020	[27]
Construction of targeted and integrative promoter-reporter plasmids pDK-K and pDK-G to measure gene expression activity in *Haemophilus parasuis*.	Ke Dai, Zheng Yang, Yung-Fu Chang, Lvquin él, Sanjie Cao, Qin Zhao, Xiaobo Huang, Rui Wu, Yong Huang, Qigui Yan, Han Xinfeng, Xiaoping Mama, Xintian Wen, and Yiping Wen	2019	[28]

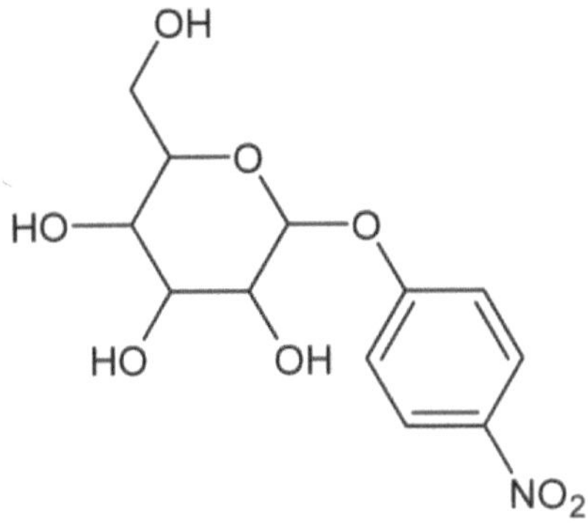

FIGURE 14.3 Chemical structure of the p-nitrophenyl-β-D-galactopyranoside (*p*NPβGal).

The p-nitrophenyl-β-D-galactopyranoside has been applied in the environmental area to purify and characterize the gene β-galactosidase (GH2 family) present in the marine bacterium in the deep sea, namely, in *Alteromonas* sp. ML52. The authors mention that relative percentual activity was determined for 2-nitrophenyl-β-D-galactopyranoside (100 ± 4.2) and 4-nitrophenyl-β-D-galactopyranoside (12.8 ± 0.5), concluding that the evidence shows efficient hydrolysis of lactose in milk at cold temperatures [29].

Zhang et al. reported the catalytic evaluation of β-galactosidase BGL0224 from *Oenococcus oeni* SD-2a in seven different substrates as a p-nitrophenyl-β-D-glucopyranoside, p-nitrophenyl-β-D-galactopyranoside, p-nitrophenyl-β-D-xylopyranoside,p-nitrophenyl-β-D-cellobioside,p-nitrophenyl-β-D-glucuronide, p-nitrophenyl-α-D-glucopyranoside, p-nitrophenyl-α-D-galactopyranoside in which they have analyzed the effect of temperature and pH in the catalytic activity and fluorescence quenching analysis [30].

The authors reported that BGL0224 has a heightened catalytic effect on p-nitrophenyl-β-D-glucopyranoside, p-nitrophenyl-β-D-cellobioside, p-nitrophenyl-β-D-galactopyranoside, and p-nitrophenyl-β-D-xylopyranoside but that the catalytic effect is present in all the analyzed substrates. The authors concluded that there exists a potent catalytic effect present in all the analyzed substrates and higher catalytic activity on substrates containing β-bonds in comparison with α-bonds (p-nitrophenyl-α-D-glucopyranoside and p-nitrophenyl-α-D-galactopyranoside); the latter having a weaker catalytic effect. p-nitrophenyl-β-D-glucopyranoside was considered the most suitable substrate, indicating a substrate affinity of BGL0224, and the authors mentioned that higher catalytic temperatures and lower pH values were more desirable for BGL0224 [30].

14.1.3 p-NITROPHENYL-α-D-GLUCOPYRANOSIDE (pNPαG) AND 2-NITROPHENYL-β-D-GLUCOPYRANOSIDE (oNPβG)

14.1.3.1 p-NITROPHENYL-α-D-GLUCOPYRANOSIDE (pNPαG)

The p-nitrophenyl-α-D-glucopyranoside in Figure 14.4, is a monosaccharide derivative of α-glucopyranoside with attached p-nitrophenol employed to produce α-glucoside hydrolysates and α-glucosidase. This substrate has been instrumental in determining α-glucosidase activity by a simple colorimetric assay, with applications in biochemistry, genetics, and sugar fermentation [9, 10].

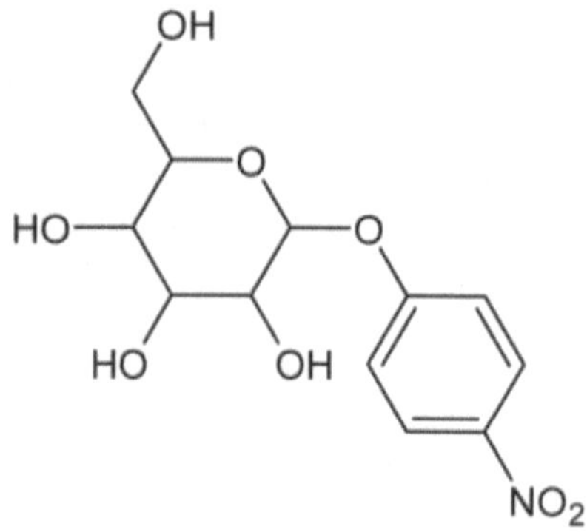

FIGURE 14.4 Chemical structure of the p-nitrophenyl-α-D-glucopyranoside (*p*NPαG).

p-nitrophenyl-α-D-glucopyranoside has been applied to immobilize α-glucosidase in thin films of SnO_2:Sb/ITO thin films employing organic p-nitrophenyl-α-D-glucopyranoside as a substrate for their immobilization [31]. p-nitrophenyl-α-D-glucopyranoside has also been reported for antibacterial activity, antioxidant, and α-glucosidase inhibition analysis in algal biomass with *Chroococcus minutus*. The determination was made in a mixture of α-glucosidase in algal extract, exhibiting a decrease in α-glucosidase activity. The authors commented that this enzyme is more effective than acarbose and presents antibacterial activity in Gram-positive and Gram-negative bacteria. p-nitrophenyl-α-D-glucopyranoside has also been employed to analyze the (2,2-diphenyl-1-picrylhydrazyl) DPPH radical scavenging activity of *Chroococcus minutus* with ascorbic acid [32].

14.1.3.2 2-NITROPHENYL-β-D-GLUCOPYRANOSIDE (oNPβG)

2-nitrophenyl-β-D-glucopyranoside in Figure 14.5, is a monosaccharide derivative with a substituted 2-nitrophenyl in the carbon #6 present in β-D-glucopyranoside. It has been employed to evaluate the activity of β-glucosidase, and it has been used for colorimetric determinations of free o-nitrophenol. β-glucosidase is commonly employed to degrade soil organic material and indicates soil quality [33, 34].

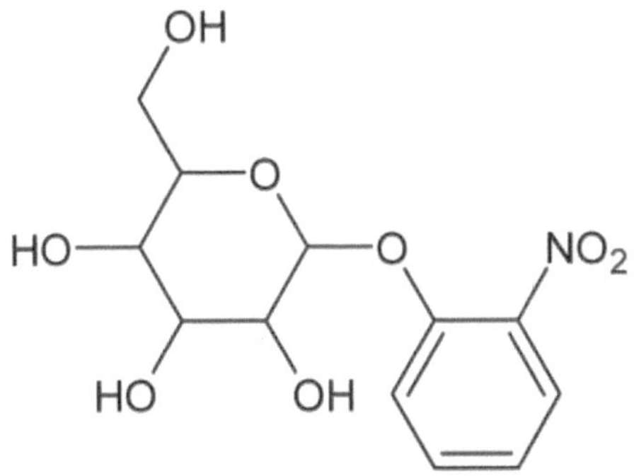

FIGURE 14.5 Chemical structure of the 2-nitrophenyl-β-D-glucopyranoside (*o*NPβG).

Trichoderma reesei β-glucosidase (Bg 11) was employed to degrade cellulose, and the enzymatic activity of protein Bgl1H was active with 2-nitrophenyl-β-D-glucopyranoside and 4-methylumbelliferyl-β-D-glucopyranoside. The specific activity of the enzyme using 2-nitrophenyl-β-D-glucopyranoside as substrate was 11 u/mg at 50°C (pH 5.0). According to the authors, the results obtained were that *Trichoderma* β-glucosidase is partially hydrolyzed hemicelluloses, β-glucans, and xylans [35].

The 2-nitrophenyl-β-D-glucopyranoside was employed to synthesize 3′-acetylated testosterone 17-O-β-glucoside (3′-AT-G) to determine OcUGT1, a flavonoid glycosyltransferase from *Ornithogalum caudatum* that catalyzes the glycosylation of testosterone with Uridine diphosphate glucose, forming testosterone 17-O-glucoside (T-17-G). The authors mentioned that the generation of T-17-G was achieved with 2-nitrophenyl-β-D-glucopyranoside or p-nitrophenyl-β-D-glucopyranoside and had an enhanced glycosylation efficiency with 2-nitrophenyl-β-D-glucopyranoside compared to uridine diphosphate glucose, heightening its potential as an alternative in the glycosylation of testosterone [36].

14.1.4 p-NITROPHENYL ACETATE (pNPA)

p-nitrophenyl acetate in Figure 14.6, this substrate is produced from the reaction between the p-nitrophenol carbon-1 hydroxyl group and the carboxyl group of acetic acid. p-nitrophenyl acetate has been employed for protein acetylation, demonstrating fast reaction times in serum albumin due to its nucleophilic affinity to the groups in the lateral chain of albumin amino-acids, increasing its reactivity with the proteins [37, 38].

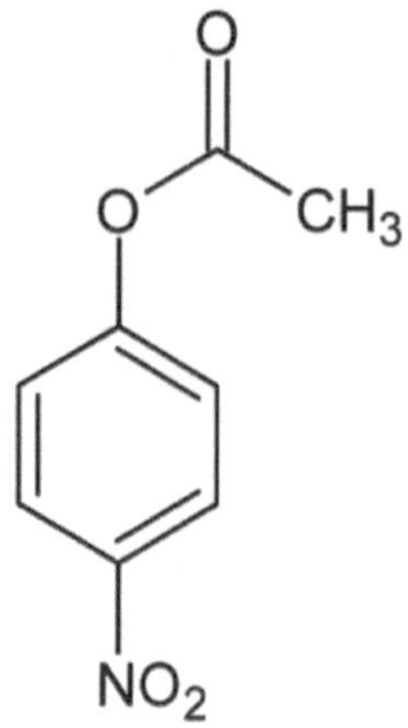

FIGURE 14.6 Chemical structure of the p-nitrophenyl acetate (*p*NPA).

The environmental application of p-nitrophenyl acetate was reported by Cortés-Gómez et al. where it was employed for the determination and evaluation of the relation of the activity esterase with cortisol and esterase with metals such as Cd, Pb, Ti, Sr, Se, Al, As, and Zn in the serum of marine turtles (*Lepidochelys olivacea*). Hydrolysis of p-nitrophenyl acetate to p-nitrophenol and esterase was analyzed in the presence of these inorganic elements and is directly related to the variation of the activity esterase with metals Cd, Pb, and Ti. Consequently, the authors consider such compounds practical biomarkers correlated to different pathologies in the Olive Ridley Turtles (Lepidochelys olivacea) [39].

In another application, p-nitrophenyl acetate activity was tested in substrates with different chain sizes employed by the thermophilic bacterium *Geobacillus thermodenitrificans* PS01, a few described in the literature. Chen et al. conclude that esterase Est1 is a thermophilic hydrolase that prefers short-chain C6 to C12 triglycerides or substrates as p-nitrophenyl acetate. The authors mention that rEst1 has ample thermal and pH stability, highlighting its potential industrial applications [40].

14.1.5 p-NITROPHENYL-α-L-ARABINOFURANOSIDE (pNPαARA)

The substrate p-nitrophenyl-α-L-arabinofuranoside in Figure 14.7 has been employed to determine α-L-arabinofuranoside activity. It has also been featured in the hydrolysis of hemicelluloses in the presence of the enzyme α-L-arabinofuranoside [42].

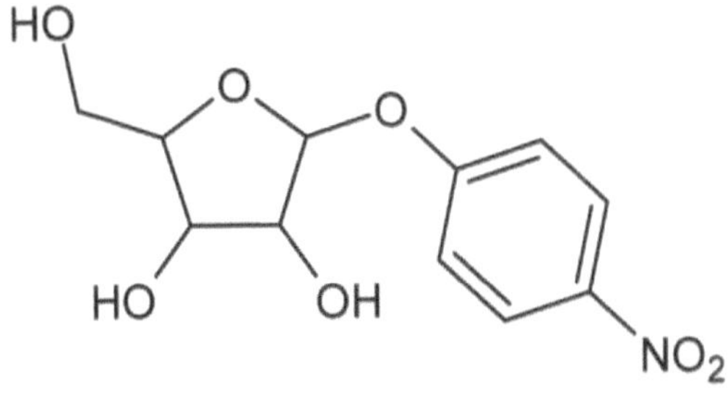

FIGURE 14.7 Chemical structure of the p-nitrophenyl-α-L-arabinofuranoside (*p*-NPαAra).

The application of p-nitrophenyl-α-L-arabinofuranoside for the characterization of β-xilosidase produced for *Aspergillus niger* for bioethanol obtained through solid-state fermentation was evaluated through p-nitrophenyl-α-L-arabinofuranoside activity and p-nitrophenyl-β-D-glucopyranoside, showing high activity for the p-nitrophenyl-β-D-xylopyranoside. Therefore, using β-xylosidase is desirable for bioethanol production through lignocellulose hydrolysis, in which great interest has been shown [43].

Gao et al. have reported the process optimization for the obtention of α-L-arabinofuranoside of the family 51 GH from *Paenibacillus polymyxa*, getting a maximum concentration of the enzyme α-L-arabinofuranoside, denominating it as PpAFase-1. The authors show high activity for arabinofuranosidase compared to xylosidase and arabinose activities. The specific activity (U/mg) was determined on different substrates such as the p-nitrophenyl-β-D-glucopyranoside 2.5 ± 0.09, p-nitrophenyl-α-L-arabinofuranoside 76.7 ± 1.2, and p-nitrophenyl-α-D-glucopyranoside not detected. The enzyme α-L-arabinofuranoside of family 51 GH is thermostable and stable over a wide pH range, and your tolerance to arabinose and xylose could be used for hemicellulose degradation, as mentioned by the authors [44].

14.2 EXPERIMENTAL: MATERIALS AND METHODS

5-bromo-4-chloro-3-indolyl-β-D-galactopyranoside (X-Gal) (Purity HPLC ≥99%) was acquired Apex Bioresearch products (APExBio), p-nitrophenyl-β-D-galactopyranoside (pNPβGal) and 2-nitrophenyl-β-D-glucopyranoside (oNPβG) were acquired by goldbiochtecnology, p-nitrophenyl-α-D-glucopyranoside (pNPαG) (Purity ≥99%) and p-nitrophenyl acetate (pNPA) (Purity HPLC ≥98%) were acquired in Sigma-Aldrich, p-nitrophenyl-β-L-arabinofuranoside (p-NPβAra) (Purity HPLC >99%) was acquired Research Products International (RPI). All reagents were stored in a freezer. The spectra were analyzed in the absorbance range 600 to 4000 cm^{-1} with 35 scans in equipment supplied with ATR-diamond crystal (PerkinElmer, USA). The spectra were baseline corrected and normalized to an absorbance of 1.

14.3 RESULTS AND DISCUSSION

14.3.1 CHARACTERIZATION FOR ATR-FTIR SPECTROSCOPY

14.3.1.1 5-BROMO-4-CHLORO-3-INDOLYL-β-D-GALACTOPYRANOSIDE (X-GAL)

The analysis of the 5-bromo-4-chloro-3-indolyl-β-D-galactopyranoside in Figure 14.8 signals are observed in 3421 cm^{-1} corresponding to O-H stretching, 2872 cm^{-1} for C-H stretching that is attributed 4-nitrophenyl-β-D-galactopyranoside, the signal in 1609 cm^{-1} for C=C aromatic groups stretching, 1557 cm^{-1} to C-N tension, in 1456 cm^{-1} C=C aromatic group tension, in 1421 cm^{-1} flexion of O-H in the plane, in 1265 cm^{-1} C-N tension, 1140 cm^{-1} stretching of the C-O and 938 cm^{-1} corresponding flexion of the aromatic C-H bonds.

14.3.1.2 p-NITROPHENYL-β-D-GALACTOPYRANOSIDE (pNPβGAL)

The analysis of the for p-nitrophenyl-β-D-galactopyranoside in Figure 14.9 the following signals are observed: 3323 cm^{-1} corresponding to stretching of O-H, 2886 cm^{-1} for stretching of C-H bonds, in 1609 cm^{-1}, stretching of aromatic C=C. Signals of tension are found in 1595 cm^{-1} for C-N groups, 1509 cm^{-1} corresponding asymmetric stretching of NO_2, in 1496 cm^{-1} tension

C=C aromatic groups, 1351 cm^{-1} corresponds with symmetric stretching of NO_2 and stretching C-O in 1180 cm^{-1}.

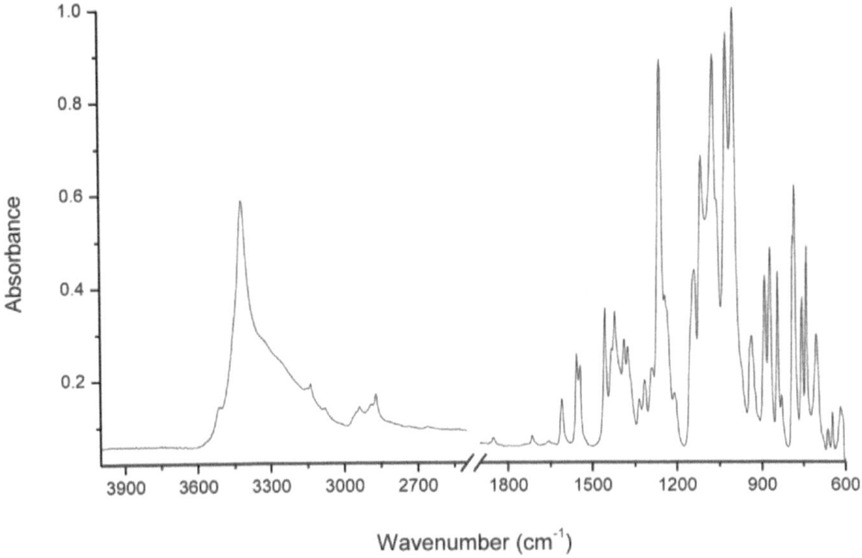

FIGURE 14.8 FTIR spectra for 5-bromo-4-chloro-3-indolyl-β-D-galactopyranoside (X-Gal).

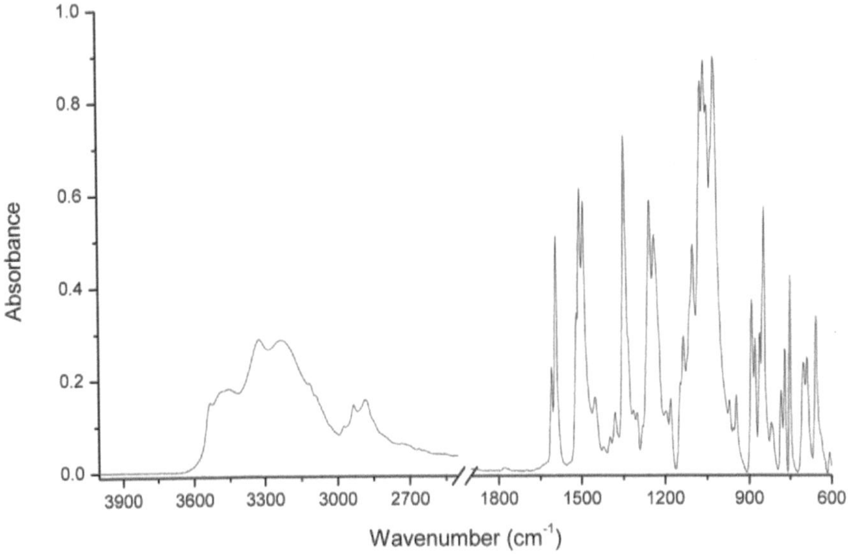

FIGURE 14.9 FTIR spectra for p-nitrophenyl-β-D-galactopyranoside (*p*NPβGal).

14.3.1.3 p-NITROPHENYL-α-D-GLUCOPYRANOSIDE (pNPαG) AND 2-NITROPHENYL-β-D-GLUCOPYRANOSIDE (oNPβG)

The analysis of the p-nitrophenyl-α-D-glucopyranoside and 2-nitrophenyl-β-D-glucopyranoside in Figure 14.10, observed signal corresponding stretching of O-H in 3473 cm^{-1} for p-nitrophenyl-α-D-glucopyranoside and 3588 cm^{-1} for 2-nitrophenyl-β-D-glucopyranoside, the tension of O-H in 3305 cm^{-1} for p-nitrophenyl-α-D-glucopyranoside and 3341 cm^{-1} for oNPβG, in this bands there are differences in shifts are observed due to the effect of substituents in ortho- or para-orientation.

There are signals corresponding to stretching C-H in 2943 cm^{-1} for p-nitrophenyl-α-D-glucopyranoside and 2937 cm^{-1} for 2-nitrophenyl-β-D-glucopyranoside, stretching C=C aromatic groups in 1609 cm^{-1} for p-nitrophenyl-α-D-glucopyranoside and 1607 cm^{-1} for 2-nitrophenyl-β-D-glucopyranoside, tension of groups C-N in 1591 cm^{-1} for *p*-nitrophenyl-α-D-glucopyranoside and 1583 cm^{-1} for 2-nitrophenyl-β-D-glucopyranoside, asymmetric stretching of the NO_2 groups in 1513 cm^{-1} for p-nitrophenyl-α-D-glucopyranoside and 1522 cm^{-1} for 2-nitrophenyl-β-D-glucopyranoside, aromatic C=C tension in 1489 cm^{-1} for p-nitrophenyl-α-D-glucopyranoside and 1485 cm^{-1} for 2-nitrophenyl-β-D-glucopyranoside, symmetric stretching of NO_2 in 1340 cm^{-1} for p-nitrophenyl-α-D-glucopyranoside and 1337 cm^{-1} 2-nitrophenyl-β-D-glucopyranoside, stretching C-O in 1173 cm^{-1} for p-nitrophenyl-α-D-glucopyranoside and 1174 cm^{-1} for 2-nitrophenyl-β-D-glucopyranoside, 749 cm^{-1} flexion C-H group aromatic for p-nitrophenyl-α-D-glucopyranoside and 745 cm^{-1} for 2-nitrophenyl-β-D-glucopyranoside. In these signals, is observable a minor difference in their displacement.

14.3.1.4 p-NITROPHENYL ACETATE (pNPA)

For the analysis of the p-nitrophenyl acetate in Figure 14.11, signs were observed in 3114 cm^{-1} corresponding to stretching of C-H, 1758 cm^{-1} for C=O tension, 1615 cm^{-1} for stretching of C=C aromatic groups, signals at 1591 cm^{-1} correspond to C-N tension, 1516 cm^{-1} matches asymmetric NO_2 stretching, tension of C=C aromatic groups is shown in 1487 cm^{-1}, 1428 cm^{-1} shows flexion in the plane, 1343 cm^{-1} shows symmetric of NO_2, stretching C-O is found at 1187 cm^{-1}, 1154 cm^{-1} equals stretching of $COCH_3$, 1102 cm^{-1} shows tension C-O-C, 1040 cm^{-1} stretching C-O-C, 1007 cm^{-1} stretching

C-OH, 913 cm^{-1} flexion C-H group aromatic and 866 cm^{-1} attributed phenyl *p*-substituted.

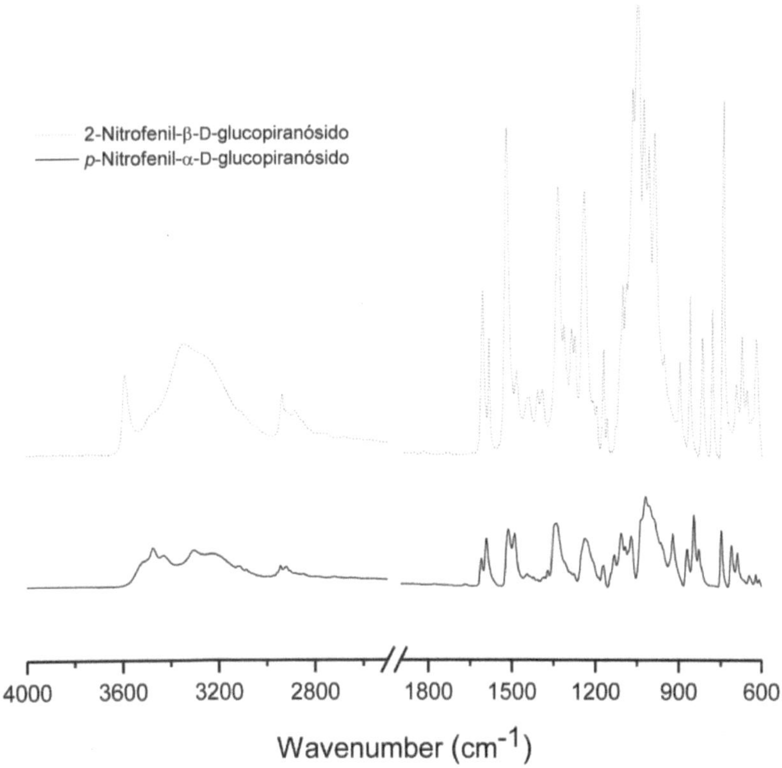

FIGURE 14.10 FTIR spectra for p-nitrophenyl-α-D-glucopyranoside (*p*NPαG) and 2-nitrophenyl-β-D-glucopyranoside (*o*NPβG).

In the spectrum in the database, broadband signals are observed in 3429 cm^{-1} and 3439 cm^{-1} that do not correspond to the characteristic signs of the p-nitrophenyl acetate, so the spectrum obtained shows free p-nitrophenol signals, where these signals are not observed [36].

14.3.1.5 p-NITROPHENYL-α-L-ARABINOFURANOSIDE (pNPαARa)

The analysis of the p-nitrophenyl-α-L-arabinofuranoside in Figure 14.12 signals are observed at 3468 cm^{-1} corresponding to stretching of O-H,

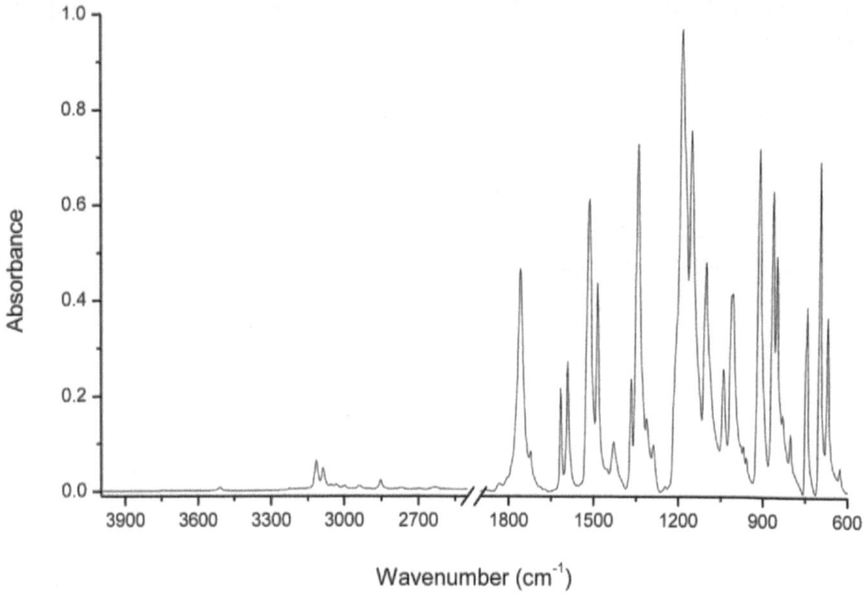

FIGURE 14.11 FTIR spectra for p-nitrophenyl acetate (*p*NPA).

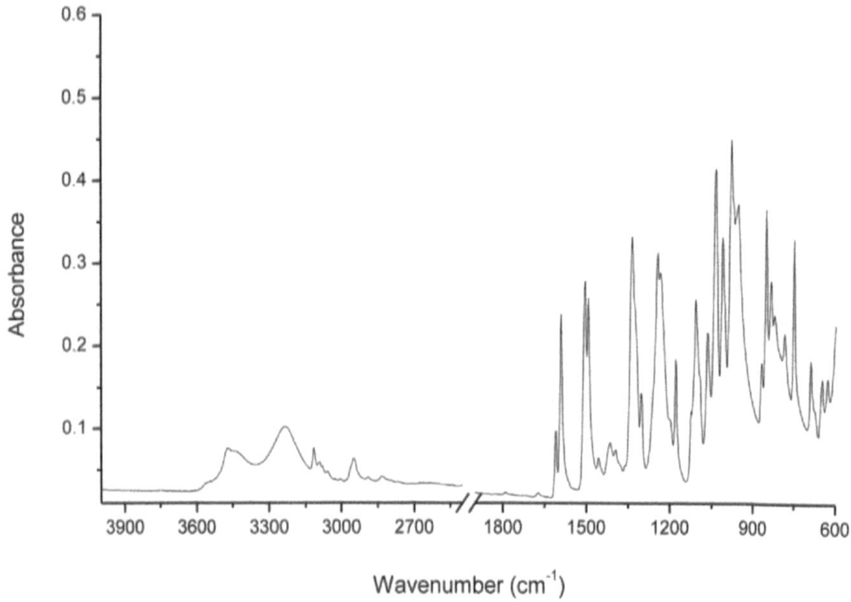

FIGURE 14.12 FTIR spectra for p-nitrophenyl-α-L-arabinofuranoside (*p*NPαAra).

3234 cm^{-1} corresponding to tensioned O-H, 2951 cm^{-1} for stretching of C-H aromatic groups, a signal of stretching in 1609 cm^{-1} of C=C aromatic groups, 1591 cm^{-1} corresponding to tension of C-N, in 1506 cm^{-1} asymmetric stretching of the NO_2 link, in 1494 cm^{-1} tension of aromatic C=C and at 1337 cm^{-1} corresponding with symmetric stretching of NO_2, respectively, the signal in 1246 cm^{-1} for the tension of C-O-C bonds and 1179 cm^{-1} of the stretching C-O.

14.4 CONCLUSION

The current text discusses the Fourier transform infrared characterization of organic substrates used for carbohydrate hydrolysis and commonly employed in colorimetric assays for detachment of free molecules, which reacts with several compounds to generate a quantifiable color change.

We also mentioned the applications and the results obtained in different research articles where authors employ these substrates to release colorimetric molecules that are quantifiable for analytical techniques and spectrophotometry. Additionally, this research addresses the characterization of such compounds through Fourier transform infrared spectrum to provide information of the signals obtained during this analysis. This document contributes with information on the characterization of these substrates that are widely used to detect of enzymes: α-glucosidase, β-glucosidase, β-galactosidase, and β-arabinofuranosidase with various applications reported in the literature.

KEYWORDS

- **Fourier transform infrared spectrum**
- ***p*NPA**
- ***p*NPαG**
- ***p*NPβAra**
- ***p*NPβGal**
- **X-Gal**
- **β-galactosidase**

REFERENCES

1. Alhifthi, A., & Williams, S., (2021). Unimolecular, biomolecular, and intramolecular hydrolysis mechanisms of 4-nitrophenyl β-D-glucopyranoside. *The Journal of Organic Chemistry*. doi.org/10.1021/acs.joc.1c00836.
2. Jéquier, E., (1994). Carbohydrates as a source of energy. *American Journal of Clinical Nutrition, 59*, 682S–685S.
3. Mentesana, P. E., Dosil, M., & Konopka, J. B., (2002). Functional assays for mammalian G-protein-coupled receptors in yeast. *Activation of G Proteins by Regulators* (Vol. 344, pp. 92–111). Academic Press.
4. MacGregor, E. A., (2003). Biopolymers, *Encyclopedia of Physical and Technology* (3rd edn.). ISBN 9780122274107.
5. Xie, K., Zhang, X., Sui, S., Ye, F., & Dai, J., (2020). Exploring and applying the substrate promiscuity of a C-glycosyltransferase in the chemo-enzymatic synthesis of bioactive C-glycosides. *Nature Communications*. doi.org/10.1038/s41467-020-18990-9.
6. Cardelli, R., Becagli, M., Marchini, F., & Saviozzi, A., (2019). Biochar impact on the estimation of the colorimetric-based enzymatic assays of soil. *Soil Use and Management, 35*, 478–481.
7. Liu, J., Wu, F., Liu, C., Bao, H., & Fu, T., (2020). "Turn-on" fluorometric probe for α-glucosidase activity using red fluorescent carbon dots and 3,3′,5,5′-tetramethylbenzidine. *Microchimica Acta*. https://doi.org/10.1007/s00604-020-04479-1.
8. Chojnowska, S., Ptaszyńska-Sarosiek, I., Kępkac, A., Dariusz-Szajdad, S., Waszkiewiczd, N., & Zwierz, K., (2018). Optimization of the method for α-L-fucosidase, β-D-galactosidase and β-D-glucuronidase determination in serum from hemolyzed blood. *Advances in Medical Sciences, 63*, 306–311.
9. Hollatz, C., & Stambuk, B. U., (2001). Colorimetric determination of active α-glucoside transport in *Saccharomyces cerevisiae*. *Journal of Microbiological Methods, 46*, 253–259.
10. Guerin, J. F., Ali, H. B., Jacques, R., Souchier, C., & Czyba, J. C., (1986). α-glucosidase as a specific epididymal enzyme marker its validity for the etiologic diagnosis of azoospermia. *Journal of Andrology, 7*, 56–162.
11. Bernard, H., Terrapon, N., Coutinho, P. M., Lombard, V., Drula, E., Garron, M. L., & Hornung, B., (2022). *Carbohydrate active Enzymes Database*. [Online] http://www.cazy.org/ (accessed on 10 July 2023).
12. Elakiya, N., & Arulmozhiselvan, K., (2021). Characterization of substrates of growing media by Fourier transform infrared (FT-IR) spectroscopy for containerized crop production. *Journal of Applied and Natural Science, 13*, 35–42.
13. Burn, S. F., (2012). *Detection of β-Galactosidase Activity: X-gal Staining* (Vol. 886, pp. 241–250). Humana Press.
14. Pati, S., Jainista, S., Behera, M., Acharya, P. A., Panda, S. K., & Senapati, S., (2014). X-gal staining of canine skin tissues: A technique with multiple possible applications. *Journal of Natural Science, Biology, and Medicine, 5*, 245–249.
15. Cai, X., Li, L., Krumholz, A., Guo, Z., Erpelding, T. N., Zhang, C., Zhang, Y., et al., (2012). Multi-scale molecular photoacoustic tomography of gene expression (Vol. 7, Issue 8, e43999). *Plos One,* doi.org/10.1371/journal.pone.0043999.
16. Matchynski, J., Manwar, R., Kratkiewicz, K. J., Mandangopal, R., Lennon, V. A., Makki, K. M., Reppen, A. L., et al., (2021). Direct measurement of neuronal ensemble

activity using photoacoustic imaging in the stimulated Fos-LacZ transgenic rat brain: A proof-of-principle study. *Photoacoustics, 24*, 10297.
17. Wei, L., & Yao, J., (2018). Photoacoustic microscopy: Principles and biomedical applications, *Biomedical Engineering Letters, 8*, 203–213.
18. Green, M. R., & Sambrook, J., (2019). Screening bacterial colonies using x-gal and IPTG: Α-complementation. *Cold Spring Harbor Protocols*. doi: 10.1101/pdb.prot101329.
19. Banovac, I., Grgurevic, L., Rumenovic, V., Vukicevic, S., & Erjavec, I., (2022). BMP3 affects cortical and trabecular long bone development in mice. *International Journal of Molecular Sciences, 23*. doi.org/10.3390/ijms23020785.
20. Lee, S. K., Shen, Z., Han, M. S., & Tung, C. H., (2022). Developing a far-red fluorogenic beta-galactosidase probe for senescent cell imaging and photoablation. *RSC Advances, 12*, 4543–4549.
21. Naeem, M. A., (2022). CRISPR/Cas 9 gene editing in bacteria: Leading to hematopoietic stem cell editing in Pakistan, *Journal of Haematology & Stem Cell Research, 2*, 4–7.
22. Watanabe-Takano, H., Fukumoto, M., Fukuhara, S., & Mochizuki, N., (2022). Protocol for whole-mount X-gal staining combined with tissue clearing in embryo and adult mouse using CUBIC. *Star Protocols, 3*. doi.org/10.1016/j.xpro.2022.101127.
23. Klimova, N., Chu, S., & Turcotte, B., (2021). A lacZ reporter with high activity in the human fungal pathogen *Candida albicans*. *FEMS Yeast Research, 21*. doi.org/10.1093/femsyr/foab013.
24. Kramer, M. S., Feil, R., & Schmidt, H., (2021). Analysis of gene expression using LacZ reporter mouse lines. *Mouse Genetics*, 29–45.
25. Sah, R. K., Bahadar, N., Bah, F. B., Adlat, S., Oo, Z. M., Zhang, L., Alí, F., et al., (2021). Analysis of Dip2B expression in adult mouse tissues using the LacZ reporter gene. *Current issues in Molecular Biology, 43*, 529–542.
26. Álvarez-Rivera, F., Rey-Rico, A., Venkatesan, J. K., Díaz-Gomez, L., Cucchiarini, M., Concheiro, A., & Alvarez-Lorenzo, C., (2020). Controlled release of rAAV vectors from APMA-functionalized contact lenses for corneal gene therapy. *Pharmaceutics, 12*, 335.
27. Wang, X. D., (2020). Single-stranded adeno-associated virus serotype 9 mediated gene expression in astrocyte via intravenous delivery. *Journal of Shanghai Jiaotong University (Medical Science)*, 1036–1040.
28. Dai, K., Yang, Z., Chang, Y. F., He, L., Cao, S., Zhao, Q., Huang, X., et al., (2019). Construction of targeted and integrative promoter-reporter plasmids pDK-K and pDK-G to measure gene expression activity in *Haemophilus parasuis*. *Microbial Pathogenesis*, doi.org/10.1016/j.micpath.2019.103565.
29. Sun, J., Congyu, Y., Wang, W., Zhuang, Z., Liu, J., Dai, F., & Hao, J., (2018). Cloning, expression and characterization of a novel cold-adapted β-galactosidase from the deep-sea bacterium *Alteromonas sp*. ML52. *Marine Drugs, 16*, 469.
30. Zhang, J., Zhao, N., Xu, J., Qi, Y., Wei, X., & Fan, M., (2021). Exploring the catalytic mechanism of a novel β-glucosidase BGL0224 from *Oenococcus oeni* SD-2a: Kinetics, spectroscopic and molecular simulation. *Enzyme and Microbial Technology, 148*. doi.org/10.1016/j.enzmictec.2021.109814.
31. Gur, B., (2020). Determination of the pH-dependent immobilization efficacy of α-glycosidase and its catalytic performance on SnO_2:Sb/ITO thin films. *Biochemical Engineering Journal, 163*. doi.org/10.1016/j.bej.2020.107758.
32. Arguelles, E., (2022). Total phenolic content and *in vitro* analysis of antioxidante, antibacterial, and alpha-glucosidase inhibition properties of *Chroococcus minutus*

(Kutzing) Nageli (chroococcales, cyanobacteria). *Journal of Faculty of Pharmacy of Ankara University, 46*, 170–181.

33. De Almeida, F. R., Naves, E. R., & Da Mota, R. P., (2015). Soil quality: Enzymatic activity of soil β-glucosidase. *Global Journal of Agricultural Research and Reviews, 3*, 146–150.
34. Turner, B. L., Hopkins, D. W., Haygarth, P. M., & Ostle, N., (2002). β-glucosidase activity in pasture soils. *Applied Soil Ecology, 20*, 157–162.
35. Brumm, P., Xie, D., Allen, L., & Mead, D. A., (2019). *Trichoderma reesei* (*Hypocrea jecorina*) Bgl1 is a novel, exo-acting glucanase and xylanase. *Journal of Enzymes, 1*, 19–30.
36. Yuan, S., Yin, S., Liu, M., He, J. M., & Kong, J. Q., (2019). OcUGT1-catalyzed glycosylation of testosterone with alternative donor substrates. *Process Biochemistry, 73*, 82–85.
37. Means, G. E., & Bender, M. L., (1975). Acetylation of human serum albumin by *p*-nitrophenyl acetate. *Biochemistry, 14*, 4989–4994.
38. Awad-Elkarim, A., & Means, G. E., (1988). The reactivity of *p*-nitrophenyl acetate with serum albumins. *Comparative Biochemistry and Physiology Part B: Comparative Biochemistry, 91*, 267–272.
39. Cortés-Gómez, A. A., Tvarijonaviciute, A., Teles, M., Cuenca, R., Fuentes-Mascorro, G., & Romero, D., (2018). *p*-nitrophenyl acetate esterase activity and cortisol as biomarkers of metal pollution in blood of olive ridley turtles (*Lepidochelys olivacea*). *Archives of Environmental Contamination and Toxicology, 75*, 25–36.
40. Chen, P. T., Liu, C. H., Chen, Y. T., Hsu, F. Y., & Shaw, J. F., (2020). Isolation, expression and characterization of the thermophilic recombinant esterase from *Geobacillus thermodenitrificans* PS01. *Applied Biochemistry and Biotechnology, 191*, 112–124.
41. *Spectral Database for Organic Compounds SDBS,* (2022). National Institute of Advanced Industrial Science and Technology (AIST). [Online] https://sdbs.db.aist.go.jp/sdbs/cgi-bin/landingpage?sdbsno=6108 (accessed on 10 July 2023).
42. Zheng, F., Liu, J., Basit, A., Miao, T., & Jiang, W., (2018). Insight to improve α-L-arabinofuranosidase productivity in *Pichia pastoris* and its application on corn stover degradation. *Frontiers in Microbiology*. doi.org/10.3389/fmicb.2018.03016.
43. Boyce, A., & Walsh, G., (2018). Purification and characterization of a thermostable β-xylosidase from *Aspergillus Niger* van Tieghem of potential application in lignocellulosic bioethanol production. *Applied Biochemistry and Biotechnology, 186*, 712–730.
44. Gao, J., Zhao, Y., Zhang, G., Li, Y., & Li, Q., (2018). Production optimization, purification, expression, and characterization of a novel α-L-arabinofuranosidase from *Paenibacillus polymyxa*. *Electronic Journal of Biotechnology, 36*, 24–33.

CHAPTER 15

Biotechnological Advances in Obtaining Phytochemical Compounds with Biological Properties from *Opuntia ficus-indica* and Their Potential Applications: A Review

SERGIO ARTURO CORONADO-CONTRERAS,[1]
XOCHITL RUELAS-CHACÓN,[2] MIRIAM DESIRÉE DÁVILA-MEDINA,[1]
YADIRA KARINA REYES-ACOSTA,[1] JUAN A. ASCACIO-VALDÉS,[1] and
LEONARDO SEPÚLVEDA-TORRE[1]

[1]*Bioprocess and Microbial Biochemistry Group, School of Chemistry, Autonomous University of Coahuila, Saltillo, Coahuila, Mexico*

[2]*Department of Food Science and Technology, Autonomous Agrarian University Antonio Narro, Saltillo, Coahuila, Mexico*

ABSTRACT

Plants belonging to the *Cactaceae* family are xerophytes, distributed in arid and semi-arid zones. *Opuntia ficus-indica* is a perennial cactus native to the American continent, adapted to desert, and semi-desert climates. It has flattened fleshy cladodes with thorns, flowers, and fruits known as tuna. This cactus presents a great amount of phytochemicals with biological activity, being this plant of great interest for research, since the method of obtaining these compounds and their possible applications are topics of interest for scientific and industrial development *O. ficus-indica* include polyphenols, betalains, colorants, and vitamins, since these compounds have different properties such as being anticarcinogenic and antioxidants.

Bioresources and Bioprocess in Biotechnology for a Sustainable Future. Leonardo Sepúlveda Torre, Juan Carlos Contreras-Esquivel, Ann Rose Abraham, & A. K. Haghi (Eds.)

15.1 INTRODUCTION

Plants belonging to the *Cactaceae* family are xerophytic plants, mostly distributed in arid and semi-arid regions [1]. They also have a great variety of shapes and sizes, succulent stems and in some cases edible fruits. In addition to having an important socioeconomic importance, since they are used as ornamental plants, for human food and as fodder for animals, thus being important for the economic development of the regions [2].

There are studies that have shown that there are species of the *Cactaceae* family with nutrients and bioactive compounds [3, 4] being also this family of interest for new research due to their potential as sources of natural dyes, agents for the control of diseases such as obesity and diabetes, as well as in the development of edible films [5–7].

Specifically referring to the nopal, this is a perennial cactus native to the American continent that belongs to the cactaceae, so it is adapted to the semi-desert climates. As it is a drought-resistant plant, this cactus has flattened fleshy cladodes with thorns, which are oval and flat shoots with thick bark, without leaves, also being cultivated in Mexico due to its wide gastronomic use, the most common species is found in 29 states of the Mexican Republic is *Opuntia ficus-indica* [8, 36]. Due to the favorable climate present in these states and the cultural background, Mexico are home to the most extensive cultivation, with an area of between 50,000 and 70,000 ha, but it is also cultivated in other countries in South America and Spain [9].

In arid areas, prickly pear cactus spreads easily, it can be considered a renewable lignocellulosic biomass to generate reducing sugars [10], being also considered an option to prevent desertification [11], in addition, this plant is used in the preparation of various foods, including its fruit known as tuna, this is a berry with a thick shell and full of seeds with a particular and mild flavor, its pulp is sweet and with little acidity, it is also used for the preparation of sweets and liquors. There are four types of prickly pears divided by color: green, red, yellow, and purple, whose color depends on the species and ripening stage of the fruit [12]. Due to its use, proximate studies of prickly pear have been carried out, determining that they contain approximately 85% water, 15% sugar, 0.3% ash and less than 1% protein [13].

Opuntia ficus-indica is exploited due to the production of prickly pear cactus and parts of the cactus such as nopal, but even so there are parts that are undervalued being considered waste [14] and some are used as feed for livestock [15–17] therefore, alternatives are sought to take advantage of the waste by obtaining compounds or materials of interest such as paper [18], since the industrial production of nopalitos and their pruning implies the generation of waste that is abandoned on site, thus being a source of severe contamination and being able to spread diseases to healthy plants [19].

The following review will focus on the main phytochemicals reported in *Opuntia ficus-indica*, the extraction methods used to obtain them and their possible application in different areas, thus promoting research on the total utilization of this plant, such as its cultivation.

15.2 PHYTOCHEMICALS OF *OPUNTIA FICUS-INDICA*

The chemical composition of this cactus (*Opuntia ficus-indica*) varies depending on the species, the age of the cladodes, the soil, the region where it is grown, as well as the season of the year [20, 21]. The peel of this fruit is acidic, contains polysaccharides, lipids, sterols, fat-soluble vitamins as well as pigments such as betalains, chlorophylls, carotenoids, and coumarins [22].

Secondary metabolites of *Opuntia ficus-indica* extracts have been determined using different methodologies such as LC-MS/MS, HPLC-DAD-MS, HPLC-UV, demonstrating the presence of different phenols and other compounds of interest, which are shown in Table 15.1, where we can see that different acids and phenolic molecules are found.

There are also studies showing that the cladodes of this plant are rich in carbohydrates such as rhamnose, fructose, arabinose, xylose, galactose, glucose, and galacturonic acid [27, 28], Likewise, proteins, carbohydrates, fiber, minerals, vitamins, and pigments such as chlorophylls, carotenoids, betalains, and phenolic compounds have been reported in *Opuntia ficus-indica* [29, 30].

Other compounds of interest are fatty acids, lipids [31] sterols and fat-soluble vitamins have been reported in *O. ficus-indica*, being also a rich source of vitamin E, being approximately 80% the α-tocopherol constituent of the total vitamin E, in turn vitamin K was also found present in this plant [32].

Several studies have been carried out to identify compounds of interest in *O. ficus-indica* and to see the changes in their concentration among the different parts of the plant. In studies that analyze the profile of these compounds, xanthophylls, and carotenoids have been identified, the main carotenoid found in *O. ficus-indica* prickly pears being lutein, followed by β-carotene, also reporting (all-E)-violaxanthin, (all-E)-neoxanthin, (9Z)-violaxanthin, (all-E)-anteraxanthin, concluding that prickly pear peel has the highest content of phytochemicals, being of great value for the development of nutraceuticals and functional foods [33], the presence of coumarins in *O. ficus-indica*, such as dihydroxy psoralen-O-hexoside, has also been reported, but these compounds are found in lower proportion in the plant [34].

TABLE 15.1 Compounds Frequently Found in *Opuntia ficus-indica*

Opuntia ficus-indica	Extraction Method	Molecule References	References
Prickly pear peel, prickly pear pulp, cladiodes	Acidified methanol and different solvents	Piscidic acid	[23–25]
Prickly pear peel, prickly pear pulp, cladiodes	Methanol and formic acid/different solvents	Eucomic acid	[23–25]
Prickly pear peel, cladiodes	Acidified methanol	p-coumaric acid 3-O-glucoside	[9, 23, 26]
Prickly pear peel, cladiodes	Methanol	Isorhamnetin 3-O-glucoside	[9, 23]
Cladiodes	Metanol 80% and acidified methanol	Gallic acid	[9, 26]
Cladiodes, prickly pear peel, prickly pear pulp	Methanol 80% and formic acid	Protocatechuic acid	[9, 25]
Cladiodes	Methanol 80% and acidified methanol	4-hidroxybenzoic acid	[9, 26]
Cladiodes	Methanol 80% and acidified methanol	Caffeic acid	[9, 26]
Cladiodes	Methanol 80% and acidified methanol	Ferulic acid	[9, 26]

Among the most studied compounds are betalains [35, 36], since they are the ones that give the characteristic color to the fruits, likewise they exert specific physiological functions, mainly related to plant pollination and protection against external agents [37], and have a high affinity to water because they contain hydroxyl groups in their structures, which form hydrogen bonds giving them this property [38], they are classified in two: betacyanins and betaxanthins [39].

If we focus on phenolic compounds, they are a family of organic molecules distributed in the plant kingdom, which, as their name indicates, are characterized by the presence of phenolic groups, which may be associated with other chemical groups; they are by-products of plant metabolism that are synthesized in response to external stimuli such as stress [40], both fruits and cladodes of *O. ficus-indica* contain them, and even though flowers have been little studied compared to the rest of the plant parts it has been reported that they can be considered the most important source of these compounds [41], as well as the source of other metabolites such as acids and condensed tannins [42] likewise, because *O. ficus-indica* flowers have been reported to be used as traditional medicine in different parts of the world, *in vivo* studies have been performed demonstrating that extracts rich in phenolic compounds possess bioactive activities [43], flavonols with bioactive effects have also been isolated in previously reported works [44].

On the other hand, in the studies carried out by different researchers and of which some molecules are shown in Table 15.1, we can also mention different compounds such as quercetin or products derived from this molecule (Quercetin pentosyl-rutinoside, quercetin rutinoside (rutin), Quercetin 7-O-rutinoside, Quercetin-3-rhamninoside, Quercetin-malonyl-hexoside, Quercetin-rhamnose-hexoside-rhamnose, Quercetin-hexoside-pentoside) as well as different organic acids (quinic acid, malic acid, diferuloyl-syringsic acid, dicaffeoylferulic acid, sinapic acid 3-O-glucoside, sinapic acid 3-O-galactoside, behenic acid, vanillic acid, syringic acid, chorogenic acid, sinapic acid, salicylic acid, coumaric acid, citric acid) [9, 23–26, 45], of which some molecules are presented in Figure 15.1.

Other phytochemical compounds have also been reported as phytosterols, terpenoids [29, 46] likewise it has been reported that in the fatty acid profile of *O. ficus*-indica shows an incidence of PUFA omega-3 series which are essential components for humans as they cannot be synthesized by the body and must be supplied by the diet [47], polyunsaturated fatty acids such as linolenic and palmitic acid whose concentrations vary depending on the age of the plant [48], lignin has also been reported present in this plant being this natural polymer a good candidate for the development of new materials due to the presence of phenolic and hydroxyl groups in its structure [49, 50].

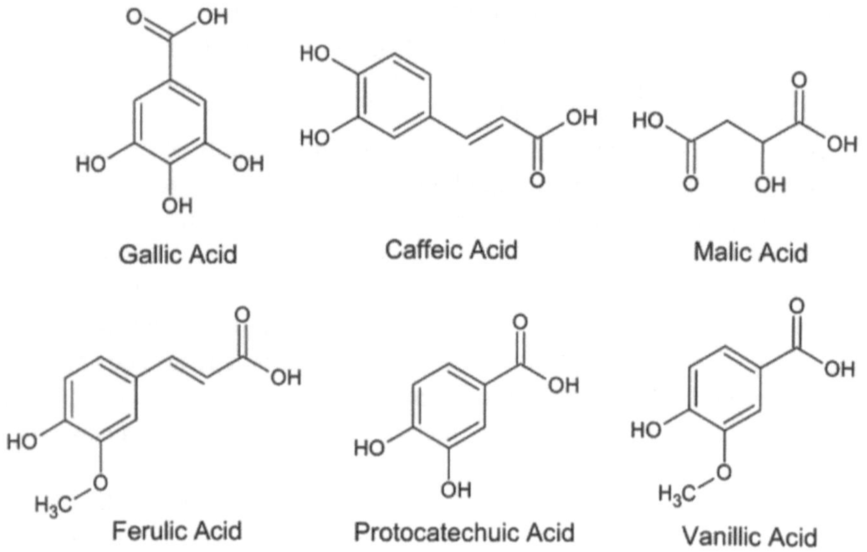

FIGURE 15.1 Polyphenol molecules reported in *Opuntia ficus-indica.*

Finding these compounds in *O. ficus-indica* makes this plant of interest for the possible application of these compounds due to their bioactive properties, thus contributing to different branches such as medicine, the food industry, etc. That is, why different methods to obtain these compounds have also been investigated [51].

15.3 METHODS OF EXTRACTION OF PHYTOCHEMICALS

Among the methods that have been reported for obtaining and quantifying the phytochemicals present in Opuntia ficus-indica are the classic methods, such as extraction with solvents such as methanol, acetone, ethyl acetate, chloroform, and water [24, 52], among the most used are mentioned the extractions with methanol at different percentages [23] which in some cases acidified [26], extraction with ethanol and temperature have also been performed [27], in terms of temperature the nutritional and antioxidant properties of *O. ficus-indica* after heat treatment have been measured, showing that the antioxidant and phenolic capacity increases after these treatments, in addition that microwave and iron heating are the preferable methods to improve the amount of extractable compounds, because the heat treatment causes the rupture of the cell wall thus releasing the phenols from the matrix [53].

To obtain betalains, due to their hydrophilic properties, they are mostly extracted with solvents such as ethyl acetate, ethanol, methanol, and water, so currently alternatives have been studied to obtain them through emerging technologies such as ultrasound-assisted extraction, microwaves, pulsed electric fields, supercritical fluids, these alternatives are studied because they can be applied as natural additives in food [54, 55]. Finally, for obtaining oils from the seeds of this fruit, the use of traditional methods such as pressing has been reported [29].

Due to the growing need for alternatives to obtain compounds of interest, studies have already been conducted with ultrasound-assisted extraction, considered a green technology, to obtain polyphenols with antioxidant activity from the flowers of *Opuntia ficus-indica* using 36% ethanol for 60 minutes at 53°C, and the extracts were also analyzed by DPPH, ABTS, and FRAP tests to demonstrate their antioxidant activity [56], another example of the application of green technologies in the utilization of this plant is the application of pulsed electric fields to improve the extraction of compounds of interest (betalains and polyphenols) in the fruits of *O. ficus-indica*, thus increasing the antioxidant capacity of these juices [57], also in the extraction of flavonoids have been applied emerging technologies, it has been reported the use of supercritical fluids for obtaining compounds of interest, supercritical carbon dioxide (CO_2) has been used to evaluate the influence of pressure, temperature, and flow rate in the extract yield and recovery of flavonoids such as Isorhamnetin-3-O-glucosyl-rhamnosyl-rhamnoside, reporting that in the use of this technology the pressure has a significant effect on the extraction yield, in addition the extraction using this method obtained a flavonoid profile similar to those reported for extracts with acidified methanol [58].

Because of the properties of the compounds present in *O. ficus-indica* it has been reported that they confer a nutraceutical value [59] having a potential in the food industry with health benefits [60], due to the little research in obtaining compounds of interest using emerging technologies to give value to the residues that should be given a greater focus on the search for ecological alternatives for obtaining these compounds, as well as giving an added value to the residues that can generate the use of this plant (seeds and peel).

15.4 BIOLOGICAL PROPERTIES OF *OPUNTIA FICUS-INDICA*

The *Opuntia ficus-indica* cactus has been traditionally used to treat different conditions, such as hypertension, asthma, edema, hyperglycemia, burns, and indigestion, in Mexico the flat and spineless stems are known as "nopalitos" or "pencas" and have been used for centuries to treat inflammation [12, 23, 61], because of this the application of nopal parts as an additive in the preparation of cookies has already been investigated, increasing the fiber value and betalain content to the cookie [62].

It is for this reason that the phytochemical composition of *Opuntia ficus-indica* has been studied by obtaining different extracts, which are rich in phytochemicals to perform different studies to determine their bioactivity in different aspects, which will be addressed in subsections.

15.4.1 ANTICANCER ACTIVITY

Studies have reported the potential of extracts of *O. ficus-indica* extracts in cancer prevention, several examples can be mentioned, all performing extracts and in some the purification of certain compounds to perform their tests, as in the case of the study of opuntiol, which is a flavonol present in OFI with an effect on oral carcinoma KB cells, They performed MTT tests, revealing that opunthiol decreased the viability of KB cell lines in a dose-dependent manner, and reported that opunthiol did not cause any significant cell death in keratinocytes of normal human skin [44] another example can be mentioned using an *OFI* extract with lung squamous cell (A549) and colon cell carcinoma (Caco-2) demonstrating inhibition of cell proliferation, and positive regulation of casp-3 genes, and negative regulation of Bcl-2, also an antiproliferative potential against liver (HepG2), colorectal adenocarcinoma (Caco-2) and breast (MCF-7) cancer cells [63], a similar work, using *OFI* extracts was shown to have antiproliferative potential against liver (HepG2), colorectal adenocarcinoma (Caco-2) and breast (MCF-7) cancer cells, decreasing the cell viability of these cells as the concentration of the extract used increased [64], of the compounds that *OFI* extracts may contain is kaempferol, a flavonol that has been found in different cacti, a study showed that extracts containing kaempferol decrease the viability of prostate carcinoma cells (PC3) and mammary carcinoma cells (Mcf7-7), whose results showed that in terms of cytotoxicity, it increased depending on the dose used of these extracts [65], so *OFI* compounds have potential in the development of cancer treatments, but this is not the only application of these compounds; more may be mentioned and investigated.

15.4.2 ANTI-INFLAMMATORY ACTIVITY

Among the compounds that can be mentioned with biological properties present in OFI are betalains, which have been shown to have anti-inflammatory properties, so they could have therapeutic applications, promoting health [66], another example of the applications of this plant as an anti-inflammatory agent is the study of the methanolic extract of the flowers of *O. ficus*-indica in which its *in vivo* activity was measured in rats, giving positive results in that test, this methanolic extract was analyzed phytochemically concluding that this activity is due to its content of flavonoids and phenolic compounds together, so there may come to be a synergy between its compounds [67], another example is the use of polyphenol-rich infusions of Opuntia ficus-indica cladodes to evaluate their anti-inflammatory effect, this on inflammation associated with obesity and ulcerative colitis induced in Swiss mice, whose results were also positive in this study [68].

Work has also been done previously to evaluate the effect of extracts of this plant on dermal wound healing, being used topically, being also evaluated the antioxidant and antibacterial activity, giving satisfactory results [69], so in addition to being a good anti-inflammatory agent, it can get to have more uses dermally, that is why more studies have been conducted to evaluate the compounds present in OFI for possible application in dermatology, it has been reported the use of extracts of *O. ficus*-indica (peel and fruit) rich in polyphenols and with antioxidant activity for the formulation of stable phytocosmetic emulsions, showing that the formulation did not induce toxicity, although in this work the soxhlet method was used to obtain the compounds [70], but these technologies can be displaced by other greener technologies, on the other hand it is also suggested that lignins from *O. ficus*-indica can be used for the development of new studies related to wound repair, antiviral use and immunological targets [49].

15.4.3 ANTIOXIDANT ACTIVITY

Extracts of *Opuntia ficus-indica* can be beneficial for human health and metabolism, due to the properties of the compounds present in them, among the properties they present it is noteworthy to mention the antioxidant activity, being applicable for the inhibition of the generation and oxidation of fat, so its effect has been studied against animal and human obesity, but this application as a nutraceutical drug against obesity requires further studies [71], likewise, extracts of this plant have been used to test its antioxidant activity and immunoprotective property in rats, evaluating the protective effect of extracts of this plant against immunotoxicity induced by chlorpyrifos (insecticides), showing positive results because these extracts have polyphenols and flavonoids that eliminate free radicals [72].

On the other hand, it has been reported that some extracted betalains *O. ficus-indica* have thermoprotective properties in ovine lymphocytes, being betanin more efficient than α-tocopherol to maintain the viability of lymphocytes under heat stress conditions, this again due to its antioxidant activity [73], in addition this thermoprotective activity is also present for lipid oxidation, since *O. ficus-indica* extracts shows a high inhibition of lipid oxidation being considered as a promising tool as a natural preservative for hamburgers since studies have been conducted that showed that hamburgers added with nopal extract show a good sensory acceptance [74], so it could also be used this plant as a food additive.

At the same time, hydroethanolic extracts have been tested to determine their antioxidant capacity, tests against hydrogen peroxide (H_2O_2), hydroxyl (-OH), peroxyl (ROO-) and ABTS radicals, giving positive results in these tests, so OFI besides being beneficial to health, is a promising source of natural antioxidants, varying in composition due to the time and place of collection [75] and studies have even been conducted with extracts from different parts of *O. ficus-indica*, reporting that the extract from prickly pear peel showed a reduction in the levels of hepatic cholesterol and an increase in its excretion, these effects being related to the phytochemical content and its antioxidant activity [76], Such antioxidant activity is strongly related to the content of carbohydrates and polyphenols present in the plant, and these in turn depend on the place of planting and time of collection of the plant [77], likewise, tests have been conducted with seed extracts, which showed a decrease in lipids and an increase in their metabolism, also improved the antioxidant system of the liver so it may have application in medicine the use of these extracts [78], this has also been demonstrated for betacyanins extracted from other *Opuntia* that have been shown to therapeutically protect against acute liver damage, reducing the gene expression of the cellular stress sensor Gadd45b modulating the expression of antioxidant enzymes that are key to protect the liver [79], which is why antioxidant activity is one of the most studied bioactive properties and with the greatest possible area of application of extracts and compounds from *Opuntia ficus-indica*, but not the only one, so more tests have been conducted to observe what other applications these compounds may have.

15.4.4 ANTIBACTERIAL AND ANTIFUNGAL ACTIVITY

As previously mentioned, the compounds present in different fractions (oily, ethanolic, methanolic, and aqueous) and parts of the plant (cladiodes, flowers, and fruits) present biological activities, if we focus on the activity against different microorganisms we can mention different works, an example of this is the work where it was reported that the oil extracted from *O. ficus*-indica showed an antimicrobial effect against Enterobacter cloacae and *Candida parapsilosis* and antifungal activity against *Penicillium* spp., *Aspergillus* spp., and *Fusarium* spp. as well as a positive effect on skin healing and a potential to prevent infections, so it can have dermatological application, as had been mentioned in a few paragraphs above [80].

Likewise, the effect of different fractions of Opuntia cactus with different microorganisms has been tested, measuring antimicrobial activity against different strains of bacteria, in a work carried out in 2020 it was reported that there is a different behavior in the percentage of inhibition against different bacteria (Acenitobacter, Escherichia coli, Klebsiella pneumonia, Streptococcus, staphylococcus aureus and Candida albicans) demonstrating that the extracts obtained with ethyl acetate, chloroform, and 96% ethanol showed the best inhibitory effect against the microorganisms studied [81], in a similar work hydroethanolic extracts were made with three cacti, among them Opuntia ficus-indica, determining minimum inhibitory concentrations, minimum bactericidal concentrations and minimum fungicidal concentrations, the antibacterial activity in that study was performed against the bacteria mentioned below: *Staphylococcus aureus* (ATCC 6538), *Bacillus cereus*, *Micrococcus flavus* (ATCC 10240), *Listeria monocytogenes* (NCTC7973), *Escherichia coli* (ATCC 35210), *Pseudomonas aeruginosa* (ATCC 27853), *Salmonella typhimurium* (ATCC 13311), *Enterobacter cloacae* (ATCC 35030). For the antifungal assays they report were as follows: *Aspergillus fumigatus* (ATCC 1022), *Aspergillus ochraceus* (ATCC 12066), *Aspergillus versicolor* (ATCC11730), *Aspergillus niger* (ATCC 6275), *Penicillium funiculosum* (ATCC 36839), *Penicillium ochrochloron* (ATCC 9112), *Penicillium verrucosum* var. *Cyclopium* and *Trichoderma viride* (IAM 5061), used ampicillin, bifonazole, and ketoconazole as positive controls, it is reported in this work that the hydroethanolic extract has effect against 6 of the 8 strains of bacteria, (*Micrococcus flavus* and *E. coli* were resistant) and even showed better results than ampicillin, as for the fungicidal activity of peel extracts showed better or equal activity compared to ketoconazole or bifonazole fungicides, it was concluded that the inhibitory effect of the extracts is due to phenolic compounds present in them, due to adsorption on cellular membranes interacting with the enzymes of these or the deprivation of substrate and metal ions [82].

Finally, it has been reported that *O. ficus*-indica extracts showed antiviral activity, recent studies report antiviral activity of an aqueous extract of this plant against cucumber mosaic virus, this thanks to an isolated protein (Opuntin B), which also showed activity against pumpkin and tobacco mosaic virus [83], thus Opuntia ficus-indica extracts can be an encouraging source for the development of new, more environmentally friendly phytosanitary products [84].

15.4.5 CARDIOVASCULAR DISEASES

With each of the compounds present in Opuntia ficus-indica biological properties can be mentioned, such is the case of betalains, which have been shown to have anti-inflammatory properties, so they could have therapeutic applications, promoting health as they can also help in diseases such as hypertension [66], in addition the seed oil of this plant has also shown to have a protective effect against diabetes induced in mice, this effect could be due to the synergism of the antioxidant compounds present [85].

15.4.6 ANTIPARASITIC ACTIVITY

Finally, another possible application is the use of extracts obtained from *O. ficus*-indica cladodes as an agent to control pests in crops, a 2021 study reported embryotoxic, larvicidal, and oviposition deterrent activity against the moth *Plutella xylostella* in an *in vivo* assay, being considered a source of insecticidal agents as it generates damage at the midgut level of these insects [86].

15.5 CONCLUSION

Opuntia ficus-indica has a wide variety of applications, which have been studied previously in different works, but it is still relevant to conduct more research for the integral application of this plant, generating the least amount of waste and taking advantage of these as well, on the other hand, it is important to seek more environmentally friendly alternatives for obtaining these compounds of interest, such is the case of the fruit of *O. ficus-indica*, which has a low protein content. *O. ficus-indica* fruit, which has a low protein content, so studies have been carried out previously to improve its protein content by using FES biotechnology with yeasts such as *Saccharomyces cerevisiae* [15] and *Kluyveromyces marxianus* [87] thus improving its nutritional quality, even so, more studies can be conducted for the identification of bioactive compounds that could be produced by this fermentation, in addition, studies have been conducted on viability for the production of biogas and fertilizer [88], as well as the use of its residues for the production of lactic acid by submerged fermentation using *Lactobacillus plantarum* [89], these studies support the integral use of the plant, but even more can be done to make the most of the whole plant and help our ecosystem in the process.

ACKNOWLEDGMENTS

The authors thank CONACyT for grant 1150316, and the Universidad Autónoma de Coahuila for access to the digital library.

CONFLICT OF INTEREST

The authors declare that there are no conflicts of interest in the preparation of this chapter.

KEYWORDS

- **cardiovascular diseases**
- **cladodes**
- **mesophilic**
- **nopal**
- ***Opuntia ficus-indica***
- **phytochemicals**
- **polyphenols**
- **semidesert plants**

REFERENCES

1. Abidi, S., Ben, S. H., Vasta, V., & Priolo, A., (2009). Supplementation with barley or spineless cactus (*Opuntia Ficus indica* F. inermis) cladodes on digestion, growth and intramuscular fatty acid composition in sheep and goats receiving oaten hay. *Small Ruminant Research, 87*(1–3), 9–16. https://doi.org/10.1016/J.SMALLRUMRES.2009.09.004.
2. De Araújo, F. F., De Paulo, F. D., Neri-Numa, I. A., & Pastore, G. M., (2021). Underutilized plants of the *Cactaceae* family: Nutritional aspects and technological applications. *Food Chemistry, 362*, 130196. https://doi.org/10.1016/J.FOODCHEM.2021.130196.
3. Muhammad, K., Nur, N. I., Gannasin, S. P., Mohd. Adzahan, N., & Bakar, J., (2014). High methoxyl pectin from dragon fruit (*Hylocereus polyrhizus*) peel. *Food Hydrocolloids, 42*(P2), 289–297. https://doi.org/10.1016/J.FOODHYD.2014.03.021.
4. Salehi, E., Emam-Djomeh, Z., Askari, G., & Fathi, M., (2019). *Opuntia Ficus indica* fruit gum: Extraction, characterization, antioxidant activity and functional properties. *Carbohydrate Polymers, 206*, 565–572. https://doi.org/10.1016/J.CARBPOL.2018.11.035.
5. Song, H., Chu, Q., Xu, D., Xu, Y., & Zheng, X., (2016). Purified betacyanins from *Hylocereus undatus* peel ameliorate obesity and insulin resistance in high-fat-diet-fed mice. *Journal of Agricultural and Food Chemistry, 64*(1), 236–244. https://doi.org/10.1021/acs.jafc.5b05177.
6. Alvarez-Bayona, R. A., Cortez-Valadez, M., Martínez-Suárez, F., Cruz-Rivera, J. J., & Flores-Acosta, M., (2019). Green synthesis approximation of Au/Li nanoparticles with opuntia *Ficus-indica* extract. *Physica E: Low-dimensional Systems and Nanostructures, 108*, 169–173. https://doi.org/10.1016/J.PHYSE.2018.12.008.
7. Del-Valle, V., Hernández-Muñoz, P., Guarda, A., & Galotto, M. J., (2005). Development of a cactus-mucilage edible coating (*Opuntia Ficus indica*) and its application to extend strawberry (*Fragaria ananassa*) shelf-life. *Food Chemistry, 91*(4), 751–756. https://doi.org/10.1016/J.FOODCHEM.2004.07.002.
8. Pérez-Cadena, R., Espinosa-Solares, T., Medina-Moreno, S. A., Martínez, A., Lizardi-Jiménez, M. A., & Téllez-Jurado, A., (2021). Effect of the age of opuntia *Ficus-indica* cladodes on the release of simple sugars. *Biocatalysis and Agricultural Biotechnology, 33*, 102010. https://doi.org/10.1016/J.BCAB.2021.102010.
9. Di Bella, G., Vecchio, G. L., Albergamo, A., Nava, V., Bartolomeo, G., Macrì, A., Bacchetta, L., Lo Turco, V., & Potortì, A. G., (2022). Chemical characterization of Sicilian dried nopal [*Opuntia Ficus-indica* (L.) Mill.]. *Journal of Food Composition and Analysis, 106*, 104307. https://doi.org/10.1016/J.JFCA.2021.104307.
10. Yang, L., Lu, M., Carl, S., Mayer, J. A., Cushman, J. C., Tian, E., & Lin, H., (2015). Biomass characterization of agave and opuntia as potential biofuel feedstocks. *Biomass and Bioenergy, 76*, 43–53. https://doi.org/10.1016/J.BIOMBIOE.2015.03.004.
11. Albuquerque, T. G., Pereira, P., Silva, M. A., Vicente, F., Ramalho, R., & Costa, H. S., (2020). Prickly pear. *Nutritional Composition and Antioxidant Properties of Fruits and Vegetables*, 709–728. https://doi.org/10.1016/B978-0-12-812780-3.00044-1.
12. Feugang, J. M., Konarski, P., Zou, D., Stintzing, F. C., & Zou, C., (2006). Nutritional and medicinal use of cactus pear (*Opuntia* Spp.) cladodes and fruits. *Frontiers in Bioscience*, 2574–2589. https://doi.org/10.2741/1992.
13. Mohamed-Yasseen, Y., Barringer, S. A., & Splittstoesser, W. E., (1996). A note on the uses of *Opuntia* spp. in Central/North America. *Journal of Arid Environments, 32*(3), 347–353. https://doi.org/10.1006/JARE.1996.0028.

14. Snyman, H. A., (2013). Growth rate and water-use efficiency of cactus pears *Opuntia Ficus-indica* and *O. robusta*. *Arid Land Research and Management, 27*, 337–348. https://doi.org/10.1080/15324982.2013.771232.
15. Araújo, L. De. F., Medeiros, A. N., Perazzo, N. A., Oliveira, L. De. S. C., & Silva, F. L. H. Da., (2005). Protein enrichment of cactus pear (*Opuntia Ficus – indica* Mill) using *Saccharomyces cerevisiae* in solid-state fermentation. *Brazilian Archives of Biology and Technology, 48*(spe). https://doi.org/10.1590/S1516-89132005000400020.
16. Herrera-Torres, E., Murillo, M., Berumen, L., Soto-Cruz, N., & Paez-Lerma, J., (2017). Enriquecimiento de proteínas de *Opuntia Ficus-indica* utilizando kluyveromyces marxianus en fermentación en estado sólido. *Ciencia e investigación agraria, 44*(2). https://doi.org/10.7764/rcia.v44i2.1767.
17. Dubeux, J. C. B., Santos, M. V. F. D., Cunha, M. V. Da., Santos, D. C. D., Souza, R. T. De. A., Mello, A. C. L. De., & Souza, T. C. De., (2021). Cactus (*Opuntia* and *Nopalea*) nutritive value: A review. *Animal Feed Science and Technology, 275*, 114890. https://doi.org/10.1016/J.ANIFEEDSCI.2021.114890.
18. Sottile, F., Modica, A., Rosselli, S., Catania, C. A., Cavallaro, G., Lazzara, G., & Bruno, M., (2021). Hand-made paper obtained by green procedure of cladode waste of *Opuntia Ficus indica* (L.) Mill. from Sicily. *Natural Product Research, 35*(3), 359–368. https://doi.org/10.1080/14786419.2019.1631820.
19. Pérez-Cadena, R., Espinosa-Solares, T., Medina-Moreno, S. A., Martínez, A., Lizardi-Jiménez, M. A., & Téllez-Jurado, A., (2021). Effect of the age of *Opuntia Ficus-indica* cladodes on the release of simple sugars. *Biocatalysis and Agricultural Biotechnology, 33*, 102010. https://doi.org/10.1016/J.BCAB.2021.102010.
20. Vargas, R. L., Arroyo, F. G., Herrera, M. C. H., Pérez, N. A., García, V. M. I., & Rodríguez, N. J. R., (2016). Physical properties of mucilage prickly pear. *Acta Universitaria, 26*(NE-1). https://doi.org/10.15174/au.2016.839.
21. Agostini-Costa, T. Da. S., (2022). Genetic and environment effects on bioactive compounds of opuntia cacti: A review. *Journal of Food Composition and Analysis, 109*, 104514. https://doi.org/10.1016/J.JFCA.2022.104514.
22. Kuti, J. O., (2004). Antioxidant compounds from four opuntia cactus pear fruit varieties. *Food Chemistry, 85*(4), 527–533. https://doi.org/10.1016/S0308-8146(03)00184-5.
23. El-Hawary, S. S., Sobeh, M., Badr, W. K., Abdelfattah, M. A. O., Ali, Z. Y., El-Tantawy, M. E., Rabeh, M. A., & Wink, M., (2020). HPLC-PDA-MS/MS profiling of secondary metabolites from *Opuntia Ficus-indica* cladode, peel and fruit pulp extracts and their antioxidant, neuroprotective effect in rats with aluminum chloride induced neurotoxicity. *Saudi Journal of Biological Sciences, 27*(10), 2829–2838. https://doi.org/10.1016/J.SJBS.2020.07.003.
24. Aruwa, C. E., Amoo, S. O., & Kudanga, T., (2019). Extractable and macromolecular antioxidants of opuntia *Ficus-indica* cladodes: Phytochemical profiling, antioxidant and antibacterial activities. *South African Journal of Botany, 125*, 402–410. https://doi.org/10.1016/J.SAJB.2019.08.007.
25. Mena, P., Tassotti, M., Andreu, L., Nuncio-Jáuregui, N., Legua, P., Del, R. D., & Hernández, F., (2018). Phytochemical characterization of different prickly pear (*Opuntia Ficus-indica* (L.) Mill.) cultivars and botanical parts: UHPLC-ESI-MSn metabolomics profiles and their chemometric analysis. *Food Research International, 108*, 301–308. https://doi.org/10.1016/J.FOODRES.2018.03.062.

26. Guevara-Figueroa, T., Jiménez-Islas, H., Reyes-Escogido, M. L., Mortensen, A. G., Laursen, B. B., Lin, L. W., De León-Rodríguez, A., et al., (2010). Proximate composition, phenolic acids, and flavonoids characterization of commercial and wild nopal (*Opuntia* Spp.). *Journal of Food Composition and Analysis, 23*(6), 525–532. https://doi.org/10.1016/J.JFCA.2009.12.003.
27. Ginestra, G., Parker, M. L., Bennett, R. N., Robertson, J., Mandalari, G., Narbad, A., Lo Curto, R. B., et al., (2009). Anatomical, chemical, and biochemical characterization of cladodes from prickly pear [*Opuntia Ficus-indica* (L.) Mill.]. *Journal of Agricultural and Food Chemistry, 57*(21), 10323–10330. https://doi.org/10.1021/jf9022096.
28. Blando, F., Russo, R., Negro, C., De Bellis, L., & Frassinetti, S., (2019). Antimicrobial and antibiofilm activity against *Staphylococcus aureus* of *Opuntia Ficus-indica* (L.) Mill. cladode polyphenolic extracts. *Antioxidants, 8*(5), 117. https://doi.org/10.3390/antiox8050117.
29. Brahmi, F., Haddad, S., Bouamara, K., Yalaoui-Guellal, D., Prost-Camus, E., De Barros, J. P. P., Prost, M., et al., (2020). Comparison of chemical composition and biological activities of Algerian seed oils of *Pistacia lentiscus* L., *Opuntia Ficus indica* (L.) Mill. and *Argania spinosa* L. skeels. *Industrial Crops and Products, 151*, 112456. https://doi.org/10.1016/J.INDCROP.2020.112456.
30. Bakar, B., Çakmak, M., Ibrahim, M. S., Özer, D., Saydam, S., & Karatas, F., (2020). Investigation of amounts of vitamins, lycopene, and elements in the fruits of *Opuntia Ficus-indica* subjected to different pretreatments. *Biological Trace Element Research, 198*(1), 315–323. https://doi.org/10.1007/s12011-020-02050-w.
31. Ramadan, M. F., & Mörsel, J. T., (2003). Oil cactus pear (*Opuntia Ficus-indica* L.). *Food Chemistry, 82*(3), 339–345. https://doi.org/10.1016/S0308-8146(02)00550-2.
32. Ramadan, M. F., & Mörsel, J. T., (2003). Recovered lipids from prickly pear [*Opuntia Ficus-indica* (L.) Mill] Peel: A good source of polyunsaturated fatty acids, natural antioxidant vitamins and sterols. *Food Chemistry, 83*(3), 447–456. https://doi.org/10.1016/S0308-8146(03)00128-6.
33. Cano, M. P., Gómez-Maqueo, A., García-Cayuela, T., & Welti-Chanes, J., (2017). Characterization of carotenoid profile of Spanish *Sanguinos* and *Verdal* prickly pear (*Opuntia Ficus-indica*, Spp.) Tissues. *Food Chemistry, 237*, 612–622. https://doi.org/10.1016/J.FOODCHEM.2017.05.135.
34. Farag, M. A., Sallam, I. E., Fekry, M. I., Zaghloul, S. S., & El-Dine, R. S., (2020). Metabolite profiling of three *Opuntia Ficus-indica* fruit cultivars using UPLC-QTOF-MS in relation to their antioxidant potential. *Food Bioscience, 36*, 100673. https://doi.org/10.1016/J.FBIO.2020.100673.
35. Yao, X., Hu, H., Qin, Y., & Liu, J., (2020). Development of antioxidant, antimicrobial and ammonia-sensitive films based on quaternary ammonium chitosan, polyvinyl alcohol and betalains-rich cactus pears (*Opuntia Ficus-indica*) extract. *Food Hydrocolloids, 106*, 105896. https://doi.org/10.1016/J.FOODHYD.2020.105896.
36. García-Cayuela, T., Gómez-Maqueo, A., Guajardo-Flores, D., Welti-Chanes, J., & Cano, M. P., (2019). Characterization and quantification of individual betalain and phenolic compounds in Mexican and Spanish prickly pear (*Opuntia Ficus-indica* L. Mill) tissues: A comparative study. *Journal of Food Composition and Analysis, 76*, 1–13. https://doi.org/10.1016/J.JFCA.2018.11.002.
37. Pereira, C., Dias, M. I., Pinela, J., Roriz, C. L., Ferreira, I. C. F. R., & Barros, L., (2022). Betalains. *Nutraceutical and Functional Food Components*, 461–507. https://doi.org/10.1016/B978-0-323-85052-0.00005-2.

38. Fathordoobady, F., Mirhosseini, H., Selamat, J., & Manap, M. Y. A., (2016). Effect of solvent type and ratio on betacyanins and antioxidant activity of extracts from hylocereus polyrhizus flesh and peel by supercritical fluid extraction and solvent extraction. *Food Chemistry, 202*, 70–80. https://doi.org/10.1016/J.FOODCHEM.2016.01.121.
39. Silva, J. P. P., Bolanho, B. C., Stevanato, N., Massa, T. B., & Silva, C., (2020). Ultrasound-assisted extraction of red beet pigments (*Beta vulgaris* L.): Influence of operational parameters and kinetic modeling. *Journal of Food Processing and Preservation, 44*(10). https://doi.org/10.1111/jfpp.14762.
40. Messina, C. M., Renda, G., La Barbera, L., & Santulli, A., (2013). By-products of farmed european sea bass (*Dicentrarchus labrax* L.) as a potential source of n-3 PUFA. *Biologia (Bratisl), 68*(2), 288–293. https://doi.org/10.2478/s11756-013-0148-8.
41. El-Mostafa, K., El Kharrassi, Y., Badreddine, A., Andreoletti, P., Vamecq, J., El Kebbaj, M., Latruffe, N., et al., (2014). Nopal cactus (*Opuntia Ficus-indica*) as a source of bioactive compounds for nutrition, health and disease. *Molecules, 19*(9), 14879–14901. https://doi.org/10.3390/molecules190914879.
42. Ouerghemmi, I., Harbeoui, H., Aidi, W. W., Bettaieb, R. I., Hammami, M., Marzouk, B., & Saidani, T. M., (2017). Phytochemical composition and antioxidant activity of Tunisian cactus pear (*Opuntia Ficus indica* L.) flower. *Journal of Food Biochemistry, 41*(5), e12390. https://doi.org/10.1111/jfbc.12390.
43. Alimi, H., Hfaiedh, N., Bouoni, Z., Sakly, M., & Ben, R. K., (2011). Evaluation of antioxidant and antiulcerogenic activities of *Opuntia Ficus indica* F. inermis flowers extract in rats. *Environmental Toxicology and Pharmacology, 32*(3), 406–416. https://doi.org/10.1016/J.ETAP.2011.08.007.
44. Veeramani, K. P., Dhineshkumar, E., Karthikeyan, R., Anbuselvan, C., Maqbool, I., Kanimozhi, G., Arul, P. B., & Rajendra, P. N., (2021). Isolation and characterization of opuntiol from *Opuntia Ficus indica* (L. Mill) and its antiproliferative effect in KB oral carcinoma cells. *Natural Product Research, 35*(18), 3146–3150. https://doi.org/10.1080/14786419.2019.1690484.
45. Amaya-Cruz, D. M., Pérez-Ramírez, I. F., Delgado-García, J., Mondragón-Jacobo, C., Dector-Espinoza, A., & Reynoso-Camacho, R., (2019). An integral profile of bioactive compounds and functional properties of prickly pear (*Opuntia Ficus indica* L.) peel with different tonalities. *Food Chemistry, 278*, 568–578. https://doi.org/10.1016/J.FOODCHEM.2018.11.031.
46. Jihane, O., Zrira, S., Petretto, G. L., Saidi, B., Salaris, M., & Pintore, G. A. M., (2016). Volatile constituents and polyphenol composition of *Opuntia Ficus-indica* (L.) Mill from Morocco. *Revue Marocaine des Sciences Agronomiques et Vétérinaires, 4*.
47. Messina, C. M., Arena, R., Morghese, M., Santulli, A., Liguori, G., & Inglese, P., (2021). Seasonal characterization of nutritional and antioxidant properties of *Opuntia Ficus-indica* [(L.) Mill.] Mucilage. *Food Hydrocolloids, 111*, 106398. https://doi.org/10.1016/J.FOODHYD.2020.106398.
48. Andreu-Coll, L., Cano-Lamadrid, M., Sendra, E., Carbonell-Barrachina, Á., Legua, P., & Hernández, F., (2019). Fatty acid profile of fruits (pulp and peel) and cladodes (young and old) of prickly pear [*Opuntia Ficus-indica* (L.) Mill.] from six Spanish cultivars. *Journal of Food Composition and Analysis, 84*, 103294. https://doi.org/10.1016/J.JFCA.2019.103294.
49. Da Cruz, F. I. J., Da Silva, B. B. R., De Souza, A. L. M., Navarro, C. D. C., Ruas, J. S., De Lorena, V. M. B., De Moraes, R. G. J., et al., (2019). Lignins isolated from prickly

pear cladodes of the species *Opuntia Fícus-indica* (Linnaeus) Miller and *Opuntia cochenillifera* (Linnaeus) Miller induces mice splenocytes activation, proliferation and cytokines production. *International Journal of Biological Macromolecules, 123*, 1331–1339. https://doi.org/10.1016/J.IJBIOMAC.2018.09.120.

50. Jablonsky, M., Nosalova, J., Sladkova, A., Haz, A., Kreps, F., Valka, J., Miertus, S., et al., (2017). Valorization of softwood bark through extraction of utilizable chemicals: A review. *Biotechnology Advances, 35*(6), 726–750. https://doi.org/10.1016/J.BIOTECHADV.2017.07.007.
51. Gil-Martín, E., Forbes-Hernández, T., Romero, A., Cianciosi, D., Giampieri, F., & Battino, M., (2022). Influence of the extraction method on the recovery of bioactive phenolic compounds from food industry by-products. *Food Chemistry, 378*, 131918. https://doi.org/10.1016/J.FOODCHEM.2021.131918.
52. Abd El-Moaty, H. I., (2020). Structural elucidation of phenolic compounds isolated from opuntia littoralis and their antidiabetic, antimicrobial and cytotoxic activity. *South African Journal of Botany, 131*, 320–327. https://doi.org/10.1016/J.SAJB.2020.03.005.
53. De Santiago, E., Domínguez-Fernández, M., Cid, C., & De Peña, M. P., (2018). Impact of cooking process on nutritional composition and antioxidants of cactus cladodes (*Opuntia Ficus-indica). Food Chemistry, 240*, 1055–1062. https://doi.org/10.1016/J.FOODCHEM.2017.08.039.
54. Calva-Estrada, S. J., Jiménez-Fernández, M., & Lugo-Cervantes, E., (2022). Betalains and their applications in food: The current state of processing, stability and future opportunities in the industry. *Food Chemistry: Molecular Sciences,* 100089. https://doi.org/10.1016/J.FOCHMS.2022.100089.
55. Salehi, E., Emam-Djomeh, Z., Askari, G., & Fathi, M., (2019). *Opuntia Ficus indica* fruit gum: Extraction, characterization, antioxidant activity and functional properties. *Carbohydrate Polymers, 206*, 565–572. https://doi.org/10.1016/J.CARBPOL.2018.11.035.
56. Brahmi, F., Blando, F., Sellami, R., Mehdi, S., De Bellis, L., Negro, C., Haddadi-Guemghar, H., Madani, K., & Makhlouf-Boulekbache, L., (2022). Optimization of the conditions for ultrasound-assisted extraction of phenolic compounds from *Opuntia Ficus-indica* [L.] Mill. flowers and comparison with conventional procedures. *Industrial Crops and Products, 184*, 114977. https://doi.org/10.1016/J.INDCROP.2022.114977.
57. Surano, B., Leiva, G., Marshall, G., Maglietti, F., & Schebor, C., (2022). Pulsed electric fields using a multiple needle chamber to improve bioactive compounds extraction from unprocessed *Opuntia Ficus-indica* fruits. *Journal of Food Engineering, 317*, 110864. https://doi.org/10.1016/J.JFOODENG.2021.110864.
58. Antunes-Ricardo, M., Gutiérrez-Uribe, J. A., & Guajardo-Flores, D., (2017). Extraction of isorhamnetin conjugates from *Opuntia Ficus-indica* (L.) Mill using supercritical fluids. *The Journal of Supercritical Fluids, 119*, 58–63. https://doi.org/10.1016/J.SUPFLU.2016.09.003.
59. Moussa-Ayoub, T. E., Youssef, K., El-Samahy, S. K., Kroh, L. W., & Rohn, S., (2015). Flavonol profile of cactus fruits (*Opuntia Ficus-indica*) enriched cereal-based extrudates: Authenticity and impact of extrusion. *Food Research International, 78*, 442–447. https://doi.org/10.1016/J.FOODRES.2015.08.019.
60. Ramírez-Rodríguez, Y., Martínez-Huélamo, M., Pedraza-Chaverri, J., Ramírez, V., Martínez-Tagüeña, N., & Trujillo, J., (2020). Ethnobotanical, nutritional and medicinal properties of Mexican drylands cactaceae fruits: Recent findings and research opportunities. *Food Chemistry, 312*, 126073. https://doi.org/10.1016/J.FOODCHEM.2019.126073.

61. Barba, F. J., Putnik, P., Bursać, K. D., Poojary, M. M., Roohinejad, S., Lorenzo, J. M., & Koubaa, M., (2017). Impact of conventional and non-conventional processing on prickly pear (*Opuntia* Spp.) and their derived products: From preservation of beverages to valorization of by-products. *Trends in Food Science & Technology, 67*, 260–270. https://doi.org/10.1016/J.TIFS.2017.07.012.
62. Bouazizi, S., Montevecchi, G., Antonelli, A., & Hamdi, M., (2020). Effects of prickly pear (*Opuntia Ficus-indica* L.) peel flour as an innovative ingredient in biscuits formulation. *LWT, 124*, 109155. https://doi.org/10.1016/J.LWT.2020.109155.
63. Fatma, A. H., Mohamed, G., & Aly, F. M., (2019). Evaluation of anti-cancer potential of proteinacious *Euphorbia tirucalli* and *Opuntia Ficus-indica* cactus extracts: *In vitro* study. *Paripex Indian Journal of Research, VIII* (IX).
64. El-Beltagi, H., Mohamed, H., Elmelegy, A., Eldesoky, S., & Safwat, G., (2019). Phytochemical screening, antimicrobial, antiaxidant, anticancer activities and nutritional values of cactus (*Opuntia Ficus indicia*) pulp and peel. *Fresenius Environmental Bulletin, 28*, 1534–1551.
65. Heikal, A., Abd El-Sadek, M. E., Salama, A., & Taha, H. S., (2021). Comparative study between *in vivo*- and *in vitro*-derived extracts of cactus (*Opuntis Ficus-indica* L. Mill) against prostate and mammary cancer cell lines. *Heliyon, 7*(9). https://doi.org/10.1016/J.HELIYON.2021.E08016.
66. Moreno-Ley, C. M., Osorio-Revilla, G., Hernández-Martínez, D. M., Ramos-Monroy, O. A., & Gallardo-Velázquez, T., (2021). Anti-inflammatory activity of betalains: A comprehensive review. *Human Nutrition & Metabolism, 25*, 200126. https://doi.org/10.1016/J.HNM.2021.200126.
67. Ammar, I., Ben, S. M., Harrabi, B., Mzid, M., Bardaa, S., Sahnoun, Z., Attia, H., & Ennouri, M., (2018). Anti-inflammatory activity and phenolic composition of prickly pear (*Opuntia Ficus-indica)* flowers. *Industrial Crops and Products, 112*, 313–319. https://doi.org/10.1016/J.INDCROP.2017.12.028.
68. Aboura, I., Nani, A., Belarbi, M., Murtaza, B., Fluckiger, A., Dumont, A., Benammar, C., et al., (2017). Protective effects of polyphenol-rich infusions from carob (*Ceratonia siliqua*) leaves and cladodes of *Opuntia Ficus-indica* against inflammation associated with diet-induced obesity and DSS-induced colitis in Swiss mice. *Biomedicine & Pharmacotherapy, 96*, 1022–1035. https://doi.org/10.1016/J.BIOPHA.2017.11.125.
69. Ammar, I., Bardaa, S., Mzid, M., Sahnoun, Z., Rebaii, T., Attia, H., & Ennouri, M., (2015). Antioxidant, antibacterial and *in vivo* dermal wound healing effects of opuntia flower extracts. *International Journal of Biological Macromolecules, 81*, 483–490. https://doi.org/10.1016/J.IJBIOMAC.2015.08.039.
70. Atiya, A., Majrashi, T., Yasmin, B. M., Abdul, Q. S. F., Alqahtani, A. S., Ali Alosman, A. S., Alahmari, A. A., Mesfer Al, A. A. N., Alshahrani, A. T., & Alshahrani, R. R. M., (2022). Influence of solvent selection and extraction methods on the determination of polyphenols, antioxidant, lipoxygenase and tyrosinase inhibition activities of *Opuntia Ficus-indica* Fruits peel and pulp collected from the Kingdom of Saudi Arabia (KSA). *Natural Product Research*. https://doi.org/10.1080/14786419.2021.1983571.
71. Sirotkin, A. V., (2022). Can nopal cactus (*Opuntia Ficus-indica* L. Miller) treat obesity? *Obesity Medicine, 30*, 100390. https://doi.org/10.1016/J.OBMED.2022.100390.
72. Smida, A., Ncibi, S., Taleb, J., Ben, S. A., Ncib, S., & Zourgui, L., (2017). Immuno-protective activity and antioxidant properties of cactus (*Opuntia Ficus indica*) extract against chlorpyrifos toxicity in rats. *Biomedicine & Pharmacotherapy, 88*, 844–851. https://doi.org/10.1016/J.BIOPHA.2017.01.105.

73. Alimi, H., Hfaeidh, N., Bouoni, Z., Sakly, M., & Ben, R. K., (2012). Protective effect of *Opuntia Ficus indica* F. Inermis prickly pear juice upon ethanol-induced damages in rat erythrocytes. *Alcohol, 46*(3), 235–243. https://doi.org/10.1016/J.ALCOHOL.2011.09.024.
74. Parafati, L., Restuccia, C., Palmeri, R., Fallico, B., & Arena, E., (2021). Impact of prickly pear extract on the quality parameters of beef burger patties after cooking. *Food Bioscience, 42*, 101146. https://doi.org/10.1016/J.FBIO.2021.101146.
75. Camboim, R. F., De Oliveira, S. H., Schmidt, L., Rodrigues, E., Tischer, B., Ruffo De, O. V., Lima Da, S. V., et al., (2020). Phenolic compounds and antioxidant activity *in vitro* and *in vivo* of *Butia* and *Opuntia* fruits. *Food Research International, 137*, 109740. https://doi.org/10.1016/J.FOODRES.2020.109740.
76. Milán-Noris, A. K., Chavez-Santoscoy, R. A., Olmos-Nakamura, A., Gutiérrez-Uribe, J. A., & Serna-Saldívar, S. O., (2016). An extract from prickly pear peel (*Opuntia Ficus-indica*) affects cholesterol excretion and hepatic cholesterol levels in hamsters fed hyperlipidemic diets. *Current Bioactive Compounds, 12*(1), 10–16. https://doi.org/10.2174/1573407212666151231183553.
77. Messina, C. M., Arena, R., Morghese, M., Santulli, A., Liguori, G., & Inglese, P., (2021). Seasonal characterization of nutritional and antioxidant properties of *Opuntia Ficus-indica* [(L.) Mill.] mucilage. *Food Hydrocolloids, 111*, 106398. https://doi.org/10.1016/J.FOODHYD.2020.106398.
78. Kang, J. W., Shin, J. K., Koh, E. J., Ryu, H., Kim, H. J., & Lee, S. M., (2016). *Opuntia Ficus-indica* seed attenuates hepatic steatosis and promotes M2 macrophage polarization in high-fat diet–fed mice. *Nutrition Research, 36*(4), 369–379. https://doi.org/10.1016/J.NUTRES.2015.12.007.
79. González-Ponce, H. A., Martínez-Saldaña, M. C., Tepper, P. G., Quax, W. J., Buist-Homan, M., Faber, K. N., & Moshage, H., (2020). Betacyanins, major components in opuntia red-purple fruits, protect against acetaminophen-induced acute liver failure. *Food Research International, 137*, 109461. https://doi.org/10.1016/J.FOODRES.2020.109461.
80. Khémiri, I., Essghaier, H. B., Sadfi, Z. N., Ben, G. N., & Bitri, L., (2019). The antimicrobial and wound healing potential of *Opuntia Ficus Indica L. Inermis* extracted oil from Tunisia. *Evidence-Based Complementary and Alternative Medicine, 2019*, 1–10. https://doi.org/10.1155/2019/9148782.
81. Abd El-Moaty, H. I., (2020). Structural elucidation of phenolic compounds isolated from *Opuntia littoralis* and their antidiabetic, antimicrobial and cytotoxic activity. *South African Journal of Botany, 131*, 320–327. https://doi.org/10.1016/J.SAJB.2020.03.005.
82. Melgar, B., Dias, M. I., Ciric, A., Sokovic, M., Garcia-Castello, E. M., Rodriguez-Lopez, A. D., Barros, L., & Ferreira, I., (2017). By-product recovery of *Opuntia* Spp. peels: Betalainic and phenolic profiles and bioactive properties. *Industrial Crops and Products, 107*, 353–359. https://doi.org/10.1016/J.INDCROP.2017.06.011.
83. Rasoulpour, R., Afsharifar, A., & Izadpanah, K., (2018). Antiviral activity of prickly pear (*Opuntia Ficus-indica* (L.) Miller) extract: Opuntin B, a second antiviral protein. *Crop Protection, 112*, 1–9. https://doi.org/10.1016/J.CROPRO.2018.04.017.
84. Rasoulpour, R., Izadpanah, K., & Afsharifar, A., (2020). Opuntin B, the antiviral protein isolated from prickly pear (*Opuntia Ficus-indica* (L.) Miller) cladode exhibits ribonuclease activity. *Microbial Pathogenesis, 140*, 103929. https://doi.org/10.1016/J.MICPATH.2019.103929.
85. Berraaouan, A., Abderrahim, Z., Hassane, M., Abdelkhaleq, L., Mohammed, A., & Mohamed, B., (2015). Evaluation of protective effect of cactus pear seed oil (*Opuntia*

Ficus-indica L. MILL.) against alloxan-induced diabetes in mice. *Asian Pacific Journal of Tropical Medicine, 8*(7), 532–537. https://doi.org/10.1016/J.APJTM.2015.06.013.

86. Batista, F. E. C., Vila, N. I. C., De Almeida, W. A., Dos, S. A. F. M., Dos Santos, C. G., Da Costa, H. N., Procópio, T. F., et al., (2021). *Opuntia Ficus-indica* cladode extract is an embryotoxic, larvicidal, and oviposition-deterrent agent for the diamondback moth, *Plutella xylostella. Crop Protection, 139*, 105351. https://doi.org/10.1016/J.CROPRO.2020.105351.
87. Herrera, E., Murillo, M., Berumen, L., Soto-Cruz, N. O., & Páez-Lerma, J. B., (2017). Protein enrichment of *Opuntia Ficus-indica* using kluyveromyces marxianus in solid-state fermentation. *Ciencia e Investigacion Agraria, 44*(2), 113–120. https://doi.org/10.7764/rcia.v44i2.1767.
88. Quiroz, M., Varnero, M. T., Cuevas, J. G., & Sierra, H., (2021). Cactus pear (*Opuntia Ficus-indica*) in areas with limited rainfall for the production of biogas and biofertilizer. *Journal of Cleaner Production, 289*, 125839. https://doi.org/10.1016/J.JCLEPRO.2021.125839.
89. Derabli, B., Nancib, A., Nancib, N., Aníbal, J., Raposo, S., Rodrigues, B., & Boudrant, J., (2022). *Opuntia Ficus indica* waste as a cost effective carbon source for lactic acid production by *Lactobacillus plantarum. Food Chemistry, 370*, 131005. https://doi.org/10.1016/J.FOODCHEM.2021.131005.

Index

G

H

M

N

Q

R

S

Y

Z